VOLUME FORTY TWO

ADVANCES IN CHEMICAL ENGINEERING

Chemical Engineering for Renewables Conversion

ADVANCES IN CHEMICAL ENGINEERING

VOLUME FORTY TWO

ADVANCES IN CHEMICAL ENGINEERING

Chemical Engineering for Renewables Conversion

Edited by

DMITRY YU. MURZIN

Laboratory of Industrial Chemistry and Reaction Engineering
Abo Akademi University
Turku/Åbo, Finland

ELSEVIER

AMSTERDAM • BOSTON • HEIDELBERG • LONDON
NEW YORK • OXFORD • PARIS • SAN DIEGO
SAN FRANCISCO • SINGAPORE • SYDNEY • TOKYO
Academic Press is an imprint of Elsevier

Academic Press is an imprint of Elsevier
525 B Street, Suite 1900, San Diego, CA 92101–4495, USA
225 Wyman Street, Waltham, MA 02451, USA
32, Jamestown Road, London NW1 7BY, UK
The Boulevard, Langford Lane, Kidlington, Oxford, OX51GB, UK
Radarweg 29, PO Box 211, 1000 AE Amsterdam, The Netherlands

First edition 2013

ISBN: 978-0-12-386505-2
ISSN: 0065-2377

For information on all Academic Press publications
visit our website at www.store.elsevier.com

Printed and bound by CPI Group (UK) Ltd, Croydon, CR0 4YY
Transferred to digital print 2013

CONTENTS

CONTRIBUTORS

V. Eta
Industrial Chemistry and Reaction Engineering, Process Chemistry Centre, Åbo Akademi University, Biskopsgatan 8, Åbo-Turku, Finland

Jari Heinonen
Lappeenranta University of Technology, Department of Chemical Technology/Laboratory of Industrial Chemistry, Lappeenranta, Finland

S. Hyvärinen
Industrial Chemistry and Reaction Engineering, Process Chemistry Centre, Åbo Akademi University, Biskopsgatan 8, Åbo-Turku, Finland

L.J. Jönsson
Technical Chemistry, Department of Chemistry, Chemical-Biological Center, Umeå University, Umeå, Sweden

David Kubička
Department of Renewable and Environmental Technologies (UniCRE-RENTECH), Research Institute of Inorganic Chemistry (VUANCH), Chempark Litvínov, Litvínov, Czech Republic

J.P. Mikkola
Industrial Chemistry and Reaction Engineering, Process Chemistry Centre, Åbo Akademi University, Biskopsgatan 8, Åbo-Turku, Finland, and Technical Chemistry, Department of Chemistry, Chemical-Biological Center, Umeå University, Umeå, Sweden

R.W. Nachenius
Department of Biosystems Engineering, Faculty of Bioscience Engineering, Ghent University, Ghent, Belgium

W. Prins
Department of Biosystems Engineering, Faculty of Bioscience Engineering, Ghent University, Ghent, Belgium

T. Riitonen
Industrial Chemistry and Reaction Engineering, Process Chemistry Centre, Åbo Akademi University, Biskopsgatan 8, Åbo-Turku, Finland

F. Ronsse
Department of Biosystems Engineering, Faculty of Bioscience Engineering, Ghent University, Ghent, Belgium

Tuomo Sainio
Lappeenranta University of Technology, Department of Chemical Technology/Laboratory of Industrial Chemistry, Lappeenranta, Finland

Tapio Salmi
Department of Chemical Engineering, Process Chemistry Centre, Laboratory of Industrial Chemistry and Reaction Engineering, Åbo Akademi, Turku, Finland

Vratislav Tukač
Department of Organic Technology, Institute of Chemical Technology Prague, Technická 5, Praha, Czech Republic

R.H. Venderbosch
BTG Biomass Technology Group BV, Enschede, The Netherlands

PREFACE

This volume of *Advances in Chemical Engineering* is devoted to some engineering aspects of biomass processing. The topic of biomass utilization became extremely important in chemical community not only because of exciting chemistry but also due to a vision that renewable raw materials might be able to replace at least partially fossil resources such as oil. Hundreds or even thousands of original scientific papers and many dozens of reviews have appeared on biomass processing, addressing various issues, such as the structure of biomass, various transformation routes, and required catalysis.

It seems, however, that methodological and engineering aspects are very seldom discussed. Model compounds are used in academic studies, although with biomass as the feedstock, its location, age, method of storage and processing, and presence of impurities can influence significantly thermal and catalytic transformations.

Analytical procedures described in the literature are often incomplete; moreover, information about mass balance closures is often missing. In addition, issues of mass transfer, external and internal diffusion, not to mention kinetic analysis, are not discussed in numerous biomass-related research papers, even those published in top scientific journals with very high impact factors.

Deactivation of traditional heterogeneous catalysts used in biomass processing could be also rather severe and should be properly considered.

Separation or fractionation of biomass is a non-trivial issue, and selective extraction should be used which is much different in concept compared to oil fractionation. After fractionation, lignocellulosic biomass fractions should be processed in different ways using heterogeneous, homogeneous, and enzymatic catalysis. In separation of reaction products, conventional methods of petrochemical and chemical industry, such as distillation, cannot be readily applied, calling for utilization of other methods, such as chromatography.

The collection of reviews in this volume addresses some of these issues, trying to fill the existing gaps in the literature. Key biomass processing technologies were selected, and the authors were asked by the quest editor to provide a deeper insight on engineering issues going beyond mere descriptions of chemistry already available in the literature.

Chapter 1, by Mikkola and coworkers, describes the technology of ethanol production and some downstream technologies. Chapter 2, by

Nachenius et al., nicely summarizes pyrolysis and torrefaction in an attempt to describe parameters having an impact on design of these units. Kubička and Tukač, in Chapter 3, give an overview of hydrotreating of triglyceride-based feedstocks in refineries. In Chapter 4, Salmi explains the basic principles of chemical reaction engineering and their applications to biomass processing, while in Chapter 5, Heinonen and Sainio describe chromatographic fractionation of lignocellulosic hydrolysates.

The hope of the guest editor is that this volume will be very useful for chemists and chemical engineers working in the area of biomass conversion.

Dmitry Yu. Murzin
Åbo Akademi University, Turlu, Finland

CHAPTER ONE

Engineering Aspects of Bioethanol Synthesis

T. Riitonen*, V. Eta*, S. Hyvärinen*, L.J. Jönsson†, J.P. Mikkola*,†

*Industrial Chemistry and Reaction Engineering, Process Chemistry Centre, Åbo Akademi University, Biskopsgatan 8, Åbo-Turku, Finland

†Technical Chemistry, Department of Chemistry, Chemical-Biological Center, Umeå University, Umeå, Sweden

Contents

Advances in Chemical Engineering, Volume 42
ISSN 0065-2377
http://dx.doi.org/10.1016/B978-0-12-386505-2.00001-8

Abstract

This chapter examines the chemistry, technology, and engineering aspects of ethanol production from different biomass sources. The occurrence, structure, and processing of natural sugars, particularly, those obtained from lignocellulose materials, into fermentable units for bioethanol are reviewed. In addition, different industrial technologies for the fractionation of lignocellulosic materials for the production of bioethanol as a fuel and platform chemical are discussed. Finally, the synthetic routes for bioethanol production processes and the catalytic upgrading processes are addressed.

LIST OF ABBREVIATIONS

4-O-MeGlcA 4-*O*-methylglucuronic acid
5-HMF 5-(hydroxymethyl)-2-furaldehyde or 5-(hydroxymethyl)furfural
AFEX ammonia fiber explosion
AG arabinogalactan
Ara arabinose
ARP ammonia recycle pretreatment
BTL biomass to liquid
BTX benzene, toluene, xylene
CI compression ignition
DFT density functional theory
E85 fuel blend (85% ethanol, 15% gasoline)
ETBE ethyl-*tert*-butyl ether
Fru fructose
F–T Fisher–Tropsch
Gal galactose
GalA galacturonic acid
GGM galactoglucomannan
Glc glucose
GlcA glucuronic acid

IL ionic liquid
Man mannose
MTBE methyl-*tert*-butyl ether
PEO polyethyleneoxide
Rha rhamnose
SynGas synthetic gas
TAME *tert*-amyl-methyl ether
TMP thermomechanical pulp
Xyl xylose

1. INTRODUCTION

Biomass sources vary from agricultural and forest residues to municipal and industrial waste. Although typically solid (plant material, wood, bark, leafs, and needles, forest residues or food waste: meat and bone meal, chicken litter), biomass can sometimes be rendered in liquid (e.g., sewage and black liquor from pulp and paper industry) or gaseous form (e.g., methane from cows).

Regarding sustainability and ecology, environmental and social issues related to the biomass harvesting and processing methods are of significant importance (Eggleston, 2010; Evans et al., 2010; Loque et al., 2011; Pfromm et al., 2011; Soimakallio et al., 2009). *Inter alia*, a lot has been discussed about edible versus nonedible biomass. In addition to many social perspectives (e.g., possible famine caused by the use of farmland or edible crops for fuels and chemicals), there are also a number of environmental aspects to be considered, like the progressing climate change. Bioethanol production based on food crops, such as corn, is ethically unfavored because corn is also an important source of nourishment. In addition, other sugar crops like sugarcane are often grown for nonfood purposes (primarily transportation blends) on farmland that competes with human related food production. For these reasons, bioethanol production from lignocellulosic materials has attracted particular interest since biomass that does not compete with the food chain can be utilized. The second-generation bioethanol production based on lignocellulosic biomass offers possibilities of bioethanol production without competing with farmland or food crops (Centi and van Santen, 2007; Minteer, 2006).

After first-generation biofuels made from edible biomass, recently the so-called second-generation biofuels produced from nonedible biomass (e.g., agricultural residues and other waste materials) have entered the market. Utilization possibilities of woody biomass or other lignocellulosic material have been under intensive investigation for some time now. Even some commercial projects around biorefineries combined with pulp and paper mills have already been established (mainly utilizing black liquor and other waste material as sources of new products). Biorefineries of that kind, combined together with other processes and utilizing the existing technology, are considered to be sustainable.

Today, biofuel products vary from wood and lignin pellets to synthesis gas (SynGas) (through gasification of biomass), bio-oils (pyrolysis of biomass), aqueous sugars (hydrolysis), and lignin (Aho et al., 2011; Grenman et al., 2011; Mäki-Arvela et al., 2011a, 2011b). Besides fermentation to bioethanol or biobutanol, the sugars obtained have several applications for further processing to aromatic hydrocarbons (dehydration) and liquid alkanes or hydrogen (by aqueous-phase processing) (Riittonen et al., 2012; Tokarev et al., 2010). Consequently, the residual lignin can besides direct combustion be upgraded to various value-added products.

Here, we concentrate on lignocellulosic biomass and especially on wood as a bioethanol source. In addition to biofuels or the traditional wood and paper products, the future forest biorefineries will be capable to produce specialty high-value biomaterials and biochemicals from forest resources.

Ethanol is a widely used and versatile organic compound which reactivity is mainly based on the incorporated hydroxyl group. Ethanol has certain characteristic physical properties, such as high water solubility. It has been used as a solvent, fuel, depressant, and chemical intermediate (Kirk-Othmer, 2007). The old industrial production methods have been based on the conversion of ethylene which is derived from petrochemical feedstocks. The older method is related to indirect hydration process which applies sulfuric acid. In the early 1970s, the process was further developed and the outcome was elimination of the acid and the process in carried out in vapor phase. However, due to sustainable development and green chemistry, the preferred synthesis routes are based on the fermentation of biomass to obtain sugars which can be transformed to ethanol (Kirk-Othmer, 2007).

2. POLYSACCHARIDES: POTENT RAW MATERIALS FOR BIOETHANOL PRODUCTION

Generally speaking, the most abundant polysaccharides in nature are cellulose, hemicelluloses, starch, and chitin. A comparison between the chemical structures (stereochemistry) of cellulose, starch and chitin, and the constituent α- and β-bonds in the polymer chains is shown in (Fig. 1.1).

The major part of both hardwoods and softwoods consists of cellulose, hemicellulose, and lignin: of which cellulose represents about 40% of the dry matter content in the wood. Hemicelluloses and pectins make up 25–30% and lignin 25–30% of the dry matter content in softwoods, while hardwoods contain 30–35% of hemicelluloses and 20–25% of lignin. Extractive content is less than 5% for both softwoods and hardwoods (Stenius, 2000; Willför et al., 2005a, 2005b).

Lignin is a complex branched polymer synthesized from monomeric aromatic alcohols, the three monolignols coniferyl alcohol, sinapyl alcohol, and *p*-coumaryl alcohol. Units derived from coniferyl alcohol, sinapyl alcohol, and *p*-coumaryl alcohol are predominantly found in lignins from softwood, hardwood, and grasses, respectively (Eklund and Lindström, 1991;

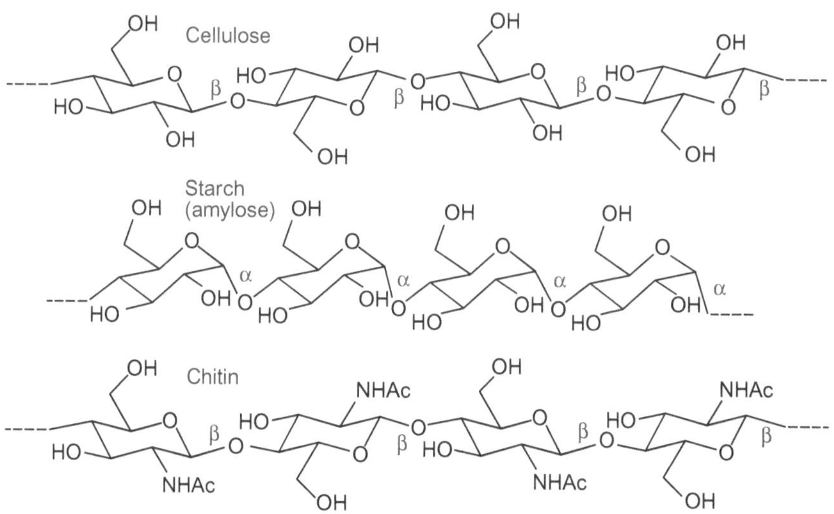

Figure 1.1 The comparison of the polysaccharide chains of cellulose, starch and chitin. (For color version of this figure, the reader is referred to the online version of this chapter.)

Stenius, 2000). Lignosulfonates obtained as by-product of sulfite pulping and kraft lignin obtained from black liquor during kraft pulping are the two types of commercialized lignin with a wide range of uses. Lignin is thought to hinder the hydrolysis of the cellulose and hemicelluloses in lignocellulosic biomass since it binds cellulose and hemicelluloses together in the living plant. This was meant from paper and pulp industry perspective, as well as concerning pretreatments when the goal is the production of other chemicals or biofuels than ethanol. Hemicelluloses are of course removed first in the traditional kraft cooking and other pulping methods. Most pretreatments affect similarly, there are some known methods (lime, ARP and AFEX) that have only minor effect on hemicelluloses but remove lignin (Mosier et al., 2005). Researches aiming to use both cellulose and hemicelluloses prefer these type of pretreatment methods.

In ethanol production, the majority of pretreatment methods target to the degradation of the hemicelluloses leaving cellulose and lignin untouched, and then letting the enzymes to depolymerize the cellulose to glucose, while lignin is left as the solid residue (referred to as "hydrolysis lignin"). On the other hand, originating from the paper and pulp industry's perspective (where the goal is to remove both hemicelluloses and lignin in order to utilize only the pure cellulose), there are nowadays other pretreatment method development projects that aim to keep the hemicelluloses inaffected despite of the removal of lignin and even decreasing the crystallinity of the cellulose: AFEX, ARP and alkali pretreatments are example of these type of methods (Lee et al., 2009; Mosier et al., 2005).

Comparison of the polysaccharide ratios (Stenius, 2000; Willför et al., 2005a, 2005b) between softwoods and hardwoods is shown in Fig. 1.2. The proportions of cellulose, hemicelluloses, glucomannans, xylans, and pectins are depicted as a percentage of the dry matter content of the wood.

2.1. Cellulose

The cell walls of plants contain cellulose, which is regarded as the most abundant biopolymer on earth (Mousdale, 2010). Lignocellulose, which is a major constituent of terrestrial biomass, has three main components: namely, cellulose, hemicellulose, and lignin (Mousdale, 2010; Sjöström and Alén, 2010). Cellulose is the most common organic matter comprising of 40% of all plant dry weight. It is often called the "backbone" of the plants, as cellulose coating protects the outer wall of cells in plants due to its chemical inertness and its complete insolubility in water. Cellulose is insoluble in conventional solvents and to some extent even resistant to acid hydrolysis, the acid-resistance depends on the crystallinity of the cellulose as well as the type

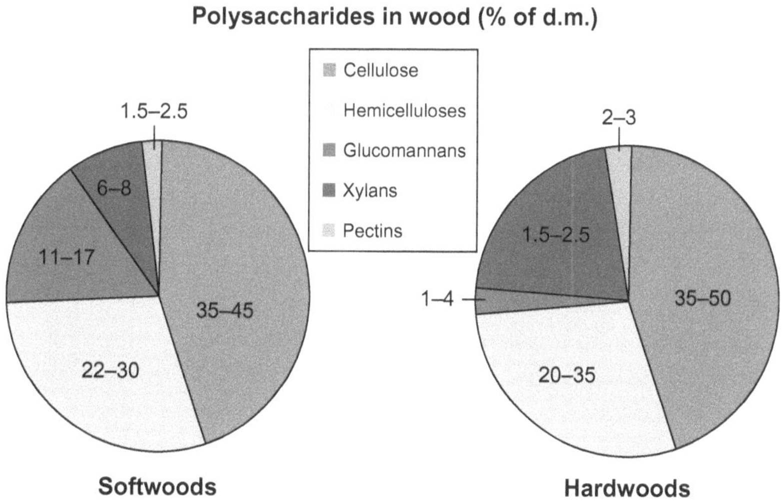

Figure 1.2 Polysaccharide ratios in softwoods and hardwoods (% of the dry matter content of the wood). Glucomannans, xylans, and pectins are usually classified to belong to hemicelluloses; thus here hemicelluloses mean other hemicelluloses excluding the mentioned major hemicelluloses groups. *Based upon data from Willför et al. (2005a, 2005b) and Stenius (2000).* (For color version of this figure, the reader is referred to the online version of this chapter.)

of the glycosidic bond, namely acid hydrolysis rate is known to be faster for glycosidic bonds with non-redusing ends–substrate accessibility, crystallinity as well as swelling of the cellulose all affect the dydrolysis rates (Mäki-Arvela et al., 2011b; Willför et al., 2009). There are numerous commercial applications utilizing cellulose, for example, paper, film, additives, adhesives, and textile fibers (Collins and Ferrier, 1995; Davis and Fairbanks, 2008; Fengel and Wegener, 1989; Sjöström, 1993; Wertz et al., 2010).

Cellulose is a linear polymer consisting of glucose units, and the degree of polymerization (DP) varies depending on the source of cellulose. There are four different known polymorphs of cellulose: cellulose I, II, III, and IV, and in addition, these polymorphs have six allomorphs: Iα, Iβ, III_1, III_2, IV_1, and IV_2 (Beckham et al., 2011; Fengel and Wegener, 1989; Sjöström, 1993; Wertz et al., 2010). Primitive organisms, such as bacteria, include Iα cellulose, while cellulose in more complicated organisms (that produce wood, cotton, and shells of certain marine animals) consist of cellulose Iβ, which is thermodynamically more stable than the allomorph Iα. Consequently, a transformation from Iα to Iβ is irreversible. The proportional division of the crystalline phases Iα and Iβ depends on the origin of the cellulose. Cellulose I is native cellulose, whereas cellulose II represents the regenerated form. The latter polymorph is energetically more stable than the first one causing the conversion reaction from the cellulose I to II to be irreversible. The preparation of cellulose III is possible either from the cellulose I (which leads to cellulose III_1) or from the cellulose II (which leads to cellulose III_2) by treating the raw material with ammonia or with certain amines (e.g., ethylene diamine). Cellulose IV can be manufactured from cellulose III by heating it at high temperature in glycerol, but in addition to this, cellulose IV can also be found with X-ray diffraction in some plants, as it is a disordered form of cellulose I (Fengel and Wegener, 1989; Gräsvik et al., 2012; Sjöström, 1993; Wertz et al., 2010).

Cellulose is a homopolymer of β-D-glucopyranose units, linked by (1 → 4)-glucosidic linkages, and the monomer unit in cellulose can also be called β-D-glucose or anhydro-β-D-glucose or simply anhydroglucose (Davis and Fairbanks, 2008; Stenius, 2000). Every other glucose (Glu) is "up-side-down," and the repeating unit including these two Glu molecules is called cellobiose (length: 1.03 nm). The cellulose chain has two types of ends: nonreducing and reducing ends. Hydrogen bonds and van der Waals' forces promote the cellulose strings to assemble into large well-ordered fibers. Interactions between functional groups, OH and H, stabilize the molecule chains in ordered systems, leading to the supramolecular structure, which enhances strength and changes the physical and chemical properties of the molecule resulting in crystalline areas in cellulose.

2.2. Heteropolysaccharides

Hemicelluloses and pectins are heteropolysaccharides, have support functions in plants, and constitute ca. 20–30% of woody biomass. These polysaccharides are easily degraded by chemical treatments during pulping. Unlike cellulose, they do not possess crystalline parts but are amorphous (Eklund and Lindström, 1991; Stenius, 2000). Typical products made of hemicelluloses and pectins besides bioethanol include emulsifiers, edible films, dietary fibers, pharmaceuticals, food additives, thickeners, gelling agents, adhesives, adsorbants, and xylitol (Willför et al., 2008).

The building blocks of the hemicelluloses and pectins are monosaccharides such as pentoses (xylose, Xyl; arabinose, Ara), hexoses (Glu, mannose, Man; and galactose, Gal), hexuronic acids (glucuronic acid, GlcA; galacturonic acid, GalA; 4-*O*-methylglucuronic acid, 4-*O*-MeGlcA), and deoxyhexoses (rhamnose, Rha). The polymer chains are often branched. The DP is lower (100–300) than that of cellulose (about 10,000 and more, depending on the source) (Collins and Ferrier, 1995; Eklund and Lindström, 1991; Fengel and Wegener, 1989; Sjöström, 1993; Stenius, 2000; Wertz et al., 2010).

Galactoglucomannan (GGM), glucomannan, arabinoglucuronoxylan, glucuronoxylan, and arabinogalactan (AG) are examples of common hemicelluloses in wood (Eklund and Lindström, 1991; Stenius, 2000; Willför et al., 2005a, 2005b, 2009). Other nonwood sources of hemicellulose are bark, foliage, peat, mosses, algae, and agricultural residue.

2.3. Mannans

Mannans (glucomannans or GGMs) are common wood hemicelluloses whose ratio varies depending on the type of wood. In softwoods (e.g., spruce, which is known to be a very rich source of GGM), mannans are mainly *O*-acetyl-galactoglucomannans, which cover 20–25% (Sjöström, 1993) or 11–17% (Willför et al., 2005a, 2005b) of dry wood. The main chains consist of mannose and glucose units, while side groups contain Gal and acetyl groups, respectively. Hemicelluloses in softwoods consist mainly of glucomannans (the main chain is built up of mannose and glucose units without any side groups), whereas only 2–5% (Sjöström, 1993) or 1–4% (Willför et al., 2005a, 2005b) dry matter of hardwoods contain mannans. The occurrence of GGMs in softwood is in the range of 15–23% (Sjöström, 1993), and the molar ratio of the Man:Glc:Gal:Acetyl units is 4:1:0.5:1. Glucomannans occurrence in hardwood, on the other hand, is within the range of 2–5% (Sjöström, 1993), and the molar Man:Glc ratio varies between 1:1 and 2:1. GGMs in TMP (thermomechanical pulping) waters can be recovered by means of ultrafiltration via direct extraction of wood (Song et al., 2011b; Willför et al., 2008; Xu and Willfor, 2012).

2.4. Xylans

In softwoods, xylans are mainly arabino-4-*O*-methylglucuronoxylans that correspond to 5–10% of the wood matter (Sjöström, 1993). They have Xyl as their main chain unit, while side groups consist of 4-*O*-MeGlcA and Ara molecules, respectively (Eklund and Lindström, 1991; Stenius, 2000). Hemicellulose xylans in hardwoods are mainly *O*-acetyl-4-*O*-methylglucuronoxylans, which is the dominating hemicellulose (15–30%) in hardwoods (Sjöström, 1993), while Xyl is the main chain unit also in softwoods. Side groups of the hardwood xylans consist also of 4-*O*-MeGlcA units, like in softwoods, but in addition, it contains acetyl groups instead of Ara (Collins and Ferrier, 1995; Fengel and Wegener, 1989; Sjöström, 1993).

Arabinoglucuronoxylan is found in softwood in the range of 5–10%, and the molar Xyl:4-*O*-meGlcA:Ara ratio is 10:2:1.3 (Sjöström, 1993). In the case of glucuronoxylan in hardwood (15–30%), the molar

Xyl:4-*O*-meGlcA:Acetyl ratio is 10:1:7 (Sjöström, 1993). Besides natural wood and other lignocellulosics, xylans can also be found in bleached kraft pulp in significant amounts (Willför et al., 2009).

2.5. Arabinogalactans

AGs that are highly water-soluble branched polymers can be found in the heartwood of larch and in compression wood (Mäki-Arvela et al., 2011b; Sjöström, 1993; Stenius, 2000). The content of AGs is about 5–35% of the dry mass. They are found in larch and reaction wood, and the molar ratio of Gal:Ara:Glc is approximately 6:3:~0 (Sjöström, 1993).

In general, 15–20% (w/w) of Siberian larch (*Larix sibirica*) heartwood consists of easily water-soluble AG. One possibility is to extract AG prior to chemical pulping or directly from saw dust. Applications of AG include products for animal feeds, fibers for skin and hair care, and product additives, for example, additives for flexographic inks (Fitzpatrick et al., 2004).

2.6. Pectins

Pectins are located in the middle lamella and primary wall of the cell wall layers (Collins and Ferrier, 1995; Eklund and Lindström, 1991; Fengel and Wegener, 1989; Sjöström, 1993; Stenius, 2000). Pectins are water soluble, and the main sugar polymer chain contains galacturonic acid and Rha units (Collins and Ferrier, 1995; Fengel and Wegener, 1989; Sjöström, 1993). Pectins are methylesterified to a high degree especially in spruce. Various alkaline treatments can be used to produce anionic pectic acids, which can be further utilized as, for example, thickeners in food (Collins and Ferrier, 1995).

2.7. Starch

Starch is the second most common polysaccharide in plants, and it is one of the main sources of D-glucose in human's diet (potato, corn, wheat, rice, tapioca) (Davis and Fairbanks, 2008). Starch is present in all green plants; thus it can also be found in trees, although in minor amounts: usually sapwood of both softwoods and hardwoods contains more starch than heartwood (Willför et al., 2005a, 2005b). Ray cells in plants include starch, and, thus, due to its hydrophilic nature, it can be extracted with polar solvents (Collins and Ferrier, 1995; Eklund and Lindström, 1991; Fengel and Wegener, 1989; Sjöström, 1993; Stenius, 2000). Besides being used as food and feed, starch is an excellent source of monosaccharides in ethanol fermentation. Furthermore, starch is widely used by paper industry as a retention agent, as a binder in pigment coatings, as an adhesive in the manufacture

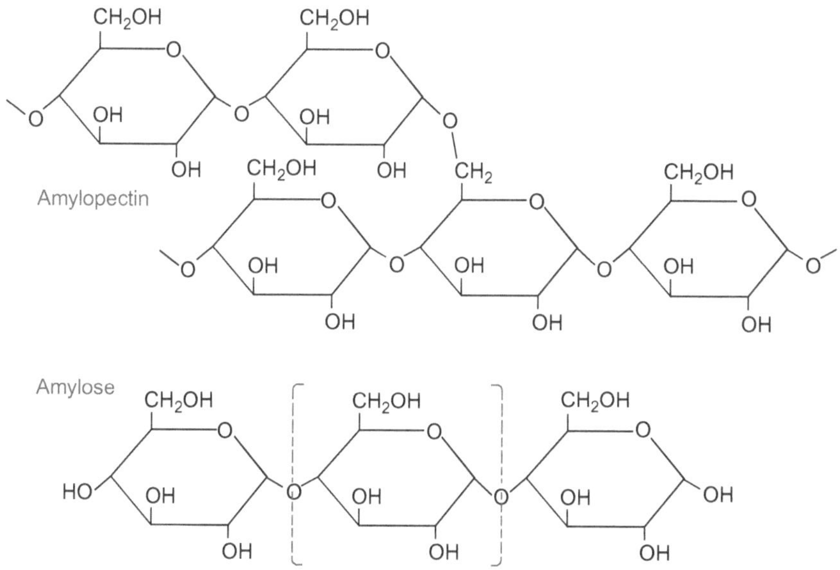

Figure 1.3 The starch components: the branched part, amylopectin, and the linear part, amylose. Amylopectin is branched at 1,6-α-glucosidic bonds. The starch chains are linked with α-1,4-bonds, which are also known as (1 → 4)-glycosidic bonds. (For color version of this figure, the reader is referred to the online version of this chapter.)

of corrugated board and for surface application in a size press, as well as for improving the interfiber bonding (Eklund and Lindström, 1991). Starch has two types of polymers consisting of D-glucose, like cellulose. The starch chains are linked with α-1,4-bonds, also known as (1 → 4)-glycosidic bonds which are known to enhance flexibility. Starch components are amylopectin and amylose, the first one being branched and the latter one being linear (Fig. 1.3). The branching of amylopectin takes place at 1,6-α-glucosidic bonds. The average distance between the branching point and the inner amylopectin chains is roughly 8 glucose units, whereas the mean distance between the branching point and the outer amylopectin chains is approximately 12 glucose units (Eklund and Lindström, 1991).

Amylopectin is usually more common in nature; for example, in potato starch, the ratio of amylopectin to amylose is 4:1, while corn (maize) and wheat contain about 2.5 times more amylopectin than amylose. Amylopectin consists of large tree-like molecule with 100–200 linear parts each consisting of 20–30 anhydroglucose units, while amylose polymer chain consists of 200–2000 anhydroglucose units. Results from X-ray diffraction and swelling patterns suggest that amylopectin is mainly crystalline, whereas a large number of intermolecular hydrogen bonds between the linear parts hold the crystalline regions together. Besides amylopectin and amylose, the starch grains (granules) also contain fat and proteins as well as some other components. Depending on the type of starch and the individual starch grain, the size of the starch grains (granules) varies quite a lot. The granule size and the composition of the mentioned additive components give different properties to the starch depending on the source from where it is originating: potato starch differs from corn and wheat starch (Eklund and Lindström, 1991).

The amorphous amylose can be dissolved from the granule without changes in its structure, but certain types of starch can also contain crystalline amylose, which makes dissolution impossible. However, the only widely accepted native crystalline form of amylose is the V-form that in addition to amylose consists of some complex forming substance present in the starch granule, for example, fatty acids or butanol (Complex-forming substances can even be utilized for fractionation of starch, as it is known that, for example, higher alcohols, fatty acids, and sulfonated oils have a special affinity for the linear part of starch.) (Eklund and Lindström, 1991).

2.8. Chitin

Chitin is a polysaccharide that is structurally close to cellulose, and it is also a very inert and strong substance. In fact, the hard exoskeletons of insects and shellfish consist of chitin. Unlike cellulose that consists of repeating cellulose units; the repeating unit of chitin is *N*-acetylglucosamine differing from cellulose only by the presence of an acetoamido group at C2 position (Davis and Fairbanks, 2008). Although originating from marine animals, chitin could be utilized as a source of fermentable sugars to bioethanol after depolymerization.

3. MONOSACCHARIDES (MONOMERIC SUGARS)

Monomeric sugars can easily be utilized for the production of 5-(hydroxymethyl)-2-furaldehyde or 5-(hydroxymethyl)furfural (5-HMF), furfural, and other specialty chemicals and can also be transformed to ethanol via fermentation. Monosaccharides (monomeric sugars) and disaccharides can be obtained as extractives (a minor constituent of wood) in the sap wood (Eklund and Lindström, 1991; Stenius, 2000; Willför et al., 2005a, 2005b). Furthermore, they can be obtained from the polysaccharides of biomass via hydrolysis. Several options exist as a depolymerization method: the conventional methods are acid methanolysis, acid hydrolysis, and enzymatic hydrolysis. Enzymatic hydrolysis is highly selective and widely used in ethanol production albeit the long operation time. The usability and benefits of different methods should be evaluated depending of the goals of the process (Fig. 1.4).

Depolymerization methods of wood polysaccharides

Method →	Product
Chemical pulping sulfate, sulfite prehydrolysis sulfate	Paper pulp, lignin, dissolving pulp
Microfluidization of pulp	Microfibrillar cellulose (nanocellulose)
Acid treatment of pulp	Cellulose nanorods
Organosolv pulping ethanol, acetic acid, etc.	Paper pulp, lignin, dissolving pulp
Supercritical fluid extraction	Extraction, biofuels, gasification
Ionic liquids, ILs	Fractionation of wood components

Figure 1.4 A typical selection of depolymerization methods (with different product aims) for the polysaccharides in wood. (For color version of this figure, the reader is referred to the online version of this chapter.)

Chemical pulping (with the contribution of the various bleaching steps) in the conventional paper mills is the leading method, resulting in almost complete delignification of the pulp (and loss of hemicelluloses) leaving the cellulose part untouched. A more rare method for industrial scale is organosolv pulping. Other methods used consist of microfluidization and acid treatment (acid hydrolysis) of pulp. However, these methods aim at different products than the two first mentioned pulping methods. Supercritical fluid extraction and treatment with ionic liquids (ILs) are currently under investigation at the academia with the goal to find out the optimized way to disassembly the polysaccharides in lignocellulosic biomass. Steam explosion/autohydrolysis, CO_2 explosion, liquid hot water, pH controlled hot water, flow-through liquid hot water, dilute acid, flow-through acid, ammonia fiber explosion (AFEX), ammonia recycle pretreatment (ARP), and lime (alkaline hydrolysis) represent some recent trials (Mosier et al., 2005; Wertz et al., 2010).

Despite challenges with respect to recovery of the IL, various IL treatments in combination with heat have been proven to function well for dissociation of wood to monomeric sugars, sugar alcohols and acids as well as their degradation products, HMR and furfural (Anugwom et al., 2012;

Maki-Arvela et al., 2012; Salmi et al., 2010). The challenges largely depend on the IL used and the process itself as well as the chemical process conditions involved. The presence of high concentration salt solutions (ion character of ILs) results in analytical challenges in traditional columns (Hyvärinen et al., 2011). Still, most ILs used are water soluble and otherwise solubilized to similar solvents as monosaccharides. Some criticism has also been targeted toward the quality of the commercially available ILs that obviously affects the results obtained. Assumedly, if a dark brown or black suspension is obtained in IL treatment of, for example, wood, the reaction conditions have been too harsh leading to degradation of the obtained monosaccharides originating from polysaccharides, hemicelluloses, and cellulose as a result of depolymerization. The dark solution indicates formation of tar and other phenolic components as well as possible IL complexes formed together with lignin, sugars, or degradation products of sugars (furfural, HMF, etc.). Nevertheless, standard quality ILs are available and provided that the analysis method is robust, an *industry* might not care if the effect was rather caused by impurities than the IL itself. An important advantage of an IL treatment is that often no pretreatment of biomass is needed.

High efficiency for the solvation of cellulose, lignin, and even wood has been proved in an increasing range of dialkyl imidazolium-based ILs (King et al., 2011). However, some recent papers (King et al., 2009a, 2009b) have also shown that one should be careful when using the term "cellulose or wood dissolution." It is known as a fact that in soda, sulfite, or kraft pulping (commercial wood fractionation processes in paper and pulp industry), wood chips are converted to fibers (fibrillation) and nearly all lignin in wood is fragmented and dissolved (delignification), while cellulose remains (i.e., cellulose is not dissolved). Similarly, it should be noted that despite the efficient solvent properties of some ILs, wood is not soluble in ILs under mild dissolution conditions (e.g., 80 °C, 18 h) (King et al., 2009a, 2009b). Usually, in the IL treatment of wood, nonfibrous pulp is obtained, lignin is not efficiently separated, and wood components are selectively precipitated. In addition, also the effects of mechanical treatments should be noted, for example, the milling/pulverization degree of the sawdust affects on the lignin linkages and solubilization of wood (King et al., 2009a, 2009b).

Furthermore, the chemical stability of ILs and solutes under process should also be taken into consideration—concerning both the degradation conditions and properties of the IL and possible reactions between the used IL and the biomass material. There are some indications that ILs such as 1-ethyl-3-methylimidazolium acetate, [emim][OAc], which is a highly effective solvator, react chemically with lignocellulosic solutes forming lignin–carbohydrate complexes, for example, phenylglycoside, benzyl esters, and ethers. Therefore, the choice of the suitable IL-based treatment method for lignocellulosic biomass or wood depends on the products that are desired. If one wants to obtain the fiber-based value chains, IL-fractionation should be based on selective extraction of biomass instead of the reactive dissolution, followed by selective precipitation of low molar weight components. Wood can be "fibrillated" with certain ILs using mild treatments. For biofuel production, efficient degradation/depolymerization of polysaccharides in wood can be useful, while in order to obtain high molar weight, polymeric components mild dissolution is recommended (King et al., 2009a, 2009b, 2011).

In recent catalytic transformation research of AG and its monomers—aiming to obtain valuable products of its monomers—it was shown that optimum pH and temperature (pH 1 at 90 °C) are crucial in order to succeed in the acid hydrolysis of AG to Ara and Gal (galactose) without further degradation of the monomers (Kusema et al., 2010). The utilization of homogenous and heterogeneous catalysts was also noticed to differ in their advantages and disadvantages, for example, the latter ones catalyzed better the hydrolysis of the AG side groups than that of the main AG chain, however, resulting in internal mass transfer limitations.

The main monosaccharides in biomass extractives are Glc and fructose (Fru); and a common dimeric sugar is sucrose that consists of Glc and Fru units (sucrose is also known as table sugar). Glucosides, for example, coniferin, can also be extracted from biomass. The content of simple sugars varies during the year depending on the season among other parameters (Eklund and Lindström, 1991; Stenius, 2000).

Monosaccharides can occur as D-sugars (*R*-) or L-sugars (*S*-): the prefix uncovers the absolute configuration of the secondary alcohol at the highest numbered stereogenic center (or chiral center) (Davis and Fairbanks, 2008). Aldopentoses (three chiral centers), such as xylose, exist as four diastereomeric pairs of enantiomers, while aldohexoses such as glucose, exist as eight diastomeric pairs. The D-forms of the sugars are much more common than their enantiomeric L-counterparts, excluding some deoxy sugars: namely, L-fucose (6-deoxy-L-galactose) and L-rhamnose (6-deoxy-L-mannose), which are both biologically important L-formic monosaccharides, the first being involved in cell–cell recognition processes and the latter representing a major component in plant cell wall (Davis and Fairbanks, 2008). These two L-monosaccharides are shown in Fig. 1.5 together with some important D-monosaccharides. The "zigzag" formulae for the most important hexoses and those for the two very important pentoses, Xyl and Ara, are displayed—all raw materials for many value-added products.

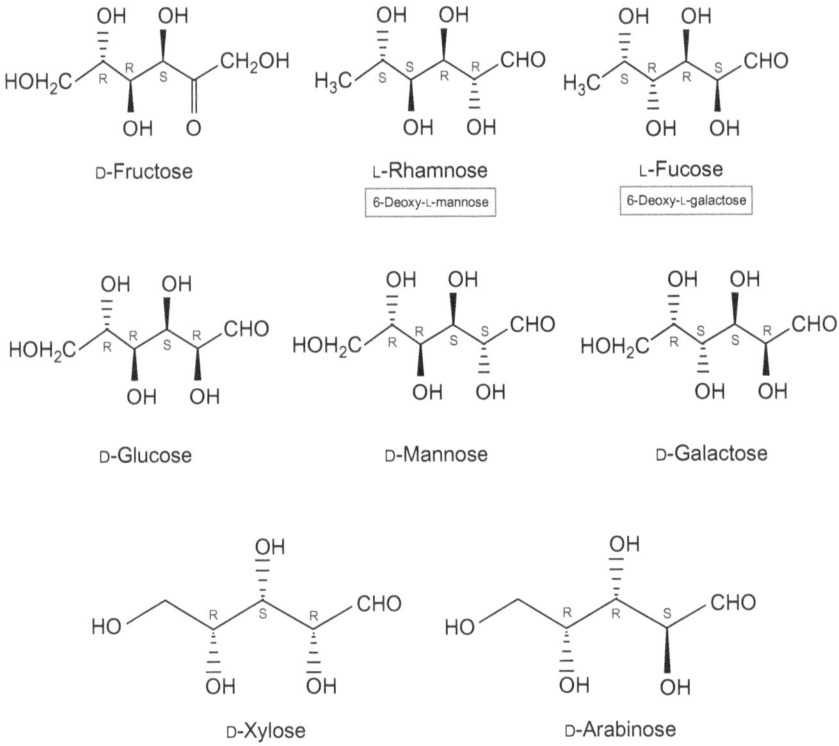

Figure 1.5 "Zigzag" formulas for some common hexoses and pentoses that can be derived from biomass. (For color version of this figure, the reader is referred to the online version of this chapter.)

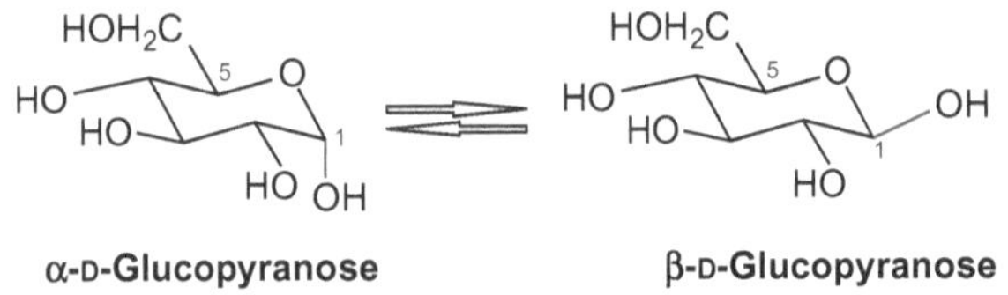

Figure 1.6 The α- or β-anomers of D-glucose. (For color version of this figure, the reader is referred to the online version of this chapter.)

In addition to D- and L-configurations, monosaccharides can be either α- or β-anomers, that is, diastereomers that differ in their stereochemistry at the anomeric position, C-1: the C-1 substituent of the α-anomer is axial, while it is equatorially arranged at the β-anomer (Fig. 1.6).

When considering chemical reactions of the monosaccharides, besides mutarotation phenomenon, it is also worth reminding of the anomeric effect: that is, electronegative substituents at the anomeric center prefer to obtain the axial orientation. Consequently, the α-anomers are usually more thermodynamically stable than β-anomers (Davis and Fairbanks, 2008). Glu is an exception to this, since in water it forms an equilibrium mixture including 62% of β-D-glucopyranose and 38% of α-D-glucopyranose. This exception, the β-anomer of Glu dominating in aqueous Glu solution, might be explained by a competition between anomeric and steric effects, where steric effects would win out in this case favoring the β-anomer (Davis and Fairbanks, 2008). Nevertheless, all sugars do display mutarotation behavior which, among other things, is strongly temperature dependent. However, at equilibrium in aqueous solution of Man, the α-anomer is the major form—an epimer of Glu (Glu and Man differ from each other only in configuration at one chiral center, C-2) (Davis and Fairbanks, 2008).

4. SUGAR ANALYSIS

When acid hydrolysis, acid methanolysis, and enzymatic hydrolysis were compared in terms of depolymerization—with GC (using both HP-1 and HP-5 capillary columns and FID and MSD detectors), HPAEC-PAD, and HPAEC-Borate techniques were compared for subsequent analysis of the released monosaccharides—it could be seen that the acid methanolysis method was generally better than the acid hydrolysis method for the samples containing xylan and uronic acid (Willför et al., 2009). Acid hydrolysis was shown to be a requirement for crystalline polysaccharide depolymerization, but the strong acid conditions evidently led to degradation of labile sugars. Acid methanolysis combined with GC analysis was proven to be a convenient method for obtaining the sugar unit composition and amount of noncrystalline polysaccharides in lignocellulosic biomass (Willför et al., 2009). However, the highest degree of recovery was observed for bleached chemical pulp samples, when the enzymatic and acid hydrolysis methods were used. The plant methanolyzates were not suitable as such for an analysis on an HPAEC–PAD system (Willför et al., 2009). For analysis of the total amount of sugar units (including cellulose, other noncrystalline hemicelluloses, and pectins), a combination of the methanolysis and hydrolysis methods is recommended (Willför et al., 2005a, 2005b, 2009).

When conventional analysis methods cannot be used, for example, due to the toleration limits exceeding contents of salts (e.g., ILs or molten salts) in the samples, the recommendations given above will not help a lot; namely the traditional GC and HPLC columns made for carbohydrate analysis do not tolerate much over 5 ppm of salt. Some reliability problems have been observed (GC) even in the case when trying a quantitative determination of monosaccharides in IL-containing samples (much diluted sample concentrations were used). Alternative analysis methods are under development, and particularly capillary electrophoresis might turn out to be the analysis method of choice in the future.

5. PRODUCTION OF ETHANOL FROM RENEWABLE FEEDSTOCKS

Ethanol from starch- and sugar-based feedstocks makes up the majority of ethanol that is produced in the world today (Mussatto et al., 2010). Important feedstocks used for ethanol production include grains, sugarcane, sugar beet, potatoes, cassava, and grapes or other fruits. When used as a component of automotive fuels, ethanol produced in this way serves as an example of a conventional biofuel (IEA Technology Roadmap, 2011), which is sometimes also referred to as a first-generation biofuel.

Production of ethanol from sugar- and starch-based feedstocks has been known for thousands of years (Mousdale, 2010). When a starch-based feedstock is utilized, the components of the starch are first hydrolyzed to sugars in an enzymatic process based on amylases. Further hydrolysis gives Glu, which is then fermented to ethanol by a microorganism, typically the budding yeast *Saccharomyces cerevisiae*. Glu is converted to pyruvate through the glycolytic pathway, which also generates ATP from ADP and NADH from NAD^+. Pyruvate is further metabolized to carbon dioxide and acetaldehyde, which is reduced to ethanol in a reaction where NAD^+ is regenerated through the oxidation of NADH. A summary of the reaction scheme for the conversion of a hexose monosaccharide, such as Glu, into ethanol is indicated in Eq. (I). The maximum yield, which is 0.51 g ethanol/g consumed Glu, was estimated by the French chemist Joseph Louis Gay-Lussac already in the beginning of the nineteenth century (Barnett and Barnett, 2011):

$$C_6H_{12}O_6 \rightarrow 2CH_3CH_2OH + 2CO_2 \qquad [I]$$

Apart from *S. cerevisiae*, there are many yeasts and bacteria that can be considered for production of bioethanol (Mousdale, 2010; Mussatto et al., 2010). Examples of microorganisms that are studied for production of bioethanol include yeasts from the genera *Saccharomyces*, *Candida*, *Kluyveromyces*, *Pichia*, *Schizosaccharomyces*, and *Zygosaccharomyces*, as well as many varieties of bacteria, for example, from the genera *Zymomonas*, *Clostridium*, *Escherichia*, and *Thermoanaerobacterium*. The capability of the microorganisms to utilize the pentose sugars Xyl and Ara may be relevant in some cases, but not with respect to conventional sugar- and starch-based processes where the di- and polysaccharides from the feedstock become hydrolyzed to hexose monosaccharides, such as Glu and Fru. Evidently, *S. cerevisiae* continues to be important in industry, since it has many valuable properties, including well-developed capabilities to efficiently convert Glu to ethanol and to withstand high ethanol concentrations in the fermentation medium.

It is noteworthy that the production of liquid biofuels from conventional feedstocks, such as grains and sugarcane, typically results in the generation of coproducts (IEA Technology Roadmap, 2011). Coproducts could be, for example, animal feed, biogas, and energy generated from solid residues. Detailed calculations of energy return on investment and greenhouse gas (GHG) balance would always take generation of coproducts into account. A compilation of recent analyses (IEA Technology Roadmap, 2011) indicates a strongly favorable life cycle GHG balance for sugarcane ethanol, cellulosic ethanol, and sugar beet ethanol and suggests that relatively good values may be achieved also for grain ethanol.

6. CELLULOSIC ETHANOL

In biorefineries that operate sulfite-based processes, bioethanol is produced as a coproduct from sugars in spent sulfite liquor (SSL). The sugars in SSL are mainly derived from hemicelluloses, and in processes with softwood as feedstock, Man will be a quantitatively important monosaccharide (Sjöström and Alén, 2010). However, as most mills that are in operation today run other types of processes, often based on kraft (sulfate) pulping or mechanical pulping, bioethanol from SSL is not very important from a quantitative point of view, considering the total ethanol production in the world.

Present-day mills typically have high pulp yield as an aim, that is, the idea is to preserve rather than to degrade cellulose that goes into the process. However, there is a lot of lignocellulose that is not used for making wood products or pulp, for instance residues from forestry, residues from agriculture, such as straw and bagasse, and various energy crops. These resources could tentatively be utilized in biorefinery processes where the aim would be to degrade the cellulose to sugars. The sugars could then be further converted through chemical and enzymatic catalysis or, more conventionally, through a fermentation process where a microorganism would produce an alcohol, an acid, or maybe a biopolymer. Alcohols and acids produced in this way could also serve as platform chemicals in the production of liquid fuels, other chemicals, and bio-based polymers. For instance, ethanol produced in this way could tentatively serve either as an advanced biofuel (IEA Technology Roadmap, 2011) or as a green chemical in the production of polyethene. The valorization of ethanol is discussed further in a subsequent section.

A difference between lignocellulose and conventional sugar- and starch-based feedstocks used in the production of ethanol is that lignocellulose is more recalcitrant to hydrolysis (Lynd et al., 2008). Considering the fact that ethanol is a relatively inexpensive bulk product, the sugar yield of the hydrolysis becomes very important. The cellulose and the hemicellulose can be hydrolyzed in a thermochemical process, in which acid catalysts are added, or in an enzymatic process. Modern concepts often rely on a combination of thermochemical and biochemical steps. To achieve a high yield of sugar, the lignocellulose is typically pretreated thermochemically before cellulolytic enzymes are added (Galbe and Zacchi, 2007).

Existing technology for pretreatment of lignocellulose includes acid-based methods, which may involve the utilization of catalysts such as sulfuric acid and sulfur dioxide, steam explosion and hydrothermolysis, and alkaline methods, such as AFEX (Galbe and Zacchi, 2007). The method and conditions for biomass pretreatment need to be selected on the basis of the recalcitrance of the feedstock. For example, acid hydrolysis serves as an appropriate option for more recalcitrant types of lignocellulose, such as softwood (Galbe and Zacchi, 2007). Pretreatment with acid hydrolysis efficiently degrades the hemicellulose and facilitates enzymatic attack on the cellulose. However, pretreatment with acid hydrolysis will also generate by-products, such as phenolics and other aromatic compounds, furan aldehydes, and aliphatic acids. By-products may in part inhibit subsequent enzymatic hydrolysis and microbial fermentation processes (Geddes et al., 2011). However, new ways to counteract inhibition of both microorganisms and enzymes have emerged (Alriksson et al., 2011; Soudham et al., 2011). Apart from the conventional pretreatment methods discussed above, there are also emerging technologies, such as treatment of lignocellulose with ILs, which is discussed further in a previous section.

After pretreatment and adjustment of pH, complex mixtures of proteins are added to effect the hydrolysis of cellulose. Critical enzymes in this process include two groups of cellulases, namely, exo-acting cellobiohydrolases that attack the ends of cellulose molecules and produce cellobiose, and endoglucanases, which attack internal bonds of cellulose molecules (Kubicek et al., 2009; Viikari et al., 2007). Furthermore, cellobiase (β-glucosidase) is required for hydrolysis of cellobiose to Glu. The role of proteins involved in amorphogenesis of cellulose has also been given attention (Arantes and Saddler, 2010). Proteins associated with glycosyl hydrolase family 61 (GH61) have recently been shown to enhance hydrolysis of cellulose. New studies indicate that GH61 proteins cleave cellulose using a metal-dependent oxidative mechanism (Quinlan et al., 2011; Westereng et al., 2011).

If the sugars that are generated by enzymatic hydrolysis are to be used as substrate in a microbial fermentation process, the procedure can be carried out as a separate hydrolysis and fermentation (SHF). However, it is also possible to combine the enzymatic hydrolysis and the fermentation in a simultaneous saccharification and fermentation (SSF). If the fermenting microorganism can contribute to hydrolysis of cellulose by secretion of cellulolytic enzymes into the medium, it may even be possible to operate the process as a consolidated bioprocess (van Zyl et al., 2007). As the search for new microbial biocatalysts suitable for mixed sugar utilization continues and as further advances in metabolic engineering are made (Geddes et al., 2011; Kim et al., 2012; Nevoigt, 2008), cofermentation of lignocellulose-derived pentose sugars may appear as a feasible option.

7. HISTORICAL AND CURRENT CONSIDERATIONS AROUND CELLULOSIC BIOETHANOL

The era of cellulosic ethanol started in the Nordic countries already more than 150 years ago. Until the 1870s, the mechanically produced pulp and "lump pulp" (cotton pulp produced from worn-out clothes) were the dominant types of pulp produced. The interest toward chemical pulping (sulfite and sulfate) steadily increased during the latter half of the nineteenth century (Persson, 2007). The so-called sulfite method experienced its breakthrough in the 1880s when Carl Daniel Ekman founded the first factory (1874, Bergvik, Hälsningland, Sweden) to produce sulfite pulp (Spaak, 1957). In 1923/1924, the Swedish capacity of sulfite cellulose reached 900,000 tons per annum. At the same time, the interest toward constructing adjacent ethanol factories was dramatically increased and 13 ethanol-producing units were in operation in Norrland and 21 in the whole of Sweden (Persson, 2007). One of the driving forces and primary reasons to establish an ethanol plant adjacent to the pulp factory was that the sulfite process converts only roughly half of the wood to pulp, whereas the other half remains dissolved in the cooking liquor also containing the cooking chemicals (primarily sulfite and alkalihydroxides), thus creating already at that time an unbearable economic and environmental burden being dumped to the closest aquifer.

That fermentation is caused by living organisms was, in essence, simultaneously discovered by three different researchers: the chemists Edward Buchner, Justus Liebig, and Jöns Jacob Berzelius. The first mentioned obtained later (1907) the Nobel Prize in chemistry for the discovery. After many interesting fights, turns and twists around the patent right for "sulfite spirit," already in 1909, the company "AB Ethyl" was founded in Sweden. The company quickly managed to sell licenses to Switzerland and the USA and later, during the First World War, and a descendant to "AB Ethyl," "AB Sulfitsprit" sold 11 licenses to German companies during 1917–1918 (Persson, 2007). The First World War resulted in a shortage of crude oil and gasoline in Germany as in many other countries, and consequently, the demand for bioethanol primarily as a transportation fuel and as a source of chemicals was dramatically increased. Although Germany had already invented the method to prepare synthetic oil from coal and hydrogen (Friedrich Bergius, an assistant of the famous Fritz Haber and a Nobel price winner in 1931 together with Carl Bosch), the process had at the time of the end of the war still not left the laboratory development stage (Persson, 2007).

Consequently, Sweden experienced an "ethanol economy" in a scale far beyond any other nation of those times. In Germany, the importance of sulfite spirit remained rather marginal, but this was not the case in Sweden: by the time the Second World War broke away, "Motyl" 50, 85, and 98 (the number indicating the percentage of ethanol in the fuel mixture) were introduced. At the same time, a few companies started to develop and build ethanol-based chemical industry in order to compensate for the lack of oil as the raw material. In 1940, "MoDo" company started in Domsjö Örnsköldsvik (Norrland region of Sweden) the first chemical plant utilizing sulfite spirit as the feed and the production assortment steadily grew to comprise more than 50 (bio-)ethanol-based chemicals such as, for example, various glycols, acrolein, *n*-hexanol, ethyl chloride, latexes, and thiols. The production steadily increased but 1960s saw the emergence of petrochemical complexes in southern Sweden (Stenungsund) and in 1970 the gradual down sizing of the ethanol-based chemical industry was a fact. Nevertheless, still in the beginning of the 1980s around 4000 tons of sulfite spirit was sold. The combination of serious environmental problems, which previously were associated with sulfite pulping and ethanol-based production of chemicals, and increasing competition from ethylene-based petrochemistry almost forced the industry to shut down. However, from 1984, the company's internal use of ethanol again increased and the following year the "Svensk ethanolkemi" company was founded in the factory premises that in continuation would come to use most of the bioethanol produced on site

(Persson, 2007). Today, the industrial park in "Domsjö" has evolved into a cradle of new type of "biorefining" site containing several companies with a multitude of products (including ethanol-based chemicals and transportation fuels like fuel blend (85% ethanol, 15% gasoline) (E85)) and a strong drive to evolve as a profitable and sustainable producer of necessities for the society of twenty-first century.

8. THERMODYNAMIC ANALYSIS OF ETHANOL PRODUCTION FROM BIOMASS

The thermodynamic analysis of the production of ethanol from lignocellulosic biomass serves as a pragmatic tool for the selection of efficient technologies capable of competing economically with fossil fuels. Improving the various stages of bioethanol production (Fig. 1.7), the process economy, energy efficiency, and the overall performance of the process are important factors for a competitive technology. Since these process steps involve energy conversion, utilization, and generation, thermodynamic tools can be deployed to analyze the efficiency of each step as well as the overall process.

Energy, chemicals, raw materials, and machinery are used during the growth, processing, and manufacture of ethanol from biomass. The total energy consumed from all process chains during ethanol production is evaluated to justify the necessity of the production of second-generation biofuels. During ethanol production from lignocellulosic biomass, cellulose, and hemicellulose obtained from the biomass are converted to the monosaccharides and disaccharides (Glu, Fru, maltose, sucrose) prior to digestion by enzymes. Comparison of the energy input during ethanol production and the energy produced when ethanol is used as a fuel or chemical serves as a tool for the evaluation of the biofuel production efficiency and for analyzing each step in the production process.

The law of conservation of energy states that the total amount of energy in an isolated system remains constant over time. The conservation of energy is expressed by energy balances which, together with the corresponding mass

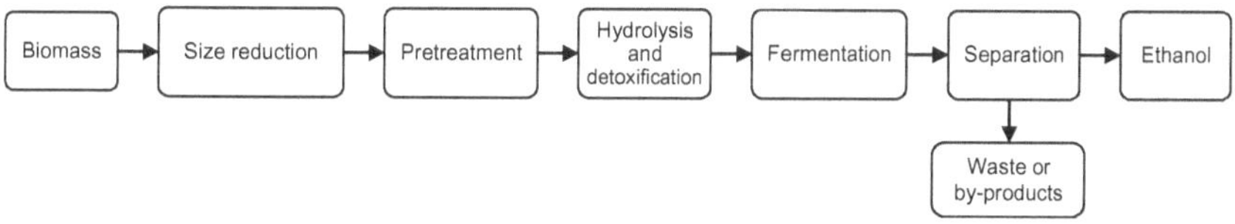

Figure 1.7 A typical process flow diagram for the conversion of lignocellulosic biomass to ethanol.

balances, are used for the modeling and analysis of energy conversion systems. Such energy balances can determine the energy supply requirements in the form of streams of matter, heat, and work. The energy balance identifies the transfer of energy to and from the system. However, energy analysis cannot identify thermodynamic inefficiencies associated with the use of energy supplied to the system (Rosen and Dincer, 2003). An account of the true thermodynamic value of an energy carrier is characterized by its potential to cause change or do meaningful work. Thus, the thermodynamic quantity, exergy, overcomes the limitations of energy analysis and also facilitates process integration.

Exergy is the maximum theoretical useful work possible during a process that brings the system into thermodynamic equilibrium with a heat reservoir (surrounding). It expresses the practical value of a substance and is defined as a maximum ability of a substance to perform work relative to the environment. The environment is a large equilibrium system containing state variables (T, P) and chemical potential of chemical components. In a thermodynamic process, heat and matter are exchanged between the system and the environment. This environment is free from irreversibility and its exergy value is zero. The thermodynamic environment is usually the reference point for calculating exergy values.

The exergy of a substance is a function of the temperature, pressure, and composition of that substance, and the temperature, pressure, and chemical form of the substance when in equilibrium with the environment. It can also be a function of the substance location and velocity. In contrast to the first law of thermodynamics, exergy accounts for the irreversibility of a process due to increase in entropy in accordance with the second law of thermodynamics. The exergy destroyed (anergy) during a process is proportional to the entropy increase of the system and that of the environment. Exergy analysis has been applied in many areas of bioethanol production (Ojeda and Kafarov, 2009; Teh and Lutz, 2010), and in the study of the performance and renewability of feedstocks (Arredondo et al., 2009).

Since energy analysis for the production technologies based only on the first law of thermodynamics (conservation of energy) is unreliable and erroneous, exergy analysis has been applied to a wide range of manufacturing processes, such as biofuel production (Sohel and Jack, 2011; Ojeda and Kafarov, 2009) and fermentation (Teh and Lutz, 2010), for efficient energy management and as a measure of the environmental impacts of manufacturing processes. Exergy analysis identifies the location, quantity and quality of thermal, and material losses in chemical and material processing.

8.1. Energy and exergy balances

The general energy balance can be expressed in terms of energy input and output as follows:

$$\sum E_{in} = \sum E_{out} \qquad [1.1]$$

In the absence of nuclear, magnetic, electrical, and surface tension effects, the total specific exergy of a system consists of four components: the physical, chemical, kinetic, and potential exergy (Hepbasli, 2008):

$$\mathrm{Ex_T} = \mathrm{Ex_{ph}} + \mathrm{Ex_{ch}} + \mathrm{Ex_{kin}} + \mathrm{Ex_{pot}} \qquad [1.2]$$

The general exergy balance can be written as follows:

$$\sum \mathrm{Ex_{in}} - \sum E_{out} = \sum E_{dest} \qquad [1.3]$$

This may also be written in the form:

$$\mathrm{Ex_{heat}} - \mathrm{Ex_{work}} + \mathrm{Ex_{mass,in}} - \mathrm{Ex_{mass,out}} = \mathrm{Ex_{dest}} \qquad [1.4]$$

The terms are defined as follows:

$$\mathrm{Ex_{heat}} = \sum Q\left(1 - \frac{T_0}{T}\right) \qquad [1.5]$$

$$\mathrm{Ex_{work}} = W \qquad [1.6]$$

$$\mathrm{Ex_{mass,\mathit{z}}} = \sum m_z \mathrm{Ex}_z \qquad [1.7]$$

where Q is the heat transfer through the boundary at temperature T, W is the work done, and m_z is the mass transfer (in/out). The specific exergy, Ex, for statically flowing open systems can be defined as follows:

$$\mathrm{Ex} = (H - H_0) - T_0(S - S_0) \qquad [1.8]$$

where the enthalpy and entropy are H and S, respectively, and the subscript zero represents properties at the restricted dead end T_0 and P_0.

For an ideal gas at constant heat capacity, Eq. (1.8) becomes

$$\mathrm{Ex} = C_p(T - T_0) - T_0\left(C_p \ln\frac{T}{T_0} - R\ln\frac{P}{P_0}\right) \qquad [1.9]$$

The exergy arising from chemical composition relative to the environment can be accounted for by defining a reference environment possessing zero exergy. Therefore, an absolute value can be assigned for every material stream within the process. Equation (1.8) can thus be written as

$$\mathrm{Ex} = (H - H_{00}) - T_0(S - S_{00}) \tag{1.10}$$

where 00 refers to the conditions in the reference environment.

For a statically flowing multicomponent material stream, the physical, chemical, and mixing terms can be merged to give the total exergy of the system (Hinderink et al., 1996).

$$\mathrm{Ex_T} = \mathrm{Ex_{ph}} + \mathrm{Ex_{ch}} + \Delta_{\mathrm{mix}}\mathrm{Ex} \tag{1.11}$$

Taking into account the total molar flow rate F, Eq. [1.11] becomes

$$\mathrm{Ex_T} = F\left(\mathrm{Ex_{ph}} + \mathrm{Ex_{ch}} + \Delta_{\mathrm{mix}}\mathrm{Ex}\right) \tag{1.12}$$

where $\mathrm{Ex_{ph}}$, $\mathrm{Ex_{ch}}$, and $\Delta_{\mathrm{mix}}\mathrm{Ex}$ are the physical, chemical, and mixing exergy, respectively.

Each exergy term must be evaluated separately in order to calculate the total exergy correctly.

Physical exergy involves the exergy of the material stream arising from the change in actual (initial) conditions (T, P) to the environment (reference) conditions (T_0, P_0), representing the thermomechanical or physical part of the total energy. Physical exergy is used for analysis of physical processes, in which the substance is considered unchanged.

8.1.1 Physical exergy

The physical exergy of a system can be written in the form:

$$\mathrm{Ex_{ph}} = S(T - T_0) - V(p - p_0) + \sum_i n_i(\mu_i - \mu_{i0}) \tag{1.13}$$

where the entropy S, volume V, and number of moles of i n_i are the extensive parameters. The intensive parameters are the temperature T, pressure p, and the chemical potential of i, μ_i (Ojeda et al., 2011). The subscript 0 describes the state when thermodynamic equilibrium is established with the environment.

The exergy of a flow may be written in the form of enthalpy H, as follows:

$$Ex_{\mathrm{ph}} = (H - H_0) - T_0(S - S_0) + \sum_i \mu_{i0}(n_i - n_{i0}) \tag{1.14}$$

8.1.2 Chemical exergy

Chemical exergy is the maximum theoretical useful work possible during a process that brings the system from environmental conditions (T, P) to the dead state (T_0, P_0, μ_0). The chemical exergy is used for analysis of chemical processes, in which the chemical elements remain unchanged. The chemical exergy corresponds to the substance calorific value. The chemical exergy of biomass is calculated from the correlation of technical fuels using low heating value (LHV) and mass fractions of organic material, sulfur, water, and ash in the biomass (Ptsainski et al., 2007):

$$\mathrm{Ex}_{\mathrm{ch,total}} = m_{\mathrm{org}}\left(\beta_{\mathrm{LHV\,org}}\right) + m_{\mathrm{s}}(\mathrm{Ex}_{\mathrm{s}} - C_{\mathrm{s}}) + m_{\mathrm{water}}\mathrm{Ex}_{\mathrm{water}} + m_{\mathrm{ash}}\mathrm{Ex}_{\mathrm{ash}} \qquad [1.15]$$

where the factor β is the ratio of the chemical exergy to the LHV of the organic fraction of biomass. The ratio is calculated from statistical correlations (Szargut and Styrylska, 1964) as follows:

$$\beta = \frac{1.0412 + 0.216\frac{m_{\mathrm{H_2}}}{m_{\mathrm{C}}} - 0.2499\frac{m_{\mathrm{O_2}}}{m_{\mathrm{C}}}\left(1 + 0.7884\frac{m_{\mathrm{H_2}}}{m_{\mathrm{C}}}\right) + 0.045\frac{m_{\mathrm{N_2}}}{m_{\mathrm{C}}}}{1 - 0.3035\frac{m_{\mathrm{O_2}}}{m_{\mathrm{C}}}} \qquad [1.16]$$

where m_i is the mass fraction of component i.

Equation (1.16) has been applied (Nilsson, 1997) for determining the specific chemical exergy of straw from the following equation:

$$\mathrm{Ex}_{\mathrm{ch,straw}} = \beta_{\mathrm{LHV}}(\mathrm{LHV} + H_{\mathrm{water}}m_{\mathrm{water}}) + \mathrm{Ex}_{\mathrm{ch,water}}m_{\mathrm{water}} \qquad [1.17]$$

Using the following data from moist straw; $m_{\mathrm{c}} = 38.9\%$, $m_{\mathrm{H_2}} = 4.9\%$, $m_{\mathrm{O_2}} = 33.8\%$, $m_{\mathrm{N_2}} = 0.4\%$, $m_{\mathrm{water}} = 18.0\%$, LHV = 13,800 kJ/kg, $H_{\mathrm{water}} = 2440$ kJ/kg, $\mathrm{Ex}_{\mathrm{ch,water}} = 50$ kJ/kg; the specific chemical exergy was calculated to be 16.1 MJ/kg.

Improvements on calculating the specific chemical energy from the elemental compositions based on structurally complicated materials, such as biomass, have been proposed (Shieh and Fan, 1982) as follows:

$$\mathrm{Ex}_{\mathrm{ch}} = 4.1868\left\{\begin{array}{l} 8177.79[\mathrm{C}] + 5.25[\mathrm{N}] + 27892.63[\mathrm{H}] - 3173.66[\mathrm{O}] \\ +0.15[\mathrm{O}](7837.67[\mathrm{C}] + 33888.89[\mathrm{H} - 4236.1[\mathrm{O}]]) \end{array}\right\} \qquad [1.18]$$

The standard chemical exergy of many compounds can be found in the literature (e.g., Ayres and Ayres, 1999). Where unavailable, the chemical exergy of a pure compound can be expressed in terms of the change standard Gibbs free energy, ΔG_{FO}, as follows:

$$\mathrm{Ex}_{\mathrm{ch}} = \Delta G_{\mathrm{FO}} + \sum_i m_i b_i \qquad [1.19]$$

where b_i is the chemical exergy of the ith pure component in the mixture.

The specific molar exergy of a specific component i having a partial pressure P_i is evaluated from the following relation:

$$\mathrm{Ex}_{\mathrm{ch,Ref},i} = RT_0 \ln \frac{P_0}{P_{\mathrm{Ref},i}} \quad [1.20]$$

where $\mathrm{Ex}_{\mathrm{ch,Ref},i}$ denotes the standard chemical exergy of reference species i, and $P_{\mathrm{Ref},i}$ refers to the partial pressure evaluated at a mean atmospheric pressure. This method has been applied to more than 80 varieties of biomass, and the results show that the specific chemical exergy of dry biomass varies between 11 and 24 MJ/kg. The average ratio of the specific chemical exergy to the biomass high heating value is 1.05 (Song et al., 2011a).

8.1.3 *Exergy of mixing*

Determination of mixing terms derived from isothermal and isobaric mixing of pure process components at actual thermomechanical conditions (T, P) is conducted through the concept of "property change of mixing." Exergy equations can be applied using enthalpy and entropy changes of mixing as follows:

$$\Delta_{\mathrm{mix}} E = \Delta_{\mathrm{mix}} H - T_0 \Delta_{\mathrm{mix}} S \quad [1.21]$$

The exergy of material streams, heat, and work flow has been evaluated. This involves determination of the mixing term resulting from isothermal and isobaric mixing of pure process components at actual thermomechanical conditions (T, P). The mixing term is described using the enthalpy and entropy changes of mixing as follows:

$$\Delta_{\mathrm{mix}} M = \Delta_{\mathrm{mix}} H - T_0 \Delta_{\mathrm{mix}} S \quad [1.22]$$

The extend of resource utilization and the thermodynamic characteristics of a system can be evaluated by accounting for the process exergy efficiency τ (Ojeda et al., 2011), given by

$$\tau = \frac{\mathrm{Ex}_{\mathrm{obtained}}}{\mathrm{Ex}_{\mathrm{consumed}}} \quad [1.23]$$

The process exergy efficiency allows for maximum improvements by limiting irreversibility.

8.2. Example of exergy analysis

Consider a simple exergy efficiency of bioethanol (Fig. 1.8):

In the bioethanol process (Fig. 1.8), the biomass introduced into the steam explosion constitutes the major exergy input, while the exergy

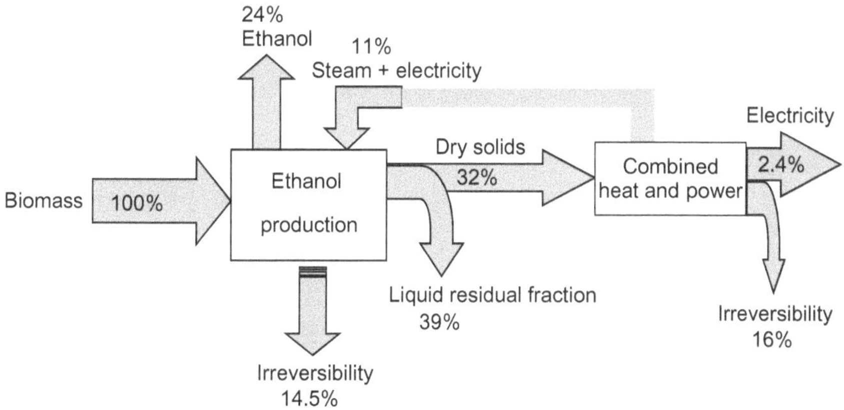

Figure 1.8 Exergy flow diagram for lignocellulose ethanol production. *Adapted from Modarresi et al. (2012).* (For color version of this figure, the reader is referred to the online version of this chapter.)

outputs are bioethanol, the residual liquid fraction, and the dry solids. The irreversibility of bioethanol production accounts for about 14–15% of the input exergy, and can be attributed to heat losses and nonreacting unknown materials. Analyses of the irreversibility during the various stages for bioethanol production have also been carried out. Bioethanol production from acid treated bagasse using different process configurations showed improvements in pretreatment and saccharification—fermentation stages (Ojeda et al., 2011). It was shown that the irreversibility generated during the acid pretreatment stage was 25.2 MJ/kg ethanol, accounting for an efficiency of 53% due to the inhibition effect of Glu and cellobiose which the enzyme velocity at low substrate concentration. Exergy analysis of the enzymatic hydrolysis step for bioethanol production from lignocellulosic biomass (Ojeda and Kafarov, 2009) also revealed that bioreactors have maximum exergetic efficiencies of 68% for the continuous stirred tank reactor (CSTR) and 72% for plug flow reactor. The exergy efficiency increases when the operation temperature increases from 40 to 50 °C which is the operative range for cellulose enzymes. The examination of the thermodynamics of fermentation and the growth of baker's yeast (Teh and Lutz, 2010) for the production of bioethanol from sugarcane bagasse showed that anaerobic respiration of yeast retains about 93% of the maximum work that could be extracted from the growth medium of a chemostat reactor. The glucose-to-ethanol conversion efficiency was 70% due to the formation of yeast cells and other metabolic by-products. When evaluating different kinds of biomass-to-biofuels processes, various exergy environmental indicators such as "renewability performance indicator" (Arredondo et al., 2009) can be defined. For examples, the reader is referred to this chapter and references therein.

9. BIOFILM REACTORS FOR BIOETHANOL PRODUCTION

Biochemical reactors (Fig. 1.9) are important in chemical processing since they affect the rate of reactions, the product yields and selectivity, and the overall process economy. A simple reactor design equipped with technique to increase the cell mass concentration is required as a biofilm reactor in order to increase the "reaction rate" (productivity) under optimized conditions, such as temperature, pH, and substrate concentration (Qureshi et al., 2005). The cell mass concentration inside the reactor increases when an attached permeable membrane (permits flow of liquid, substrate, and products) is used to retain cells, the cells are immobilized (fixing cells on supports), or they are recycled (Vega et al., 1988). The readers should realize that these kinds of reactor setups hardly operate under the conditions of

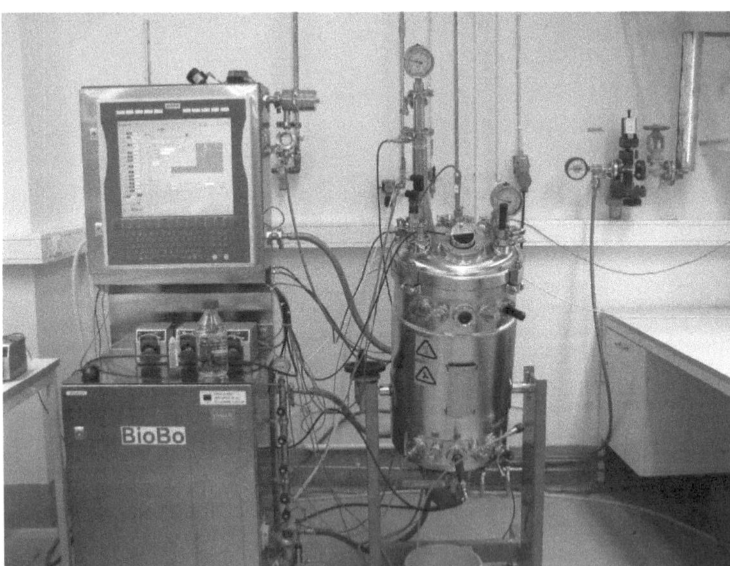

Figure 1.9 Pilot scale bioreactor at Processum Biorefinery Initiative AB, Örnsköldsvik, Sweden. *Image courtesy of Dr. Björn Alriksson.* (For color version of this figure, the reader is referred to the online version of this chapter.)

intrinsic kinetics and, therefore, the design of the reactor type influences the apparent, observed production rates.

The natural form of cell immobilization occurs via absorption or adherence of cells on the support materials resulting in cell layers or biofilm. A biofilm is defined as a structural community of bacterial cells enclosed in a self-produced polymeric matrix and adherent to a living or inert surface (Costerton et al., 1999). Biofilm formation and growth is influenced by surface roughness of the support, support porosity, hydrophobicity or flagellation of cells, the presence of nutrients, favorable temperature, etc. The lower shear forces on rough surfaces or in porous media and the presence of phosphorous provide an environment conducive for cell adherence and growth. The biofilm is used in different biological systems including wastewater treatment, gas and odor treatment, and chemicals synthesis, for example, ethanol.

The mature biofilm can be used in different processes in various reactors, such as CSTR, fluidized bed reactors (FBRs), packed bed reactors (PBR), airlift reactors, upflow anaerobic sludge reactors, expanded granular sludge reactor, and plastic composite support reactors. Reaction rates in these reactors are usually high, with achievable cell concentrations of 74 g/L (Qureshi et al., 2005). The high cell concentration leads to high activity of the bioparticles which internally contribute to the reactor productivity.

During ethanol production, the biomass previously pretreated and hydrolyzed is fed to the bioreactor containing the biofilm. Although the composition of lignocellulose materials varies in different plants, the main components (cellulose, hemicelluloses, and lignin) are fermented to sugar monomers which are subsequently transformed chemically and/or enzymatically to ethanol. High ethanol productivities have been obtained by using *Zymomonas mobilis* in an attached film expanded bed reactor using vermiculite as bed material to form an active biofilm. Productivities of 105 g/L/h were obtained at a dilution rate of 3.6 h^{-1}.

Biofilm reactor can be generally be categorized into fixed bed and expanded bed reactors. In fixed bed, the biofilm develops in a static layer, while in the expanded bed, the biofilm moves continuously driven by high air or velocity flow, or by mechanical stirring. Ethanol fermentation in different biofilm reactors has been conducted in batch and continuous operation using different microorganisms, substrates, and support materials. The low cell concentration and the inhibitory effect of ethanol have fostered the development of systems to increase ethanol productivity from lignocellulosic biomass. Ethanol productivities of 1–2 and 30 g/L/ h obtained in batch and continuous stirred tank fermentors, respectively, have been increased to about 80 g/L/h via the combined effect of vacuum fermentor and cell recycling (Cysewski and Wilke, 1977; Margaritis and Wilke, 1978). The reactors commonly used to biofilm growth and biomass processing to ethanol as described below:

9.1. Packed bed reactor

In PBR, suitable support materials are filled into tubes or cylinders followed by inoculation with the culture to form biofilm. The time for biofilm maturity depends on the culture, support, and nutrients availability. Contact between the biofilm and the substrate is typically improved when PBR is used. The biofilm is fixed and does not move with the liquid. Continuous PBR is the most widely used reactor for immobilized enzymes and immobilized microbial cells. In these systems, it is necessary to consider the pressure drop across the packed column and the residence time of the substrate in the bed. PBR is simple to construct and operate; however, these reactors are often blocked due to excessive cell growth, packing density of support, and excess nutrients. The substrate flow regimes in packed columns include downward flow, upward flow, and the recycling methods.

The recycling method is advantageous when the linear velocity of the substrate solution affects the reaction flow rate. This is because the recycling method allows the substrate solution to be passed through the column at a desired velocity. In industrial applications, upward flow is generally preferred over downward flow due to compression of enzymes or microorganisms in the latter case. When gasses are produced during enzymatic reactions, upward flow is preferred. A continuous PBR is easy to control and operate, and the operating conditions are easily stabilized and the product quality can be verified easily. The operation duration has been from 2 weeks to 3 months.

Continuous packed bed biofilm reactor filled with *S. cerevisiae* immobilized via adsorption on natural support (sugarcane bagasse) was used to produce ethanol from molasses obtained from sugarcane processing (Tyagi and Ghost, 1982). A productivity of 29 g/L/h was obtained in comparison to ca. 3 g/L/h obtained in a free cell continuous process. The reactor operated continuously for 35 days without reactor plugging.

Ethanol production in an attached film extended packed bed biofilm reactor of *Z. mobilis* using vermiculite as bed material to form an active biofilm resulted in productivities of 105 g/L/h obtained at a dilution rate of 3.6 h^{-1} (Bland et al., 1982). Absorption of *Z. mobilis* on ion exchange resins embedded in packed beds has also been used for the production of ethanol (Krug and Daugulis, 1983). Immobilized cells of *Z. mobilis* on cationic macroreticular resin formed an active biofilm and resulted in productivity of 136 g/L/h (based on the void volume). Reactor plugging occurred due to excessive cell growth after 200 h of operation.

9.2. Continuous stirred tank reactor

In a CSTR, the vessel is filled with fibrous bed support used for the absorption of cells and stirred uniformly with a mechanical device, ensuring that no variation in concentration gradients exists within the vessel. These reactors are also known as agitating or rotating continuous reactors depending on the operational mode. There may be excess growth on the surface of the fibrous bed, and the cells may be sheared off the support due to high shear forces during agitation.

In this reactor type, nutrient and substrate flow into the reactor, while there is an equal outflow of product, nutrient and microbial waste products, and microbial cells flow out. The balance between feed and discharge is maintained for long enough times to achieve steady-state operations without changes in the reactor conditions. The CSTR biofilm offers advantages of observing, separating, and evaluating the kinetics and stoichiometry of each biofilm process, the uniform bulk liquid phase, convenient steady-state reproducible conditions, and a constant shear stress of the fibrous bed.

A culture of *Z. mobilis* immobilized on plastic composite support consisting of polypropylene has been used in a CSTR and resulted in high ethanol productivity of 536 g/L/h at a dilution rate of 15 h^{-1}(Kunduru and Pometto, 1996). Conversely, *S. cerevisiae* immobilized on a plastic composite support resulted in ethanol productivity of 76 g/L/h at a dilution rate of 3 h^{-1}. An important state-of-the-art development of continuous process is to use multistage (typically four) reactors of variable sizes to maximize performance, thereby increasing the productivity.

Fermentation using steam exploded oak chips was performed in a continuous stirred tank biofilm reactor equipped with an internal membrane filtration module for cell retention. Ethanol productivity of 17 g/L/h was obtained at a dilution of 0.22 h^{-1}. No bacterial contamination occurred after 210 h with the hydrolysate sterilized at 60 °C (Lee et al., 2000).

9.3. Fluidized bed reactors

Cell growth occurs around the absorbent particles in FBR. Active biofilm formation around the particles and accumulation of sufficient biomass in the reactor take between 2 and 4 weeks and do not plug the reactor due to an excessive growth of cells. Thermophilic ethanol fermentation of wet-exploded wheat straw hydrolysate has been investigated in an FBR at 70 °C. Undetoxified wheat straw hydrolysate used (3–12% dry matter) contained sugar mixtures of Glu and Xyl ranging from 12 to 41 g/L. The anaerobic bacterium *Thermoanaerobacter* exhibited significant resistance to high levels of acetic acid (up to 10 g/L), and other metabolic inhibitors present in the hydrolysate. Although the hydrolyzate was not detoxified, ethanol yield in a range of 0.39–0.42 g/g was obtained. Overall, sugar efficiency to ethanol was 68–76%. The reactor was operated continuously for approximately 143 days, and no contamination was seen without the use of any agent for preventing bacterial infections. The tested microorganism has a considerable potential to be a novel candidate for lignocellulose bioconversion into ethanol (Georgieva et al., 2008).

Another FBR for continuous ethanol production using *S. cerevisiae* naturally immobilized on coke particles resulted in productivity of 100 g/L/h with a feed dilution rate of 2.25 h^{-1} (Demsey, 1990).

10. KINETIC ANALYSIS OF BIOETHANOL PRODUCTION

The production of ethanol from lignocellulosic biomass has emerged as an alternative method for the production of transportation fuels based on the dwindling petroleum reserves. Ethanol may be obtained via the processing of sugar-containing substrates, such as agricultural residues and food processing wastes. However, the cost of the substrate must be low in order to have a favorable impact on the economics of the ethanol production. In order to efficiently transform the sugars to ethanol, the kinetics of the reactions involved during cell growth, fermentation, inhibition, and ethanol recovery are required for optimal ethanol production. During fermentation, ethanol accumulation in the broth inhibits metabolic activities of the enzymes, thereby inhibiting cell viability, sugar consumption, and ethanol productivity. In addition to ethanol inhibition, substrate limitation and inhibition, and cell death must be described in the kinetic models. With fluctuating temperatures during fermentation, these effects may become even more severe (Phisalaphong et al., 2006). The specific growth rate in a microbial culture is usually expressed as a function of a growth-limiting substrate concentration and described by Monod-type kinetics; however, in the presence of ethanol, the specific growth rate decreases due to the inhibiting effect of ethanol and the mode of operation.

The operating modes commonly used for bioethanol production from lignocellulosic materials among others include SHF and SSF. The efficiency of operating modes for ethanol production can be evaluated using two indicators: productivity defined by the mass of ethanol produced per unit mass of dried feedstock per unit time (g/g/h) and yield defined by the mass of ethanol produced per unit mass of dried feedstock (g/g). Due to shorter operating time and less inhibition of enzymes, SSF generally has the higher productivity than SHF, but the yield depends on other factors, such as the substrate used (Shen and Agblevor, 2010).

Macroscopic approach of evaluation of the pathways and bioenergetics can be adopted under anaerobic conditions in batch fermentation processes in order to probe the product yields and develop models capable of describing process dynamics. This is applicable since batch cultures are characterized as a time dependent and irreversible process, the inhibition and deactivation of the cellular activities by ethanol. It is therefore necessary to consider that growth and ethanol production in batch cultures are influenced by ethanol concentration, deactivation of enzymes and time of culture, the mode and conditions of operation, etc., in order to determine the actual kinetics of the processes involved.

10.1. Kinetic models for bioethanol production

The quantitative analysis of complicated systems can be achieved by mathematical modeling, capable of accurately predicting reaction parameters or functions. Kinetic formulations are often used for the kinetic analyses of biochemical systems from experimental data. Such mathematical models are capable of predicting the influence of operating parameters on cell concentration, substrate utilization rate, and ethanol production rate. The model should be capable of describing substrate limitation, substrate inhibition, product inhibition, and cell death simultaneously during anaerobic fermentation to yield ethanol.

The growth kinetics of microorganisms depends on optimal maintenance of the transport of the necessary nutrients in the medium to the cell surface, rate of mass transfer from the medium into the cells and temperature, and pH. The microbial growth kinetics can be described mathematically (Ghaly and El-Taweel, 1997) as follows:

$$\frac{dX}{dt} = \mu X \tag{1.24}$$

where X is the cell concentration (g/L) and μ is the specific growth rate (h^{-1}).

The relation between specific growth rate and limiting substrate concentration S is formally described by the classical Michaelis–Menten equation, actually the Monod equation:

$$\mu = \frac{\mu_m S}{K_s + S} \tag{1.25}$$

where the saturation constant K_s is defined as the substrate concentration at half the maximum specific cell growth rate.

Increasing alcohol concentration will gradually retard microbial growth and may finally completely inhibit it. The symbol K_p denotes the concentration of ethanol at which the cell growth will be inhibited. An ethanol inhibition term P_{inh} can be defined as follows:

$$P_{inh} = \frac{K_p}{K_p + P} \tag{1.26}$$

where K_p is the ethanol inhibition concentration and P is the ethanol concentration.

A high substrate concentration inhibits cell growth due to the effect of osmotic pressure. The symbol K'_s denotes the substrate concentration at which growth is inhibited. A substrate inhibition term S_{inh} can be defined as follows:

$$S_{inh} = \frac{K'_s}{K'_s + S} \tag{1.27}$$

By combining Eqs. (1.25)–(1.27), the specific growth rate μ can be described as a function of substrate limitation, product inhibition, and substrate inhibition as follows:

$$\mu = \frac{\mu_m S}{K_s + S} \frac{K_p}{K_p + P} \frac{K'_s}{K'_s + S} \tag{1.28}$$

The Monod (for cell growth) or Michaelis–Menten (for enzymatic kinetics) rate law is often used to describe changes in biochemical systems assuming quasi-steady state. Log–linear approximations and the complete dynamic nonlinear mass balances of metabolic networks based on enzyme activities are also used, with the reference intracellular and extracellular metabolites levels (Hatzimanikatis et al., 1996). A reference elasticity coefficient is, nevertheless, necessary in order to yield approximate coefficients. The biochemical systems theory has been used as a pragmatic tool for pathway identification tasks based on time series data. Nonlinear ordinary differential equations are applied to represent the dynamics of biological systems in the following form:

$$\frac{dX_i}{dt} = \sum_{j=1}^{r} n_{ij} v_j, \quad i = 1, \ldots, n \qquad [1.29]$$

where X_i is the composition of the *i*th component, n_{ij} are stoichiometric coefficients, and v_j is the reaction rate of the *j*th pathway.

The generalized mass action and the *S* system are important variables within the biological system theory, where each rate is formulated as a power law function.

10.2. Model development

10.2.1 Batch systems

The development of a batch system for ethanol production is based on the enzymatic reaction mechanism of cellulose to Glu as follows:

$$\mathrm{Ce} \xrightarrow{r_1} \mathrm{Cb} \xrightarrow{r_2} \mathrm{G} \qquad [1.30]$$

where Ce, Cb, and G denotes the concentrations (g/L) of cellulose, cellobiose, and Glu, respectively; r_1 and r_2 are enzymatic catalytic reactions describing the reaction rates (g/L/h). The reaction rates describing the decomposition of cellulose to cellobiose and to Glu may be expressed as follows:

$$r_1 = \frac{k'_1 \mathrm{Ce}}{1 + \frac{\mathrm{G}}{K_{1\mathrm{G}}} + \frac{\mathrm{Cb}}{K_{1\mathrm{Cb}}}} \qquad [1.31]$$

$$r_2 = \frac{k_2 \mathrm{Cb}}{1 + \frac{\mathrm{G}}{K_{2\mathrm{G}}}} \qquad [1.32]$$

where k'_1 is the enzyme deactivation factor (h^{-1}); k_2 is the specific rate constant of cellobiose hydrolysis to Glu; K_{1G}, K_{2G}, and K_{1Cb} are inhibitory constants for Glu and cellobiose by their respective specific enzymes (Drissen et al., 2009; Shen and Agblevor, 2010).

The effect of enzyme deactivation can be included in the variable k'_1 and expressed as a product of the specific rate constant k and enzyme concentration e as follows:

$$k'_1 = k_1 e \qquad [1.33]$$

Substituting (1.27) into (1.25) gives

$$r_1 = \frac{k_1 e\mathrm{Ce}}{1 + \frac{\mathrm{G}}{K_{1\mathrm{G}}} + \frac{\mathrm{Cb}}{K_{1\mathrm{Cb}}}} \qquad [1.34]$$

The cell growth and transformation of glucose to ethanol can be expressed in the form:

$$G \xrightarrow{r_X, r_G, r_E} \mathrm{EtOH} + \mathrm{CO_2} + \mathrm{cell} \qquad [1.35]$$

Assuming Monod model given by

$$r_X = \mu X = \frac{\mu_m XG}{K_G + G} \qquad [1.36]$$

According to Eq. (1.36), cell growth increases exponentially when the growth-limiting substrate is sufficiently supplied. When there is limited supply of nutrients, the enzymatic reaction is slowed and the cell growth decreases. In this case, the rate of cell growth is dependent on the cell concentration.

10.2.2 Continuous stirred tank reactor

In the CSTR, the substrate is fed continuously into the culture while the outflow containing ethanol, residual cellulose, Glu, cells, by-products, and enzymes exit at the same flow rate F. Two reactors are required for prehydrolysis (reactor 1) and the simultaneous scarification and fermentation process (reactor 2) in order to achieve semi-SSF in a continuous mode (Fig. 1.10).

The change in cellobiose concentration has been proposed to be described as follows (Shen and Agblevor, 2010):

$$\frac{dB}{dt} = \frac{r_1}{0.947} - r_2 + D(B_1 - B_2) + \frac{D(C_1 - C_2)}{0.947} \qquad [1.37]$$

The dilution rate (h^{-1}) is given by

$$D = \frac{F}{V_2} \qquad [1.38]$$

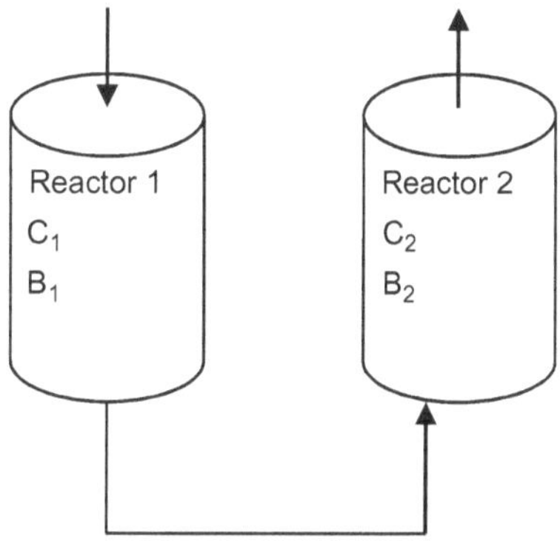

Figure 1.10 Reactors for continuous operation during semi-simultaneous saccharification and fermentation. (For color version of this figure, the reader is referred to the online version of this chapter.)

where F is the feed flow rate (L/h), V_2 is the volume of reactor 2 (L), and C and B are the cellulose and cellobiose concentrations in the fermentors (g/dm^3).

The mass flows for a continuous ethanol fermentation system can be monitored to determine the cell, substrate, and ethanol concentrations. A cell mass balance for the continuous ethanol fermentation process can be written as

$$[\text{Cells change rate}] = [\text{cell input rate}] + [\text{cell growth rate}] - [\text{cell death rate}] - [\text{cell output rate}]$$

This can be written mathematically as

$$\frac{dX}{dt}V = QX_i + \mu XV - K_d XV - QX_o \qquad [1.39]$$

where X_i and X_o are the inlet and outlet cell concentrations and Q is the flow rate (L/h).

The hydraulic retention time R can be defined as

$$R = \frac{V}{Q} \qquad [1.40]$$

Substituting yields

$$\frac{dX}{dt} = \frac{X\mu_m S}{K_s + S}\frac{K_p}{K_p + P}\frac{K'_s}{K'_s + S} + \frac{X_i}{R} - \frac{X_o}{R} - K_d S \qquad [1.41]$$

In fermentation systems, the inlet cell value X_i is usually zero and in a perfectly mixed system, $X = X_o$. Thus, Eq. (1.41) can be rewritten as

$$\frac{dX}{dt} = X\left[\frac{\mu_m S}{K_s + S}\frac{K_p}{K_p + P}\frac{K'_s}{K'_s + S} - \frac{1}{R} - K_d\right] \qquad [1.42]$$

Under steady-state conditions, $dX/dt = 0$ and the equation becomes

$$\frac{1}{R} = \frac{\mu_m S}{K_s + S}\frac{K_p}{K_p + P}\frac{K'_s}{K'_s + S} + -K_d \qquad [1.43]$$

A substrate mass balance for the continuous fermentation process can be written as

$$-\frac{dS}{dt}V = QS_i - \left(R_{sx} + R_{sm} + R_{sp}\right)V - QS_o \qquad [1.44]$$

where S and S_o are the inlet and outlet substrate concentrations, respectively, and Q is the flow rate.

The rates of substrate uptake for cell growth R_{sx}, cell maintenance R_{sm}, and product formation R_{sp} are defined as

$$R_{sx} = \frac{R_x}{Y_{s/x}} \qquad [1.45]$$

$$R_{sm} = m_s X \qquad [1.46]$$

$$R_{sp} = \frac{R_p}{Y_{p/s}} \qquad [1.47]$$

where $R_x = dX/dt$ is the cell growth rate, $Y_{x/s}$ is the growth yield coefficient (g cells/g substrate), $Y_{p/s}$ is the product yield coefficient (g product formed/g substrate), m_s is the maintenance coefficient (g substrate/g cells/h), and $R_p = dP/dt$ is the rate of product formation.

Substituting R for V/Q and using

$$-\frac{dS}{dt} = -\left[\frac{R_x}{Y_{s/x}} + m_s X + \frac{R_p}{Y_{p/s}}\right] + \frac{S_i - S_o}{R} \qquad [1.48]$$

Under steady-state conditions, $dP/dt = 0$, therefore

$$S_i - S_o = R\left[\frac{R_x}{Y_{s/x}} + m_s X + \frac{R_p}{Y_{p/s}}\right] \qquad [1.49]$$

For a given concentration S_i, the amount of substrate removed from a system $S_i - S_o$ is a function of the hydraulic retention time R, microbial cell concentration X, and the product formation R_p.

The product formation can also be described as a function of the energy coefficient for growth α as follows:

$$\frac{dP}{dt} = \alpha X \left[\frac{\mu_m S}{K_s + S} \frac{K_p}{K_p + P} \frac{K'_s}{K'_s + S} - \frac{1}{R} - K_d \right] - \frac{P_0}{R} \quad [1.50]$$

under steady state, $dP/dt = 0$ and $P_0 = P$, therefore

$$P = \alpha X R \left[\frac{\mu_m S}{K_s + S} \frac{K_p}{K_p + P} \frac{K'_s}{K'_s + S} - \frac{1}{R} - K_d \right] \quad [1.51]$$

The SSF model was apparently able to describe the (four) main and (three) by-products formed in the system. The dynamics of both continuous and fed-batch SSF operations in ethanol fermentation could be described amazingly well (experimental points vs. simulated curves). However, the authors (Shen and Agblevor, 2010) do not report the confidence intervals, relative errors for the parameters obtained and, moreover, no information is given about the degree of explanation, the mathematical methods applied, or details of the objective function. Importantly, on the basis of simulations, it was concluded that there exists an optimal dilution rate for the maximal ethanol productivity during SSF continuous process. Thus, reaction engineering still has a lot to give in rational treatment of biological systems.

11. BIOMASS-TO-LIQUID ETHANOL PRODUCTION FROM SYNTHESIS GAS

Besides biological fermentation process using sugars in biomass, ethanol can also be produced from biomass using a chemical biomass-to-liquid (BTL) processes, utilizing the "SynGas" platform. Any BTL process has the advantage of the fact that SynGas can be obtained from various biomass feedstocks by gasification and/or reforming. Thus, even lignocelluloses can be converted by the BTL process. In comparison, the fermentation process is limited in the selected biomass components (carbohydrates) for ethanol production, since lignin in the lignocellulosic biomass cannot be converted by the current fermentation process (Liu et al., 2011).

Regarding ethanol production from SynGas, the research has concentrated on comparing different catalysts for the ethanol synthesis. On the other hand, some apparatus and synthesis methods have also been patented (Stites et al., 2011) to provide a selective method for producing ethanol from SynGas and being able to achieve high yields of ethanol in the introduced processes. At the same time, they report in their previous patent a straightforward route through SynGas from cellulosic biomass via methanol which is, consequently, converted to acetic acid or acetates, thereafter being reduced to ethanol with yields of over 100 gallons per dry ton of biomass (Stites et al., 2010). Also Hurley et al. (2010) have patented a catalytic process which comprises a series of catalysts improving the selectivity and yield toward ethanol synthesis from SynGas (Hurley et al., 2010). The system is also suitable for synthesis of other alcohols. In the system, three catalysts are contacted in the reactor with SynGas: the first and third catalysts are hydrogen promoters (Cu–Zn, Mo, or Fe with an optional alkali metal additive and an optional support of aluminum oxide, zeolite, silica, or clay) and the second catalyst is a homologation promoter (containing one or more of the group VIII metals in free or combined form with cocatalyst metals consisting of lanthanide, actinide, or yttrium series metals with optional support and additives; Hurley et al., 2010). The system has the advantage of producing ethanol directly from SynGas (coproducing also methanol, acetaldehyde, acetic acid, and methane at the same time) and with a carbon selectivity toward ethanol around 69 mol%, at best compared to Stites invention comprising several process steps. The three catalysts can be applied either in a single reaction vessel or in separate reaction vessels that are connected in a series (Hurley et al., 2010). In the ethanol synthesis example provided by the authors, the selectivity toward ethanol reached 64% (mol/mol).

In the synthesis of higher alcohols from SynGas, four kinds of catalysts are typically employed: Rh-based, Mo-based, modified Fisher–Tropsch (F–T) synthesis, and modified methanol synthesis catalysts (Liu et al., 2011). Rh-based catalysts have shown the best selectivity toward ethanol among various catalysts, but due to the high cost of Rh metal, it is necessary to improve the overall activity and selectivity toward ethanol in order to achieve a commercially viable and cost-effective process. Furthermore, upon development of highly active Rh-based catalysts for the synthesis of ethanol from SynGas, the support and the promoter are very important. Selected rare earth oxides have been found as efficient promoters for Rh/SiO_2 catalysts in the synthesis of ethanol. It is also known that in the CeO_2-supported metal catalysts, there is a strong interaction between support and metal which results in a high activity in various reactions. Moreover, if Zr^{4+} ions are introduced into the CeO_2 lattices, the catalytic performance of CeO_2-supported metal catalysts is enhanced. The catalytic performance of $Rh/Ce_{1-x}Zr_xO_2$ (where x is 0–1 and Rh loading 2 wt%) catalysts was investigated for the synthesis of ethanol from SynGas in a high-pressure fixed-bed flow reactor (reaction conditions of 548 K, 2.4 MPa, the ratio of H_2/CO being 2:1, and W/F of 10 g h/mol).

Overall, the catalyst characterization results indicated that $Rh/Ce_{0.8}Zr_{0.2}O_2$ contained both acid and base sites on the surface. The main oxygenated products obtained were acetaldehyde and ethanol in case of neutral or acidic supports, whereas methanol and ethanol dominated in the case of basic supports. $Rh/Ce_{0.8}Zr_{0.2}O_2$ demonstrated the highest selectivity for ethanol among various catalysts (Liu et al., 2011).

Another study (Ngo et al., 2011) was performed over Rh catalyst in the synthesis of ethanol from SynGas, focusing on studying Li and Mn as secondary additives. In this study, Rh was used as a catalyst, TiO_2 as a support, and Fe as a promoter. Both Li and Mn (in Rh–Fe/TiO_2) slightly decreased the catalytic activity but improved the ethanol selectivity. However, at high reaction temperatures, both Mn and Li additions in Rh–Fe/TiO_2 gave high ethanol yields and selectivities due to the increased catalytic activity at high reaction temperatures (Ngo et al., 2010).

For the reference, the petrochemical route for ethanol production has been in use since the beginning of the twentieth century. The reaction is based on ethylene that is hydrated to ethanol in the so-called indirect or direct processes. The indirect process is based on the use of liquid sulfuric acid, and additionally, it was used as an initial way to produce ethanol. However, it was soon replaced by the direct process which takes advantage of solid acid catalysts. The direct hydration process has been relatively unchanged since 1940s (Nakagawa et al., 2000). The direct synthesis process has also been carried out over supported phosphoric acid catalysts but also acidic zeolites and zeolite type structures offer a modern approach (Atkins et al., 1983; Ishihara et al., 1997; Iwamoto et al., 1986). The drawback is that the supported phosphoric acid catalyst may leach and, thus, several corrosive effects can be observed in a reactor system. Moreover, the reaction produces certain by-products such as diethyl ether and side reactions, for example, oligomerization and dehydrogenation can be an issue (Nakagawa et al., 2000). Zeolites are also known to deactivate by, for example, dealumination under reaction conditions. Additionally, the process generally displays low yields and requires high pressures. The reactor technology used in the direct process is based on the continuous operation mode. The traditional fixed-bed reactor technology offers a general view on the process (Atkins et al., 1983). A more advanced approach is based on the combination of a reaction and separation (Ishihara et al., 1997). The technology in question utilizes special, porous zeolite tubes, or membranes which have an effect on the equilibrium, shifting the reaction toward products in the hydration of ethylene to ethanol (Fig. 1.11). The outcome is demonstrated by higher yields of ethanol obtained.

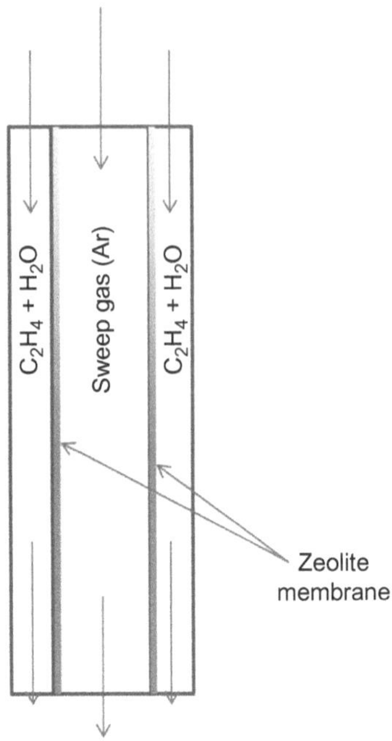

Figure 1.11 The concept of simultaneous reaction and separation. (For color version of this figure, the reader is referred to the online version of this chapter.)

12. BIOETHANOL VALORIZATION OVER INORGANIC HETEROGENEOUS CATALYSTS: CLASSICAL LIQUID AND GASEOUS PRODUCTS

Ethanol itself may contain some limitations as, for example, a gasoline replacement, and thus, scientific community is discussing the possibility of converting ethanol into more valuable chemicals. One interesting way to carry out these promising ideas is to take advantage of modern inorganic catalysis. The processes are based on the utilization of chemical reactors which convert ethanol with the aid of inorganic catalysts into products. Still, these processes are typically heterogeneously catalyzed ones which means that the reactant and the catalyst reside in different phases (Centi and van Santen, 2007; Kirk-Othmer, 2007).

Further, ethanol can be directly upgraded, without any addition of other reactants, to some interesting and simple liquid products like acetaldehyde, acetic acid, butyl alcohol, diethoxy ethane, ethyl acetate, and diethyl ether. Furthermore, some well-known gaseous products are hydrogen, methane, ethene, carbon dioxide, and carbon monoxide. The outcome of the process is depending on the choice of the catalyst and process conditions. Typically, a mixture of these compounds is obtained. Ethanol can also react with other compounds like ethylene oxide (glycol ether), amines (ethylamines), acrylic acid (ethyl acrylate), and isobutylene (ethyl-*tert*-butyl ether) (Centi and van Santen, 2007; Kirk-Othmer, 2007). Each one of these chemicals has used in some special purposes which renders the production important. In this chapter, only main reactions of ethanol without additional reactants will be discussed.

One of the most interesting topics related to ethanol is its use as a fuel additive or as a fuel itself. Interestingly, even in 1908, Henry Ford presented his famous Model T which was originally designed to run with ethanol. However, due to various reasons ethanol was then forgotten. Nevertheless, the real breakthrough occurred in the 1970s when the oil crises emerged. Nowadays, 85% ethanol containing fuel (E85) is available particularly in European markets, but a spark-ignited engine which is especially designed to run on ethanol is needed. Furthermore, it is claimed that a remarkable reduction in exhaust and GHG emissions is obtained. The fact that 15% mix with classical gasoline is needed stems from the fact that ethanol has lower vapor pressure and, consequently, problems, for example, upon engine cold-start can be regarded negligible. Ethanol cannot form satisfactory air/fuel vapor mixtures below 11 °C. Regardless, E85 is of course a marginal product due to supply and demand reasons but the car manufacturers have developed a new strategy which implements so-called flexi-fuel vehicles. These cars can be operated with any mix ranging from 100% gasoline up to 85% ratio of ethanol fuel blends. Specific precautions are needed because ethanol is corrosive for some metallic compounds (zinc, brass, aluminum) and nonmetallic compounds (natural rubber, urethane, cork gasket material, certain plastics) (Minteer, 2006). However, E85 is in wider use and has been a fully functional fuel for many years. The biggest problem with high-blend ethanol fuels is the fact that it cannot be applied directly in (nonmodified) engines which are currently in use. Also, the catalytic converters of today's cars are designed to tackle with gasoline as fuel, not ethanol. As a result, the emissions of toxic acetaldehyde are dramatically increased (Poulopoulos et al., 2001).

Another way to promote the use of ethanol in modern engines is the addition in much smaller amounts to traditional gasoline and diesel. Even legislation in many countries is supporting this strategy that is feasible in many ways. First, it supports the idea that the fuel is applicable in all engines currently in use, and second, a decrease in emissions is attained which originates from additional oxygen. Many studies have indicated that the addition of oxygen-containing compounds (e.g., ethanol) to traditional fossil-based fuels has a diminishing effect on carbon monoxide and hydrocarbon exhaust emissions (Minteer, 2006). The outcome is, consequently, cleaner combustion. Interestingly, the addition of ethanol may also have some negative consequences. Investigations have revealed that an increase in the ethanol concentration in gasoline is followed by an increase in aldehyde and formaldehyde emissions (Minteer, 2006). These chemicals are known to contribute to adverse health effects upon exposure. In addition, well-known oxygen additives in gasoline are methyl- or ethyl-*tert*-butyl ether (MTBE, ETBE) (Centi and van Santen, 2007; Minteer, 2006). However, particularly MTBE is toxic and can easily contaminate the groundwater. Thus, recent political decisions are shifting the balance toward ethanol as a better additive. Many studies confirm that ethanol can be blended with gasoline from 5% to 25% without causing any bigger problems in modern engines. Ethanol itself is not a perfect fuel additive because it possesses some negative physicochemical properties if compared to traditional gasoline. For instance, its octane number is much higher, less lubrication is obtained, and its energy content is lower (Minteer, 2006). Thus, it would be interesting to find some new ethanol derivatives which could be applied in the future.

An interesting alternative use for ethanol is to use it as ED95 in diesel engines or to blend it with diesel, which is widely used in heavy traffic and, particularly in Europe, in passenger cars. Different studies have suggested that an appropriate blending with ethanol varies between 10% and 15%, at maximum. Ethanol addition to diesel was also found to cause similar lower carbon monoxide and particulate matter emissions. The big problem with diesel engines is that ethanol blending easily lowers the viscosity, resulting in leakage in the fuel injection pump. In addition, ethanol blending stability with diesel is not at comparable levels as with gasoline. The main factors affecting blending properties are temperature and water content. Interestingly, below 10 °C these two mixtures start to separate and extra precaution is needed to countereffect these types of phenomena. It is well known that some emulsifiers and cosolvents can be added to these blends lowering the separation temperature. For instance, ethyl acetate which is one of the ethanol derivatives discussed in this chapter has been found to prevent separation and is thus one potential candidate in the future. Ethanol itself has a high octane number, but its cetane number, which is an important characterisation measure in diesel fuel, is much lower and thus limits its use as a diesel additive (Hansen et al., 2005; Minteer, 2006). Based on these observations, it would be beneficial to find some new ethanol derivatives.

12.1. Acetaldehyde

Acetaldehyde is a well-known organic chemical, and it can be found in animal and plant metabolism. Interestingly, acetaldehyde is an intermediate in ethanol fermentation process and it is presented in plant juices and tobacco smoke (Kirk-Othmer, 2007). It is well miscible in common organic solvents like water, acetone, and ethanol. In addition, it exhibits typical reactivity of aldehydes and it is mainly used as an intermediate to other valuable chemicals. For instance, acetic acid, ethyl acetate, and butyl alcohol are produced through aldehyde intermediate, but nowadays, its significance is decreasing due to the cost of ethylene feedstock and the emergence of other processes. For example, butanol can be manufactured with the oxo-process and acetic acid by applying the Monsanto process (Kirk-Othmer, 2007). However, bio-based ethanol might be a future choice for acetaldehyde and, thus, some of these valuable chemicals could be straight derived from it. As a consequence, the dependence on fossil-based resources could be diminished.

There are several possible sources available for acetaldehyde production. The most common one since 1960s has been ethylene which is oxidized in the liquid-phase process. The process is named Wacker–Hoechst process, and it was developed in the late 1950s. The reaction is based on the use of aqueous $PdCl_2$ and $CuCl_2$ catalysts (Moiseev, 2011). In addition, acetylene, lower hydrocarbons, SynGas, and methanol can be used as the raw material (Kirk-Othmer, 2007).

However, keeping in mind the bio-based needs of the future, one should take a deeper look at the bioethanol-based acetaldehyde production. Interestingly, there are many publications available which discuss the oxidation of ethanol to acetaldehyde (Batis and Najjar, 2010; Christensen et al., 2007; Hensen and Guan, 2009; Herrera et al., 2007; Idriss and Seebauer, 2000). A common feature to all these publications is that the reactions have been carried out over supported metal oxides. For instance, Christensen et al. (2007) reported ethanol oxidation via acetaldehyde intermediate up to acetic acid which is a result of further oxidation. However, their catalyst of choice was gold supported on $MgAl_2O_4$ and TiO_2 which are known to exhibit basic character. Hensen and Guan (2009) reported some results about the effect of gold particle size on the reaction. Basic nature of metal oxides is known to have a positive effect on acetaldehyde formation, whereas acidic properties normally lead to ethene and diethyl ether type products. Moreover, alumina (Avgouropoulos et al., 2006) and vanadium oxide catalysts (Herrera et al., 2007) have also shown activity to acetaldehyde formation after alkali metal addition as a result of basicity. Furthermore, perovskite-type oxide structures, which are known to contain special electric and magnetic properties, were investigated by Batis and Najjar (2010). One important aspect in future biorefinery operations will be an aqueous ethanol solution which is an outcome of fermentation and, thus, its reactivity is of great importance. From industrial viewpoint, it would be beneficial to be able to avoid the costly distillation process (Christensen et al., 2007).

A question can arise, what is actually happening on a catalyst surface? Idriss (2004), Idriss and Seebauer (2000), and Idriss et al. (2004) carried out fundamental studies of ethanol adsorption and its reactions over metal oxide surfaces. Interestingly, some common conclusions can be drawn from these results. Ethanol has been found to form an ethoxide on the catalyst surface, and furthermore, on CeO_2 surfaces, mono and bidentate ethoxides have been observed. In addition, some studies include density functional theory (DFT) calculations about the interaction of the ethoxy group with different surfaces (Dumesic et al., 2003; Illas et al., 2010; Radilla et al., 2010). It seems that the mechanism is based on O–H bond breaking upon which the surface oxygen atom forms a hydroxyl group with hydrogen which is cleaved from the adsorbed ethanol molecule. The formed ethoxide molecule binds to a metal cation on a catalyst surface which can further react to acetaldehyde through additional dehydrogenation. Interestingly, a study by Idriss (2004) demonstrates that the first dehydrogenation process is reversible, whereas the second stage is irreversible. Additionally, it is easy to conclude that the amount of ethoxides on the catalyst surface could explain the catalytic activity. However, the real factor that determines the success of a catalyst is coordinative unsaturation (Idriss, 2004). Polycrystalline surfaces are known to contain coordinatively unsaturated metal cations which are capable of reducing and reoxidizing and, thus, are applicable in dehydrogenation reactions which lead to, for example, acetaldehyde formation. Overall, we can conclude that for acetaldehyde formation, there are two crucial things: first, the catalyst should contain a crystalline structure and, second, it should be of alkaline nature. Consequently, an amorphous–crystalline structure is not suitable for this type of reaction. Hattori (1995) has reported that alcohols are typically dehydrogenated to corresponding aldehydes over basic catalysts. As previously mentioned, acetaldehyde serves as a base intermediate toward valuable products. Thus, many reactions that lead to desirable products are actually base-catalyzed reactions. Based on the information gathered, a profound view on further products would be valuable.

Ethanol oxidation is not only limited to acetaldehyde formation, but it can also be oxidized to acetic acid. One advantage of this type of processes is that acetic acid can be synthesized directly from bioethanol over a single catalyst and, thus, the production process would become much more simpler and economically feasible.

12.2. Acetic acid

Acetic acid is a common organic molecule which contains a carboxylic acid group. It has been known for over 5000 years, and its traditional name is vinegar which means its aqueous solution. It can be found everywhere including seawater and plant and animal fluids. In addition, acetic acid has a crucial role in many biological processes which are related to energy (Kirk-Othmer, 2007). Acetic acid is also soluble in common solvents like water, ethanol, and ether. Interestingly, it tends to form equilibrium between a monomer and a dimer in the gas phase, whereas in the liquid phase, equilibrium is found between a monomer and a cyclic dimer or dehydrated dimer. Acetic acid serves also as a pathway to more valuable products. For instance, its reaction with ethylene and oxygen leads to the formation of vinyl acetate which is used in textile treatments, paper coatings, and adhesives. Additionally, another important pathway is the production of acetic anhydride which is applied in the synthesis of important cellulose acetate (Kirk-Othmer, 2007).

Looking back in history, there has been an interesting development of different catalytic processes which led to the formation of acetic acid. The most common processes have been methanol carbonylation and acetaldehyde oxidation (Kirk-Othmer, 2007). Additionally, of some importance has been the oxidation of hydrocarbons and direct synthesis of acetic acid through ethylene and ethane. The carbonylation process was for the first time presented by "Badische Anilin and Soda-Fabrik," later to be restructured as the company BASF, in 1913 and it was based on a reaction between carbon monoxide and methanol. At that time, the process was carried out at high temperature and pressure. Later on, some transition metal carbonyls were found to be efficient catalysts for the reaction and cobalt iodide was chosen as the catalyst. The main idea in the development was to take advantage of nonpetroleum feedstocks. Monsanto presented a significant improvement in the process by developing a rhodium iodide promoted catalyst in the 1960s (Kirk-Othmer, 2007). The real advantage of the new process was that the reaction could be carried out at atmospheric pressure and, thus, a significant reduction in operational costs was achieved. The latest modification, named the Cativa process, was introduced in the 1990s, taking advantage of iridium metal (Kirk-Othmer, 2007). The acetaldehyde process is the second major process for acetic acid production. It is simply based on acetaldehyde oxidation, but the real reaction mechanism contains multiple steps which include free radicals. The active catalysts for the reaction have been copper, manganese, and phosphomolybdic acid. There has also been some interest in fossil-based resources which can be used in the production of acetic acid. For instance, oxidation of aliphatic hydrocarbons (*n*-butane and light naphtha) and oxidation of ethylene have been presented for the process. Overall, it would be interesting to find an alternative way of production from renewable sources.

One candidate might be bioethanol that can be oxidized through acetaldehyde to acetic acid. It is commonly agreed that acetaldehyde forms an acetate in the catalyst surface which can be further oxidized to carbon dioxide (Idriss et al., 2004). However, by proper choice of catalysts and reaction conditions, the outcome can be acetate and, thus, acetic acid. Avgouropoulos et al. (2006) have investigated ethanol reactivity on Pt/Al_2O_3 catalyst and have confirmed an advantage upon use of unmodified catalyst in the acetic acid production, whereas the addition of alkalis (Na, K) inhibits it. The explanation may be the fact that alkali metals increase the basic character of the catalyst and, thus, shift ethanol decomposition pathway. In addition, some reports (Christensen et al., 2007; Takei et al., 2011) discuss the conversion of aqueous ethanol solution to acetic acid. Interestingly, it was reported that in the liquid phase, the only possibility to carry out the reaction is the use of supported gold catalysts. This serves as an extremely interesting example of supported gold which can offer promising possibilities when present in nanoscale. In addition, the same studies also reveal that, at moderate ethanol concentrations ($<60\%$), the main product is acetic acid, whereas at higher concentrations, a shift toward ethyl acetate occurs via esterification of acetic acid. It is quite obvious that water thus inhibits strongly acetaldehyde formation and favors acetic acid. Overall, acetic acid synthesis is strictly in relation with the acetaldehyde production, and by performing some process and catalyst modifications, the production of acetic acid through bioethanol-based pathway could become feasible. It seems that supported heterogeneous metal oxide catalysts exhibit the best performance in the production of acetic acid from ethanol (acetaldehyde as an intermediate). By remembering the future needs, the advantage of straight catalytic valorization of aqueous fermented ethanol solution is important.

12.3. 1-Butanol

1-Butanol is one of the most interesting potential ethanol derivatives. For instance, it can be used as a high-blend component in gasoline in the near future. Butyl alcohols are generally soluble in all common organic solvents but not to water, the only exception being *tert*-butanol. Butyl alcohols can be found in some natural compounds, and they have typical reactivity of alcohols. For example, butanols can undergo different dehydration, oxidation, alkylation, and esterification reactions (Kirk-Othmer, 2007).

1-Butanol is a promising candidate for replacing gasoline because it exhibits similar properties. In comparison, ethanol is more corrosive, completely water soluble, and contains less energy than 1-butanol (Szwaja and Naber, 2010). However, 1-butanol is also used in many other applications like surface coatings, viscosity regulators, and plastics. Additionally, its derivatives are of considerable value including *n*-butyl acrylate and methacrylate which have been the largest volume 1-butanol derivatives in the USA. In addition, other isomers of butanol have many similar properties and in certain applications isobutanol can replace 1-butanol (Kirk-Othmer, 2007).

The first report on industrial scale production of 1-butanol already arised in 1912. During that time, the process was based on fermentation of carbohydrates with a bacterium (*Clostridium acetobutylicum*). However, from the modern point of view, the traditional way for 1-butanol production can comprise three process alternatives. The first one was in use until mid-1950s, and it was based on the hydrogenation of crotonaldehyde. This manufacturing method, the so-called aldol condensation reaction, combines two aldehyde molecules. The reaction occurs over basic catalysts, and studies have demonstrated the advantage of copper-based catalysts. Later on, the petrochemical feedstocks started to play a crucial role in the production. In 1942, Reppe presented a synthesis method where propylene, carbon monoxide, and water reacted in the presence of iron carbonyl salt catalysts, leading to the formation of 1-butanol (Kirk-Othmer, 2007). However, the Reppe process has never been taken into use due to the expensive process technology. In the mid-1950s, a new modification for the Reppe process was presented. The so-called oxo-synthesis also applies propylene which is hydroformulated with hydrogen and carbon monoxide in the presence of Co, Rh, or Ru hydrocarbonyl- or substituted hydrocarbonyl catalysts (Kirk-Othmer, 2007). Formed *n*-butyraldehyde is subsequently hydrogenated to *n*-butanol over heterogeneous catalysts. Naturally, there are many variations in the process which are related to the pressure, temperature, and catalytic systems. The process has been the main source for 1-butanol since mid-1950s, but the main disadvantage for this type of feedstock is its fossil-based origin (Kirk-Othmer, 2007). Interestingly, as we are shifting toward bio-based feedstock, the old acetaldehyde method might offer a new opportunity because it can be manufactured from bioethanol. Thus, it is worth mentioning that these types of commercial scale production routes are highly dependent on the price of the feedstock.

As we discussed earlier, ethanol tends to form acetaldehyde over certain types of compounds, for example, metal oxides. Additionally, reactivity is seldom limited to that stage and it can easily proceed further up to acetates, carbon monoxide, and carbon dioxide. However, when optimal conditions are fulfilled, two acetaldehyde molecules can combine by forming a four-carbon atom containing molecule, called crotonaldehyde (Idriss et al., 2004). The process is designated as aldol condensation, and it typically forms C_4 products like 1-butanol, 2-butenol, butanal, 1,3-butadiene, and 1-butene, depending on the catalyst used and the reaction conditions. Moreover, crotonaldehyde is known to undergo dehydration and hydrogen transfer reactions which have an effect on the final outcome (Diaz et al., 2011). The aldol condensation reaction is also present in the Guerbet chemistry. It is a well-known reaction type and typically applied in alcohol dimerization reactions, normally producing β-branched alcohols (Ishii et al., 2006; Ueda et al., 1990). Overall, the aldol condensation can occur over many metal oxides as reported by Ji et al. (1997). A question is whether the reaction is base or acid catalyzed or both. The literature indicates that base-catalyzed reaction pathway is preferred (Ji et al., 1997; Kelly and Jackson, 2003). However, depending on the type of reacting compounds, the acid sites may be beneficial. It is also known that for certain type of aldehydes and ketones, the base strength should be at an optimal level (Kelly and Jackson, 2003).

The challenge is to determine the key factors that lead to a high yield in 1-butanol production. From the mechanistic point of view, ethyl alcohol condensation reaction to 1-butanol through heterogeneous catalysis is the most challenging part of its valorization chemistry. Tsuchida et al. (2006, 2008) reported 1-butanol synthesis over hydroxyapatite catalysts and found that, at certain Ca/P ratio, the hydroxyapatite contains both acidic and basic sites. This bifunctional behavior was suggested to be the reason for successful 1-butanol production. Interestingly, similar results have been obtained over metal supported alumina (Jiang et al., 2004), Cu–Mg–Al mixed oxides (Tanchoux et al., 2009), and Mg–Al mixed oxides derived from hydrotalcites (Diaz et al., 2011). Thus, it seems quite clear that ethanol condensation for 1-butanol proceeds through bifunctional oxide-based catalytic materials which favor the aldol condensation. However, other studies have shown that the reaction can also proceed over pure basic MgO (Coville et al., 2003) and alkali cation-doped zeolite (Yang and Meng, 1993) catalysts. Consequently, now it seems that ethanol valorization to 1-butanol does not proceed through aldol condensation. The actual mechanism could be based on direct dimerization of two ethanol molecules. However, the amount of available studies is so minor that the real mechanism remains still open. Furthermore, studies by Di Cosimo et al. (1998, 2000) on Mg–Al mixed oxide catalysts suggested that ethanol condensation for 1-butanol requires not only Lewis acid–strong Brønsted base pair sites but also a high density of basic sites. This is an interesting observation, and it supports the idea of bifunctionality and the aldol condensation. Overall, the reaction seems to need at least basic sites in a metal oxide catalyst and the catalyst should be of ionic nature (Busca, 2010).

12.4. Diethyl ether

Ethers consist of a wide variety of compounds with similar ether linkages (C–O–C). There are few well-known structures which have useful properties as gasoline additives. For instance, MTBE, *tert*-amyl methyl ether (TAME), and ETBE are typical oxygenate additives. Moreover, ethers are recognized as good solvents and their inert behavior is an advantage in organic synthesis. Ethers are typically highly miscible to a large category of organic solvents but display a minimal solubility in water. They can also form so-called oxonium compounds with Lewis acids and protons and, thus, are applied as Grignard reagents and in Friedel-Crafts alkylation and acylation (Kirk-Othmer, 2007). Alkyl ethers have also been used in extractions, paint removers, resins, waxes, and lubricating oils (Kirk-Othmer, 2007).

From the process technical point of view, manufacturing methods for ethers are quite simple. First option has been based on catalytic dehydration of alcohols in the liquid or gas phase over acidic catalysts. The second approach is known as the Williamson synthesis and is mainly targeting the production of aliphatic ethers. It is based on the reaction of alkali metal alkoxides and alkyl halides. Additionally, an indirect addition of an alcohol to an olefin is also possible to carry out (Kirk-Othmer, 2007).

One of the most interesting ethanol derivatives is diethyl ether which might have an important role in the future. Diethyl ether can be mixed with ethanol and be used, for example, as a gelation agent for nitrocellulose (Kirk-Othmer, 2007), and further, it might become an important diesel additive (Dogu et al., 2009) due to its beneficial physicochemical properties. For instance, ethyl ether cetane number is almost twice the level of the traditional fossil-based diesel (45–55); its heating value is 34,000 kJ/kg which is enough for engine use and its octane number is also very high. Based on these numbers, diethyl ether should be considered as a potential fuel substituent candidate. Ethyl ethers production has mainly been based on fossil resources: ethylene is hydrated to ethanol over a supported phosphoric acid catalyst (Kirk-Othmer, 2007). Actually, diethyl ether is a minor by-product in this process, but its demand is so low that the need has been satisfied. Ethyl ether can also be produced through dehydration of ethanol solution which was discussed earlier. Additionally, the process can be homogenously or heterogeneously catalyzed. The traditional way has been to use liquid sulfuric acid as a catalyst which has been mixed with ethanol but, in the future, heterogeneous catalysis will have an increasing role (Kirk-Othmer, 2007). From the modern point of view, the most common way to produce diethyl ether is ethanol dehydration over solid acid catalysts taking advantage of continuous operating reactor technology. The process has been studied, for example, over acidic nanocomposite WO_x–silicate catalysts (Dogu et al., 2009), wide category of zeolite structures (Magnoux et al., 2009; Takahara et al., 2005), and transition metal oxides (Zaki, 2005). Additionally, solid acidic resins are also known to be effective in the reaction (Olah et al., 1997). One interesting issue in the process is that the formation of ethylene is connected with ethyl ether. In conclusion, these studies revealed that even minor modifications to the catalysts or process conditions may have a huge impact on the outcome. Overall, the parameters that favor ether over ethylene are lower temperature, lower contact time, weaker acid sites, and higher ethanol partial pressure (Magnoux et al., 2009). The discussion related to the reaction mechanism is also rather concerted. Some studies suggest that, at first, diethyl ether is formed, whereas ethylene formation proceeds through ether decomposition (Dogu et al., 2009; Magnoux et al., 2009; Takahara et al., 2005). However, Zaki (2005) suggested that decrease in surface acidity is followed by a weaker ethanol adsorption and, thus, C_2H_5–O bond becomes stronger. This could favor the so-called intermolecular reaction which leads to the formation of diethyl ether.

12.5. Diethoxy ethane

1,1-Diethoxy ethane, also known as diethyl acetal, is an important chemical which has been used as a raw material for pharmaceuticals and as an intermediate in the production of various industrial chemicals. Moreover, it is used in perfume industry where its function is mainly related to oxidation resistance and, thus, to an increase in the self-life of a perfume. It can also be used as a raw material for polyacetal resin production, as a diesel component, and even as a flavoring agent in spirit drinks (Rodrigues and Silva, 2001; Nord and Haupt, 2005; Silva and Rodrigues 2005; Tanchoux et al., 2009). One of the most modern aspects which utilize diethyl acetal is its use as a diesel additive (Centi and van Santen, 2007). Some studies have indicated certain benefits associated with its use, such as a decrease in particle number, particle mass, and to some extent in NO_x emissions. The only real negative effect was found to be a significant increase in acetaldehyde emission.

Actually, diethoxy ethane might offer a huge amount of possible applications today and in the future. However, its production is surprisingly little reported in the open literature (Abello et al., 2001; Rodrigues and Silva, 2001; Silva and Rodrigues, 2005; Tanchoux et al., 2009). Diethyl acetal is formed through acetalization reaction between one aldehyde molecule and two ethanol molecules. Two ways for the production of diethoxy ethane have been proposed: one route is based on straight upgrading of ethanol (Abello et al., 2001; Tanchoux et al., 2009), whereas the other way is to blend together aldehyde and ethanol before the reaction (Rodrigues and Silva, 2001; Silva and Rodrigues, 2005). Additionally, both directions of synthesis take advantage of solid acid catalysts as zeolites (Abello et al., 2001), ion exchange resins (Rodrigues and Silva, 2001; Silva and Rodrigues, 2005), and Cu–Mg–Al mixed oxides (Tanchoux et al., 2009). Traditionally, the reaction can be catalyzed by homogenous liquid acids but the environmental and recycling problems do not favor this approach. Interestingly, Abello et al. (2001) have suggested that the Brønsted acidity is the key factor which determines the catalytic activity toward diethoxy ethane, with higher acidity levels being more favorable. On the other hand, Tanchoux et al. (2009) suggest that aqueous ethanol solution enhances diethyl acetal formation which relies on an assumption that the strong Lewis base sites (O^{2-}) are converted to the weaker Brønsted base sites (OH^{-}). However, in this research, the catalyst is a bifunctional oxide catalyst containing both acidic and basic properties and it is not directly comparable to pure acid catalyst. Based on the minor amount of available experimental data, it is not feasible to draw any reliable conclusions about the catalyst acidity and its correlation to the catalytic activity. From industrial and economical viewpoint, 1,1-diethoxy ethane manufacturing methods should be based on simple ethanol upgrading over heterogeneous catalysts.

12.6. Ethyl acetate

Ethyl acetate, also known as ethyl ethanoate, is an important and widely used chemical in industry. It has an important role as a solvent in adhesives, paints, or coatings, and it can be used to substitute aromatic solvents such as toluene (Appel et al., 2011; Nguyen et al., 2010). Additionally, ethyl ethanoate exhibits similar physicochemical properties as some of previously discussed chemicals which may show potential as diesel additives (Ashok, 2010). In this study, a compression-ignition (CI) diesel engine was run with a blend containing ethanol and a conventional diesel. Ethyl acetate was added to the mixture and its role was to act as a surfactant. The results indicate that a clear reduction in environmentally hazardous emissions is observed such as particulate matter, smoke density, hydrocarbons, and carbon monoxide.

In fact, there are several ethyl acetate production technology alternatives. Traditionally, ethyl acetate has been manufactured from fossil-based resources but in terms of sustainability, recent research efforts have focused in the investigation of a possibility of a simple and economical one-pot synthesis strategy, taking advantage of bioethanol. The main fossil-based process comprises the esterification reaction between ethanol and acetic acid (Nguyen et al., 2010). In this process, ethanol is produced by hydration of ethylene and acetic acid by methanol carbonylation reaction. The process is known to be acid catalyzed and liquid sulfuric acid is a commonly used in the reaction. As mentioned earlier, liquid acids are not anymore suitable due to tightening environmental requirements and difficult handling problems. As a result, some researches have demonstrated the process over solid acid catalysts such as zeolites (Chen and Wu, 2004), aluminum sulfates

(Gang and Wenhui, 2010), and heteropolyacid-supported montmorillonites (Bokade and Gurav, 2010). Additionally, some other ways to produce ethyl ethanoate comprise the so-called Tishchenko reaction between two acetaldehyde molecules, liquid-phase oxidation of *n*-butane, copolymerization of butyraldehyde and polyvinyl acetate, and esterification of acetic acid and ethylene (Chen and Wu, 2004).

The simple ethanol and aqueous ethanol-based process to ethyl acetate is the most important approach today. It is also worth mentioning that the possibility to carry out the reaction without any additional reagents and nonfossil-based feedstocks is a clear benefit. Based on the research activity in this field, several studies demonstrate the process quite comprehensively. It is generally known that ethanol valorization to ethyl ethanoate can proceed via two pathways which are known as the dehydrogenative and the oxidative process (Appel et al., 2011). The difference between these two options is related to the use or absence of oxygen in the reaction. Additionally, the oxidative route is rarely reported in the scientific literature but the main advantage is the straightforward and, thus, an economically feasible acetate purification process. On the other hand, the ethanol-to-air ratio should be kept at low levels due to the explosive nature of the mixture. The oxidative route proceeds over PdO supported on silica (Appel et al., 2009), Pd supported on γ-alumina and styrene–divinylbenzene copolymer (Lin et al., 1999) as well as Au supported on $MgAl_2O_4$ and TiO_2 (Christensen et al., 2007). In general, supported metal oxides seem to be the most suitable catalysts for the reaction, while an interesting anomaly is presented by a process involving graphite nanofibers (Weinstein et al., 2011). The dehydrogenative route has gathered more interest, and even a commercial application has been presented by Davy Process Technology (Colley and Tuck, 2012). Moreover, all open literature publications (Appel et al., 2009, 2011; de la Piscina et al., 2005; Inui et al., 2004; Wang et al., 2010; Waugh et al., 2005) related to the dehydrogenative route take advantage of metal oxide catalysts (ZrO_2, CeO_2, Al_2O_3, SiO_2, Cr_2O_3, ZnO, WO_3–ZrO_2). Interestingly, some reports (Appel et al., 2010, 2011; Inui et al., 2002, 2004) have demonstrated the advantage of Cu–Zn–Zr–Al–O types of mixed metal oxides in ethyl acetate synthesis. In short, it can be concluded that basic metal oxide is an advantage. A study reported by Wang et al. (2010) proposes that copper is an important constituent in a catalyst since it can create additional Lewis acidity in the form of Cu^+ while being an activating agent for acetaldehyde and, thus, toward ethyl acetate.

When viewing the literature available, there are three possible explanations for the reaction mechanism which can occur in a one-pot system. The first is based on upgrading of ethanol over an acidic catalyst which oxidizes ethanol through acetaldehyde to acetic acid. Afterward, acetic acid reacts with another ethanol molecule by forming an ester bond (Lin et al., 1999). The second choice serves a new and more complicated strategy which takes advantage of base catalysis. In the beginning, ethanol is oxidized to acetaldehyde which then undergoes aldolization reaction with another ethanol molecule by forming a hemiacetal, followed by a dehydrogenation to ethyl ethanoate. This approach has been strongly suggested as the most reliable view (Appel et al., 2010, 2011; Bolder, 2008; de la Piscina et al., 2005; Inui et al., 2002, 2004; Waugh et al., 2005). In addition, oxides with strong basic sites and high densities are suggested to be favorable for the reaction mechanism in question (Appel et al., 2010, 2011). The third choice is related to Tishchenko reaction which is known to occur between aldehyde molecules. The reaction is known to be base catalyzed (Hattori and Seki, 2003), and additionally, it is known to compete with the aldol condensation reaction (Hattori, 1995) that leads to 1-butanol and other C_4 products. In the case of ethanol, the reaction proceeds via oxidation of two ethanol molecules to corresponding acetaldehydes which further react together by oxidation and reduction. It is known that the reaction requires hydrogen transfer properties from the oxide support (Idriss and Seebauer, 2000). Overall, the ethyl acetate production mechanism seems to be sensitive to the catalyst structure and it is interesting to note how close the acetaldehyde-related pathways are to each other. For instance, it is important to consider the effect of a crystalline structure and, thus, how it affects on the product distribution, especially in an industrial scale. In addition, even a minor increase in catalyst basicity may have an impact whether the product is, for example, ethyl acetate or 1-butanol.

12.7. Ethylene

Ethene (ethylene) entered the stage as a large-volume commodity in the 1940s when oil and chemical companies decided to collect it from different refinery compounds. It is the most basic and widely applied petrochemical compound in the world. Before that time, acetylene was an important feedstock in different chemical syntheses. Nowadays, ethene production is mainly based on thermal cracking of different hydrocarbon feeds that are of fossil-based origin. Several alternatives can be applied as the raw materials in the process. Some common compounds such as ethane, propane, butane, naphthas, and gas oil can be applied (Kirk-Othmer, 2007). Each raw material contains some special properties that have an effect on the final product distribution and the type of furnace needed. Overall, ethene is a reactive molecule and it is mainly used for the production of other petrochemical compounds. Ethylene can undergo a wide variety of reactions including polymerization, oxidation, and addition. For example, polyethylene is a widely known polymer which is found in numerous applications (Kirk-Othmer, 2007).

Interestingly, other feedstocks have risen as alternatives to the formation of ethylene. It is of utmost importance to consider renewable sources as a possible raw material for future large-volume chemical production. Suitable candidates could include feeds such as SynGas, methanol, and even ethanol. Lignocellulosic second-generation bioethanol would offer a tremendous opportunity for the process in question. As already discussed, the synthesis of ethylene is closely related to that of diethyl ether. However, it is important to note that the cost of a feedstock is of crucial importance on refinery operations and, thus, the manufacturing expenses of bio-derived ethanol remain a big question in the industrial scale.

Heterogeneous catalytic dehydration of ethanol to ethylene has been widely studied (Datta and Phillips, 1997; Rodriguez-Mirasol et al., 2011; Takahara et al., 2007; Trakarnpruk and Arenamnart, 2006; Wu et al., 2011; Xiao et al., 2009). It is well known that the catalysts need to contain acidic properties and most studies have been conducted over zeolite type structures such as ZSM-5 which are similar to those used in the synthesis of diethyl ether. Additionally, the most modern methods take advantage of metal oxide-coated zeolites (Xiao et al., 2009), SAPO-11 molecular sieves (Wu et al., 2011), and carbon-based catalysts (Rodriguez-Mirasol et al., 2011). These all are known to exhibit acidic properties. In terms of selectivity, it would be interesting to determine the parameters that have an effect on the product distribution between ethyl ether and ethylene. Some common physicochemical properties that favor ethylene formation include higher reaction temperature, higher acidity, increased contact time, and lower ethanol partial pressure. As both compounds can be produced over similar catalyst and the selectivity can be adjusted by altering the process conditions, it might be possible to combine the syntheses of both ethylene and diethyl ether in the future.

As mentioned above, ethylene is known to be the primary chemical for the production of other derivatives. Oxidation of ethylene yields ethylene oxide as a product which has a wide range of applications including sterilizing agent, ethylene glycols, polyesters, surfactants, and poly(ethylene oxide) (PEO) polymers. The early production method was called the chlorohydrin process, but due to pollution problems, the method was later on replaced by the direct oxidation process (Ullman, 2003). The direct oxidation process is known to proceed efficiently over metallic silver, but virtually, all manufacturing is carried out over supported silver catalysts, nowadays. Additionally, the reaction can proceed with air or pure oxygen, while the latter is the basis for modern methods (Kirk-Othmer, 2007). Surprisingly, recent reports described a one-pot approach for the production of ethylene oxide from ethanol (Lippits and Nieuwenhuys, 2010; Nieuwenhuys and Lippits, 2010). It was shown that alumina as a support is selective toward the desired product when coated with silver, copper, and gold nanoparticles. In general, these metals are known to exhibit oxidative behavior. Furthermore, Li_2O was found to have a suppressive effect on the formation of ethylene and diethyl ether, apparently as a result of neutralization of the acidic sites. The one-pot approach naturally requires a supply of oxygen to the reaction stream. Thus, bioethanol truly serves as an alternative raw material to produce a wide variety of vital base chemicals to the modern society.

12.8. Hydrogen

Long-term goal should be a transformation toward greener process alternating. One obvious possibility is presented by hydrogen which can be applied, for example, in fuel cells creating electric energy (Minteer, 2006). This type of setup could replace the traditional combustion engine and lead to reduced exhaust gas emissions. Still, further questions arise considering hydrogen storage, source, equipment, price, and safety. From technical point of view, it is important to evaluate usefulness of different hydrogen sources and the conversion technology. Nowadays, hydrogen can be easily obtained from different fossil-based refinery operations. However, bio-based resources such as ethanol might offer a huge potential in the future. The actual hydrogen production is based on different chemical reactions such as steam reforming, water gas shift, and partial oxidation. Additionally, even simple water electrolysis is available but the cost is currently higher in comparison to use of fossil-based sources (Kirk-Othmer, 2007).

Bioethanol steam reforming reaction (sometimes connected with the water gas shift reaction) is studied in detail in the literature due to its importance (Epron et al., 2011; Erdohelyi et al., 2010; Leong et al., 2008; Leung et al., 2007; Minteer, 2006; Shen et al., 2008). Interestingly, different catalytic systems and catalyst compositions have a significant role in the final outcome of the process. Contrary to steam reforming, also aqueous-phase reforming can be used to produce hydrogen, for example, from polyols. The presence of oxygen atoms bonded to each carbon atom renders the C–C bonds weaker; consequently, this allows for easier splitting of molecules to hydrogen and CO that via subsequent water gas shift reacts to CO_2 and an equimolar amount of hydrogen will be produced. Interestingly, mixtures of polyols with ethanol resulted enhanced yields of hydrogen (Tokarev et al., 2010).

In general, ethanol steam reforming was usually occurring over metal supported metal oxide catalysts. It is known that certain metals (e.g., Ni, Rh, Ir, Pt) are more active toward the desired reaction. Furthermore, basic metal oxide supports (MgO, ZnO, CeO_2, La_2O_3) are known to exhibit a better performance in the steam reforming conditions. Acidic properties in the catalyst support are generally the reason for dehydration and decomposition products which are not advantageous in the steam reforming of ethanol. Overall, depending on the catalyst, numerous reaction pathways can occur which clearly indicates the need for further research.

12.9. Diethyl carbonate

Dialkylcarbonates are gaining increasing interest as potential replacements for hazardous and toxic chemical, such as phosgene, alkyl halides, or alkyl sulfates used for carbonylation or alkylation. Diethyl carbonate (DEC) is a transparent, colorless liquid under normal conditions, exhibits a mild toxicology profile (Pacheco and Marshall, 1997) and is one of the most significant green chemicals among carbonate esters. DEC can, for example, be used in the manufacturing of polycarbonates, as an intermediate for various pharmaceuticals (Berridge et al., 1983) and as an alternative to ethyl halides and phosgene for ethylation and carbonylation processes. DEC is also considered as a promising replacement for MTBE as an attractive oxygen-containing fuel additive due its high oxygen content (40.6%). DEC does not persist in the environment since it biodegrades to ethanol and carbon dioxide.

Traditional methods for the production of DEC, such as the phosgene process (Muskat and Strain, 1941), oxidative carbonylation of ethanol (Dunn et al., 2002), reaction of ethanol with urea (Wang et al., 2007;

Yang and Meng, 1993), and carbonylation of ethyl nitrite (Fan et al., 2007), suffer from drawbacks including the use of poisonous phosgene and carbon monoxide, and low DEC production rates. Therefore, green alternatives utilizing ethanol and carbon dioxide could be applicable at industrial scale and would be of tremendous importance considering the large-scale availability of bioethanol and anthropogenic CO_2. Thus, a novel synthesis method of DEC starting from affordable, nontoxic, and easy to handle carbon dioxide and ethanol is a promising approach.

The novel technology for DEC synthesis from the direct reaction of bioethanol and CO_2 is an attractive approach owing to the transformation of anthropogenic carbon dioxide to value-added chemicals. However, the reaction of ethanol and carbon dioxide is both reversible and exothermic ($\Delta_r H^{\theta}_{298K} = -16.60$ kJ/mol) and does not occur spontaneously under normal conditions ($\Delta_r G^{\theta}_{298K} = +35.85$ kJ/mol>0) (Leino et al., 2011). High pressures, highly active catalysts, and promoters are required to obtain DEC in high yield.

The use of butylene oxide as a chemical dehydrating agent catalyzed by ceria has shown great improvements in the yield resulting in a ninefold enhancement of DEC yield in comparison to the synthetic method without any water removal. In addition, commercial cerium oxide appeared to be an active catalyst toward DEC formation.

On the basis of the performed kinetic studies, it has been proposed that DEC synthesis proceeds via a consecutive route including an intermediate, butylene carbonate, that undergoes transesterification with ethanol to generate DEC accompanied with 1,2-butanediol. The basicity of the catalyst plays an important role in one-pot DEC formation starting from ethanol, CO_2, and butylene oxide. Moreover, tuning of the catalyst properties toward enhancement of its basicity, surface area, and particle size could be a key for the increase of DEC yield and selectivity.

12.10. Other products

Previously, we discussed a large number of common important chemical compounds which can be derived from ethanol directly. However, from the practical point of view, the product distribution is seldom unique and a large category of (undesired) by-products are formed. Typically, these types of reactions can proceed through multiple and simultaneous reaction steps which result in demanding catalyst preparation procedures. For instance, ethanol is known to consist of two carbon atoms and it is well known that the cleavage of this bond leads to different single carbon atom products (CO, CO_2, CH_4). Different reaction mechanisms of this type have been reviewed in the literature (Erdohelyi et al., 2007; Idriss et al., 2004). For instance, methane is a well-known compound in the biogas and carbon monoxide has some importance in applications such as carbonylation agent. Additionally, CO has been used in the production of acetic acid, acetic anhydride, formic acid, and methyl formate (Kirk-Othmer, 2007).

Some rarely reported and maybe surprising final products can emerge in one-pot approach upon upgrading of bioethanol. A few studies demonstrated the production of BTX (benzene, toluene, xylenes) over H-ZSM-5 zeolite catalysts (Inaba et al., 2006, 2011; Li et al., 2010; Murata et al., 2008). These compounds are aromatic and show the clear potential of bioethanol as the feedstock of the future. In addition, low Si/Al ratio was found to result in a positive effect upon formation of aromatic hydrocarbons over H-ZSM-5 catalysts. Furthermore, some reports demonstrated the production of paraffins (Inaba et al., 2011) and propylene (Inaba et al., 2011; Inoue et al., 2010) over similar types of zeolite catalysts. It is clear that ethylene and diethyl ether are formed under milder reactions temperatures, whereas above-mentioned products require higher temperatures. Consequently, low-temperature processes should be considered due to slower catalyst deactivation and higher selectivity toward the desired reaction products.

13. SUMMARY AND CONCLUSIONS

In summary, there has recently been a rapid development of various technologies useful for the production of particularly, cellulosic ethanol. The technical progress in the area suggests that ethanol produced from cellulosic feedstocks has the potential to emerge as a quantitatively significant complement to bioethanol produced from starch- and sugar-based raw materials. Overall, both monosaccharides and polysaccharides as such, obtained from biomass (and especially from wood), create an interesting new field for whole new research topics with relatively high expectations. More environmentally friendly and more sustainable products and process methods are waiting to be born. Some high-value products utilizing biomass-derived mono- or polysaccharides (or new forest-based biorefinery type technique in its manufacturing processes) have already been successfully developed and commercialized (e.g., xylitol, furfural, and the wide range of lignin-, cellulose-, or hemicelluloses-based products). Biodegradable films and other products, which might (at least to some extent) have a chance to compete with plastic in future, are also under development and power plants utilizing renewable energy sources are built in progressive amounts. Better biochemicals and fuels and their manufacturing processes should be developed albeit the complexity of biomass yields this task anything but simple. Nevertheless, we will need greener chemicals and fuels being available in reasonably near future. Besides understanding the nature of biomass and advanced analytics, successful concepts require careful analysis and understanding of thermodynamics, development of appropriate kinetic approaches able to describe the complex biological systems in engineering terms, and optimized reactor setups.

A wide variety of different important chemical compounds which can be synthesized from bioethanol were discussed. The so-called one-pot approach is known to reduce the overall energy demand and likely render the economics of an industrial process more feasible. In terms of converting ethanol to other products, the majority of scientific reports discussing the topic are mainly concentrating on metal oxide catalysts. It is also clear that carefully selected metals and well-engineered catalyst structures can enhance the desired reaction. On the other hand, the latest scientific publications presented the use of some new materials, for example, carbon nanotubes in these types of reactions. The catalyst composition should also contain homogeneous, tailored nanostructure for a high selectivity toward a certain product and, at the same time, provide a robust material with a long lifetime. This brings along some challenges considering large-scale catalyst syntheses. Overall, amorphous crystal structures are not beneficial in these reactions. Reaction temperature should also be taken into account because it is a crucial factor that can lead to a wide product distribution and rapid catalyst deactivation. From industrial viewpoint, low-temperature synthesis technology would offer more promising possibilities in terms of energy savings and economics. As discussed, catalyst acidity and basicity are the main factors that determine the final outcome of the process. Many products (acetic acid, ethyl acetate, hydrogen, 1-butanol, and other C_4s) are formed through acetaldehyde-related pathways known to be base catalyzed. The more demanding exception is 1-butanol formation that might proceed via aldol condensation reaction, considering the acid and base dual functionality. Acid catalysis is the relevant pathway in the synthesis of diethyl ether, ethylene, 1,1-diethoxy ethane, aromatics, and possibly ethyl acetate. An overall view of these phenomena is presented in Fig. 1.12. In conclusion, bioethanol can be considered as one of the most versatile and promising feedstock candidates of the future.

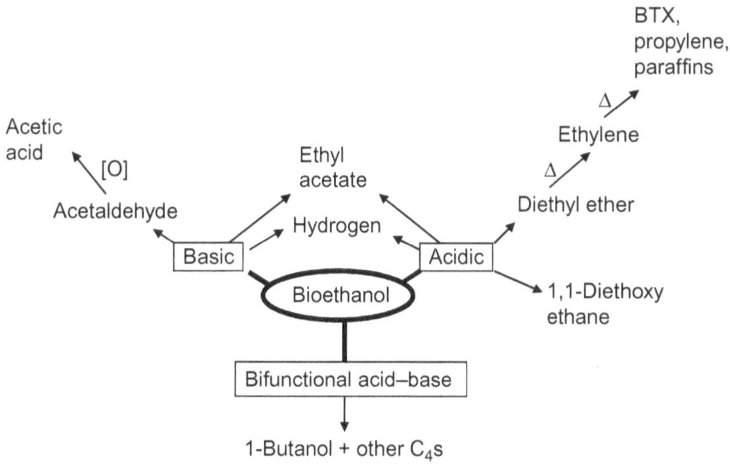

Figure 1.12 Bioethanol valorization scheme based on catalyst acidity and basicity. (For color version of this figure, the reader is referred to the online version of this chapter.)

ACKNOWLEDGMENTS

In Sweden, the Bio4Energy programme, Kempe Foundations as well as Knut and Alice Wallenberg Foundation are gratefully acknowledged. The authors are grateful to the Academy of Finland and the Finnish Agency for Innovation and Technology (TEKES) for financial support. This work is part of the activities of the Åbo Akademi Process Chemistry Centre.

REFERENCES

Abello MC, Gomez MF, Arrua LA: Synthesis of 1,1-diethoxyethane from bioethanol. Influence of catalyst acidity, *React Kinet Catal Lett* 73(1):143–149, 2001.

Aho A, Tokarev A, Backman P, et al: Catalytic pyrolysis of pine biomass over H-beta zeolite in a dual-fluidized bed reactor: effect of space velocity on the yield and composition of pyrolysis products, *Top Catal* 54:941–948, 2011.

Alriksson B, Cavka A, Jönsson LJ: Improving the fermentability of enzymatic hydrolysates of lignocellulose through chemical in-situ detoxification with reducing agents, *Bioresour Technol* 102:1254–1263, 2011.

Anugwom I, Mäki-Arvela P, Virtanen P, Willför S, Sjöholm R, Mikkola J-P: Selective extraction of hemicelluloses from spruce using switchable ionic liquids, *Carbohydr Polym* 87(3):2005–2011, 2012.

Appel LG, Gaspar AG, Esteves AML, Mendes FMT, Barbosa FG: Chemicals from ethanol—the ethyl acetate one-pot synthesis, *Appl Catal A Gen* 363:109–114, 2009.

Appel LG, Gaspar AB, Barbosa FG, Letichevsky S: The one-pot ethyl acetate syntheses: the role of the support in the oxidative and the dehydrogenative routes, *Appl Catal A Gen* 380:113–117, 2010.

Appel LG, Zonetti PC, Celnik J, Letichevsky S, Gaspar AB: Chemicals from ethanol—the dehydrogenative route of the ethyl acetate one-pot synthesis, *J Mol Catal A Chem* 334:29–34, 2011.

Arantes V, Saddler JN: Access to cellulose limits the efficiency of enzymatic hydrolysis: the role of amorphogenesis, *Biotechnol Biofuels* 3:4, 2010.

Arredondo VHI, Colorado RAA, Junior OS: Ethanol production from banana fruit and its lignocellulosic residues: exergy and renewability analysis, *Int J Thermodyn* 12:155–162, 2009.
Ashok MP: Study of the performance and emissions of the compression-ignition (CI) engine using ethyl acetate as a surfactant in ethanol-based emulsified fuel, *Energy Fuel* 24:1822–2828, 2010.
Atkins MP, Smith DJH, Westlake D: Monmorillonate catalysts for ethylene hydration, *J Clay Miner* 18:423–429, 1983.
Avgouropoulos G, Oikonomopoulos E, Kanistras D, Ioannides T: Complete oxidation of ethanol over alkali-promoted Pt/Al2O3 catalysts, *Appl Catal B Environ* 65:62–69, 2006.
Ayres RU, Ayres W: *Accounting for resources 2: The life cycle of materials,* Cheltenham, UK and Lyme, MA, 1999, Edward Elgar.
Barnett JA, Barnett L: *Yeast research—A historical overview,* Washington, DC, 2011, ASM Press.
Batis H, Najjar H: La–Mn perovskite-type oxide prepared by combustion method: catalytic activity in ethanol oxidation, *Appl Catal A Gen* 383:192–201, 2010.
Beckham GT, Matthews JF, Peters B, Bomble YaJ, Himmel ME, Crowley MF: Molecular-level origins of biomass recalcitrance: decrystallization free energies for four common cellulose polymorphs, *J Phys Chem B* 115:4118–4127, 2011.
Berridge M, Comar D, Crouzel C: A new synthesis of [2-11C] phenobarbital, *Int J Appl Radiat Isot* 34:1657, 1983.
Bland RR, Chen HC, Jewell WJ, Bellamy WD, Zall RR: Continuous high rate production of ethanol by Zymomonas mobilis in an attached film expanded bed fermentor, *Biotechnol Lett* 4:323, 1982.
Bokade VV, Gurav HJ: Synthesis of ethyl acetate by esterification of acetic acid with ethanol over a heteropolyacid on montmorillonite K10, *Nat Gas Chem* 19:161–164, 2010.
Bolder FHA: Dehydrogenation of alcohol mixtures to esters and ketones, *Ind Chem Eng Res* 47:7496–7500, 2008.
Busca G: Bases and basic materials in chemical and environmental processes. Liquid versus solid basicity, *Chem Rev* 110:2217–2249, 2010.
Centi G, van Santen RA: *Catalysis for renewables*, Weinheim, Germany, 2007, Wiley-VCH.
Chen Y-W, Wu K-C: An efficient two-phase reaction of ethyl acetate production in modified ZSM-5 zeolites, *Appl Catal A Gen* 257:33–42, 2004.
Christensen CH, Jorgensen B, Christiansen SE, Thomsen MLD: Aerobic oxidation of alcohols over gold catalysts: role of acid and base, *J Catal* 251:332–337, 2007.
Colley SW, Tuck MWM: Davy process technology, 2012. http://www.davyprotech.com/pdfs/Ethyl%20Ethanoate%20Synthesis.pdf (accessed 01.03.2012) containing Fawcett CR, Tuck MWM, Rathmell C, Colley SW: WO 0020375, 2000; Harris N, Rathmell C, Colley SW: WO 0020374, 2000; Fawcett CR, Tuck MWM, Watson DJ, Sharif CM, Colley SW, Wood MA: WO 0020373, 2000.
Collins P, Ferrier R: *Monosaccharides: their chemistry and their roles in natural products,* West Sussex, England, 1995, John Wiley & Sons Ltd.
Costerton JW, Stewart PS, Greenberg EP: Bacterial biofilms: a common cause of persistent infections, *Science* 248:1318–1322, 1999.
Coville NJ, Ndou AS, Plint N: Dimerisation of ethanol to butanol over solid-base catalysts, *Appl Catal A Gen* 251:337–345, 2003.
Cysewski GR, Wilke CR: Rapid ethanol fermentations using vacuum and cell recycle, *Biotechnol Bioeng* 19:1125, 1977.
Datta R, Phillips CB: Production of ethylene from hydrous ethanol on H-ZSM-5 under mild conditions, *Ind Eng Chem Res* 36:4466–4475, 1997.
Davis BG, Fairbanks AJ: *Carbohydrate chemistry,* Oxford, United Kingdom, 2008, Oxford Science Publications, Oxford University Press.

de la Piscina PR, Sanchez AB, Homs N, Fierro JLG: New supported Pd catalysts for the direct transformation of ethanol to ethyl acetate under medium pressure conditions, *Catal Today* 107–108:431–435, 2005.

Demsey MJ: *Physiology of immobilized cells,* Amsterdam, 1990, Elsevier Science Publisher BC pp 137–148.

Di Cosimo JI, Diez VK, Xu M, Iglesia E, Apesteguia CR: Structure and surface and catalytic properties of Mg-Al basic oxides, *J Catal* 178:499–510, 1998.

Di Cosimo JI, Apesteguia CR, Gines MJL, Iglesia E: Structural requirements and reaction pathways in condensation reactions of alcohols on MgyAlOx catalysts, *J Catal* 190:261–275, 2000.

Diaz E, Leon M, Ordonez S: Ethanol catalytic condensation over Mg-Al mixed oxides derived from hydrotalcites, *Catal Today* 164:436–442, 2011.

Dogu G, Dogu T, Varisli D: Novel mesoporous nanocomposite WOx-silicate acidic catalysts: ethylene and diethylether from ethanol, *Ind Eng Chem Res* 48(21):9394–9401, 2009.

Drissen RET, Maas RHW, Tramper J, Beeftink HH: Modelling ethanol production from cellulose: separate hydrolysis and fermentation versus simultaneous saccharification and fermentation, *Biocatal Biotransfor* 27(1):27–35, 2009.

Dumesic JA, Alcala R, Mavrikakis M: DFT studies for cleavage of C–C and C–O bonds in surface species derived from ethanol on Pt(111), *J Catal* 218:178–190, 2003.

Dunn BC, Guenneau C, Hilton SA, et al: Production of diethyl carbonate from ethanol and carbon monoxide over a heterogeneous catalyst, *Energy Fuel* 16:177, 2002.

Eggleston G: Future sustainability of the sugar and sugar-ethanol industries. In *ACS symposium series, 1058: sustainability of the sugar and sugar-ethanol industries*, 2010, pp 1–19.

Eklund D, Lindström T: *Paper chemistry: an introduction,* Grankulla, Hangö tryckeri, Finland, 1991, Dt Paper Science Publications.

Epron F, Le Valant A, Garron A, Bion N, Duprez D: Effect of higher alcohols on the performances of a 1%$Rh/MgAl_2O_4/Al_2O_3$ catalyst for hydrogen production by crude bioethanol steam reforming, *Int J Hydrogen Energy* 36:311–318, 2011.

Erdohelyi A, Dömök M, Toth M, Raskó J: Adsorption and reactions of ethanol and ethanol–water mixture on alumina-supported Pt catalysts, *J Appl Catal B Environ* 69:262–272, 2007.

Erdohelyi A, Dömök M, Oszko A, Baan K, Sarusi I: Reforming of ethanol on Pt/Al_2O_3-ZrO_2 catalyst, *Appl Catal A Gen* 383:33–42, 2010.

Evans A, Strezov V, Evans TJ: Sustainability considerations for electricity generation from biomass, *Renew Sust Energ Rev* 14(5):1419–1427, 2010.

Fan M, Zhang P, Ma X: Study on Wacker-type catalysts for catalytic synthesis of diethyl carbonate from ethyl nitrite route, *Fuel* 86:902–905, 2007.

Fengel D, Wegener G: *Wood: chemistry, ultrastructure, reactions,* Berlin, New York, 1989, Walter de Gruyter & Co.

Fitzpatrick A., Roberts A., Witherly, S: Larch arabinogalactan: A novel and multifunctional natural product, AgroFOOD industry hi-tech Jan/Feb 2004, www.nutritioninnovation.com/AGPaper.pdf (accessed Oct. 2012).

Galbe M, Zacchi G: Pretreatment of lignocellulosic materials for efficient bioethanol production, *Adv Biochem Eng Biotechnol* 108:41–65, 2007.

Gang L, Wenhui P: Esterifications of carboxylic acids and alcohols catalyzed by $Al_2(SO_4)_3 \cdot 18H_2O$ under solvent-free condition, *Kinet Catal* 51:559–565, 2010.

Geddes CC, Nieves IU, Ingram LO: Advances in ethanol production, *Curr Opin Biotechnol* 22:312–319, 2011.

Georgieva TI, Mikkelsen MJ, Ahring BK: Ethanol production from wet-exploded wheat straw hydrolysate by thermophilic anaerobic bacterium Thermoanaerobacter BG1L1 in a continuous immobilized reactor, *Appl Biochem Biotechnol* 145:99, 2008.

Ghaly A, El-Taweel A: Kinetic modelling of continuous production of ethanol from cheese whey, *Biomass Bioenerg* 12:461–472, 1997.

Gräsvik J, Raut D, Mikkola J-P: Mun J, Sim H, editors: *Handbook of ionic liquids, Properties, Applications and Hazards*, US, 2012, Nova Science Publishers, Inc.

Grenman H, Eränen K, Krogell J, Willför S, Salmi T, Murzin DYu: Kinetics of aqueous extraction of hemicelluloses from spruce in an intensified reactor system, *Ind Eng Chem Res* 50(7):3818–3828, 2011.

Hansen AC, Zhang Q, Lyne PWL: Ethanol-diesel fuel blends, *Bioresour Technol* 96:277–285, 2005.

Hattori H: Heterogeneous basic catalysis, *Chem Rev* 95:537–558, 1995.

Hattori H, Seki T: Tishchenko reaction over solid base catalysts, *Catal Surv Asia* 7:2–3, 2003.

Hatzimanikatis V, Floudas A, Bailey E: Optimization of regulatory architectures in metabolic reaction networks, *Biotechnol Bioeng* 52:485–500, 1996.

Hensen EJM, Guan Y: Ethanol dehydrogenation by gold catalysts: the effect of the gold particle size and the presence of oxygen, *Appl Catal A Gen* 361:49–56, 2009.

Hepbasli A: A key review on exergetic analysis and assessment of renewable energy resources for a sustainable future, *Renew Sust Energ Rev* 12:593, 2008.

Herrera JE, Chimentao RJ, Kwak JH, Medina F, Wang Y, Peden CHF: Oxidation of ethanol to acetaldehyde over Na-promoted vanadium oxide catalysts, *Appl Catal A Gen* 332:263–272, 2007.

Hinderink AP, Kerkhof F, Lie A, Arons J, Kooi H: Exergy analysis with a flowsheeting simulator—I. Theory; calculating exergies of material streams, *Chem Eng Sci* 51:4693, 1996.

Hurley RG, Schuetzle D, Summers MD: Patent, US 7,718,832 B1, May 18, 2010.

Hyvärinen S, Damlin P, Gräsvik J, Murzin DYu, Mikkola J-P: Ionic liquid fractionation of woody biomass for fermentable monosaccharides, *Cell Chem Technol* 45(7–8):483–486, 2011.

Idriss H: Ethanol reactions over the surfaces of noble metal/cerium oxide catalysts, *Platinum Met Rev* 48(3):105–115, 2004.

Idriss H, Seebauer EG: Reactions of ethanol over metal oxides, *Appl Catal A Gen* 152:201–212, 2000.

Idriss H, Sheng P-Y, Bowmaker GA: The reactions of ethanol over Au/CeO_2, *Appl Catal A Gen* 261:171–181, 2004.

IEA Technology Roadmap: *Biofuels for transport*, Paris, France, 2011, International Energy Agency.

Illas F, Radilla J, Boronat M, Corma A: Monitoring the interaction of adsorbates on metal surfaces by surface site engineering: the case of ethoxy on Cu, Pd, Ag and Au regular and stepped surfaces, *Phys Chem Eng Phys* 12:6492–6498, 2010.

Inaba M, Murata K, Saito M, Takahara I: Ethanol conversion to aromatic hydrocarbons over several zeolite catalysts, *React Kinet Catal Lett* 88(1):135–142, 2006.

Inaba M, Murata K, Takahara I, Inoue K-I: Production of olefins from ethanol by Fe and/or P-modified H-ZSM-5 zeolite catalysts, *J Chem Technol Biotechnol* 86:95–104, 2011.

Inoue K, Okabe K, Inaba M, Takahara I, Murata K: Metal modification effects on ethanol conversion to propylene by H-ZSM-5 with Si/Al2 ratio of 150, *React Kinet Mech Catal* 101:477–489, 2010.

Inui K, Kurabayashi T, Sato S: Direct synthesis of ethyl acetate from ethanol carried out under pressure, *J Catal* 212:207–215, 2002.

Inui K, Kurabayashi T, Sato S, Ichikawa N: Effective formation of ethyl acetate from ethanol over Cu-Zn-Zr-Al-O catalyst, *J Mol Catal A Chem* 216:147–156, 2004.

Ishihara T, Matsuo J, Ito M, Nishiguchi H, Takita Y: Porous H-ZSM-5 zeolite tube as a novel application of catalyst for the synthesis of ethanol by hydration of ethylene, *Ind Eng Chem Res* 36:4427–4429, 1997.

Ishii Y, Matsu-ura T, Sakaguchi S, Obora Y: Guerbet reaction of primary alcohols leading to alkylated dimer alcohols catalyzed by iridium complexes, *J Org Chem* 71:8306–8308, 2006.

Iwamoto M, Tajima M, Kagawa S: Gas-phase direct hydration of ethylene over proton-exchanged zeolite catalysts at atmospheric pressure, *J Catal* 101:195–200, 1986.

Ji W, Chen Y, Kung HH: Vapor phase aldol condensation of acetaldehyde on metal oxide catalysts, *Appl Catal A Gen* 161:93–104, 1997.

Jiang XZ, Yang KW, Zhang WC: One-step synthesis of n-butanol from ethanol condensation over alumina-supported metal catalysts, *Chem Lett* 15(12):1497–1500, 2004.

Kelly GJ, Jackson SD: Aldol condensation of aldehydes and ketones over solid base catalysts, *Catalysis in Application*, 129–135, 2003, Cambridge, UK.

Kim SR, Ha SJ, Wei N, Oh EJ, Jin YS: Simultaneous co-fermentation of mixed sugars: a promising strategy for producing cellulosic ethanol, *Trends Biotechnol* 30:274–282, 2012.

King AWT, Zoia L, Filpponen I, et al: In situ determination of lignin phenolics and wood solubility in imidazolium chlorides using 31P NMR, *J Agric Food Chem* 57:8236–8243, 2009a.

King AWT, Kilpeläinen I, Heikkinen S, Järvi P, Argyropoulos DS: Hydrophobic interactions determining functionalized lignocellulose solubility in dialkylimidazolium chlorides, as probed by 31P NMR, *Biomacromolecules* 10:458–463, 2009b.

King AWT, Asikkala J, Mutikainen I, Järvi P, Kilpeläinen I: Distillable acid-base conjugate ionic liquids for cellulose dissolution and processing, *Angew Chem Int Ed Engl* 50:6301–6305, 2011.

Kirk-Othmer editor: *Concise encyclopedia of chemical technology*, ed 5, Hoboken, NJ, 2007, John Wiley & Sons, Inc. vol. 1.

Krug TA, Daugulis AJ: Ethanol-production using Zymomonas mobilis immobilized on an ion-exchange resin, *Biotechnol Lett* 5:159–164, 1983.

Kubicek CP, Mikus M, Schuster A, Schmoll M, Seiboth B: Metabolic engineering strategies for the improvement of cellulase production by Hypocrea jecorina, *Biotechnol Biofuels* 2:19, 2009.

Kunduru MR, Pometto AL: Continuous ethanol production by Zymomonas mobilis and Saccharomyces cerevisiae in biofilm reactors, *J Ind Microbiol Biotechnol* 16:249–256, 1996.

Kusema BT, Xu C, Maki-Arvela P, et al: Kinetics of acid hydrolysis of arabinogalactans, *Int J Chem React Eng* 8:1–18, 2010.

Lee WG, Park BG, Chang YK, Chang HN, Lee JS, Park SC: Continuous ethanol production from concentrated wood hydrolysates in an internal membrane-filtration bioreactor, *Biotechnol Prog* 16:302–304, 2000.

Lee SH, Doherty TV, Linhardt RJ, Dordick JS: Ionic liquid-mediated selective extraction of lignin from wood leading to enhanced enzymatic cellulose hydrolysis, *Biotechnol Bioeng* 102(5):1368–1376, 2009.

Leino E, Mäki-Arvela P, Eränen K, et al: Enhanced yields of diethyl carbonate via one-pot synthesis from ethanol, carbon dioxide and butylene oxide over cerium (IV) oxide, *Chem Eng J* 176–177:124, 2011.

Leong WK, Chen L, Koh ACW, et al: Highly efficient ruthenium and ruthenium–platinum cluster-derived nanocatalysts for hydrogen production via ethanol steam reforming, *Catal Commun* 9:170–175, 2008.

Leung DYC, Ni M, Leung MKH: A review on reforming bio-ethanol for hydrogen production, *Int J Hydrogen Energy* 32:3247–3328, 2007.

Li G, Ni Y, Peng W, et al: High selective and stable performance of catalytic aromatization of alcohols and ethers over La/Zn/HZSM-5 catalysts, *J Ind Eng Chem* 16:503–505, 2010.

Lin T-B, Chang J-R, Chung D-L: Ethyl acetate production from water-containing ethanol catalyzed by supported Pd catalysts: advantages and disadvantages of hydrophobic supports, *Ind Eng Chem Res* 38:1271–1276, 1999.

Lippits MJ, Nieuwenhuys BE: Direct conversion of ethanol into ethylene oxide on copper and silver nanoparticles: effect of addition of CeOx and Li2O, *Catal Today* 154:127–132, 2010.

Liu Y, Murata K, Inaba M, Takahara I, Okabe K: Synthesis of ethanol from syngas over Rh/Ce1-xZrxO2 catalysts, *Catal Today* 164:308–314, 2011.

Loque D, Eudes A, Yang F: Biomass availability and sustainability for biofuels. In *RSC energy and environment series. 4: chemical and biochemical catalysis for next generation biofuels*, 2011, pp 5–32.

Lynd LR, Laser MS, Bransby D, et al: How biotech can transform biofuels, *Nat Biotechnol* 26:169–172, 2008.

Magnoux P, Ferreira Madeira F, Gnep NS, Maury S, Cadran N: Ethanol transformation over HFAU, HBEA and HMFI zeolites presenting similar Brønsted acidity, *Appl Catal A Gen* 367:39–46, 2009.

Mäki-Arvela P, Rozmyszowicz B, Lestari S, et al: Catalytic deoxygenation of tall oil fatty acid over palladium supported on mesoporous carbon, *Energy Fuel* 25:2815–2825, 2011a.

Mäki-Arvela P, Salmi T, Holmbom B, Willför S, Murzin DYu: Synthesis of sugars by hydrolysis of hemicelluloses—a review, *Chem Rev* 111:5638–5666, 2011b.

Maki-Arvela P, Salminen E, Riittonen T, Virtanen P, Kumar N, Mikkola J-P: The challenge of efficient synthesis of biofuels from lignocellulose for future renewable transportation fuels, *Int J Chem Eng* 10:674761, 2012.

Margaritis A, Wilke CR: The rotorfermentor. II. Application to ethanol fermentation, *Biotechnol Bioeng* 20:727–753, 1978.

Minteer S: *Alcoholic fuels,* Boca Raton, 2006, CRC Press.

Modarresi A, Kravanja P, Friedl A: Pinch and exergy analysis of lignocellulosic ethanol, biomethane, heat and power production from straw, *Appl Therm Eng* 43:20–28, 2012.

Moiseev II: Green chemistry in the bulk chemicals industry, *Kinet Catal* 52(3):337–347, 2011.

Mosier N, Wyman C, Dale B, Elander R, Lee YY, Holtzapple M, Ladisch M: Features of promising technologies for pretreatment of lignocellulosic biomass, *Bioresource Technol* 96:673–686, 2005.

Mousdale DM: *Introduction to biofuels,* Boca Raton, FL, 2010, CRC Press.

Murata K, Inaba M, Takahara I: Effects of surface modification of H-ZSM-5 catalyst on direct transformation of ethanol into lower olefins, *J Jpn Petrol Inst* 51:234–239, 2008.

Muskat IE, Strain F: US Patent 2379250, 1941.

Mussatto SI, Dragone G, Guimarães PMR, et al: Technological trends, global market, and challenges of bio-ethanol production, *Biotechnol Adv* 28:817–830, 2010.

Nakagawa Y, Tajima N, Hirao K: A theoretical study of catalytic hydration reactions of ethylene, *J Comput Chem* 21(14):1292–1304, 2000.

Nevoigt E: Progress in metabolic engineering of Saccharomyces cerevisiae, *Microbiol Mol Biol Rev* 72:379–412, 2008.

Ngo H, Liu Y, Murata K: Effect of secondary additives (Li, Mn) in Fe promoted, Rh/TiO2 catalysts for the synthesis of ethanol from syngas, *React Kinet Mech Catal* 102:425–435, 2011.

Nguyen T, Kikuchi Y, Noda M, Sugiyama H, Hirao M: *Proceedings of the 20th European symposium on computer aided process engineering—ESCAPE20, 6–9 June, Ischia, Naples, Italy,* 2010.

Nieuwenhuys BE, Lippits MJ: Direct conversion of ethanol into ethylene oxide on gold-based catalysts: effect of CeOx and Li2O addition on the selectivity, *J Catal* 274:142–149, 2010.

Nilsson D: Energy, exergy and emergy analysis of using straw as fuel in district heating plants, *Biomass Bioenerg* 13:63–73, 1997.

Nord KE, Haupt D: Reducing the emission of particles from a diesel engine by adding an oxygenate to the fuel, *Environ Sci Technol* 39:6260–6265, 2005.
Ojeda K, Kafarov V: Exergy analysis of enzymatic hydrolysis reactors for transformation of lignocellulosic biomass to ethanol, *Chem Eng J* 154:390, 2009.
Ojeda K, Sánchez E, El-Halwagi M, Kafarov V: Exergy analysis and process integration of bioethanol production from acid pre-treated biomass: comparison of SHF, SSF and SSCF pathways, *Chem Eng J* 176–177:195, 2011.
Olah GA, Shamma T, Surya Prakash GK: Dehydration of alcohols to ethers over Nafion-H, a solid perfluoroalkanesulfonic acid resin catalyst, *Catal Lett* 46:1–4, 1997.
Pacheco MA, Marshall CL: Review of dimethyl carbonate manufacture and its characteristics as a fuel additive, *Energy Fuel* 11:2, 1997.
Persson B: *Sulfitsprit: Förhoppningar och besvikelser under 100 år,* Bjästa, Sweden, 2007, DAUS Tryck & Media.
Pfromm PH, Amanor-Boadu V, Nelson R: Sustainability of algae derived biodiesel: a mass balance approach, *Bioresour Technol* 102:1185–1193, 2011.
Phisalaphong M, Srirattana N, Tanthapanichakoon W: Mathematical modeling to investigate temperature effect on kinetic parameters of ethanol production, *Biochem Eng J* 28:36, 2006.
Poulopoulos SG, Samaras DP, Philippopoulos CJ: Regulated and unregulated emissions from an internal combustion engine operating on ethanol-containing fuels, *Atmos Environ* 35:4399–4406, 2001.
Ptsainski K, Prins M, Pierik A: Exergetic evaluation of biomass gasification, *Energy* 32 (568):2007.
Quinlan RJ, Sweeney MD, Lo Leggio L, et al: Insights into the oxidative degradation of cellulose by a copper metalloenzyme that exploits biomass components, *Proc Natl Acad Sci USA* 108:15079–15084, 2011.
Qureshi N, Annous BA, Ezeji TC, Karcher P, Maddox IS: Biofilm reactors for industrial bioconversion processes: employing potential of enhanced reaction rates, *Biomed Central* 4:24, 2005.http://dx.doi.org/10.1186/1475-2859-4-24.
Radilla J, Boronat M, Corma A, Illas F: Structure and bonding of ethoxy species adsorbed on transition metal surfaces, *Theor Chem Acc* 126:223–229, 2010.
Riittonen T, Toukoniitty E, Madnani DK, et al: One-pot liquid-phase catalytic conversion of ethanol to 1-butanol over aluminium oxide—the effect of the active metal on the selectivity, *Catalysts* 2:68–84, 2012.
Rodrigues AE, Silva VMTM: Synthesis of diethylacetal: thermodynamic and kinetic studies, *Chem Eng Sci* 56:1255–1263, 2001.
Rodriguez-Mirasol J, Bedia J, Barrionuevo R, Cordero T: Ethanol dehydration to ethylene on acid carbon catalysts, *Appl Catal B Environ* 103:302–310, 2011.
Rosen MA, Dincer I: Exergy–cost–energy–mass analysis of thermal systems and processes, *Energy Convers Manage* 44:1633, 2003.
Salmi T, Murzin D, Mäki-Arvela P, et al: Catalytic engineering in the processing of biomass into chemicals. In Cybulski A, Moulijn JA, Stankiewicz A, editors: *Novel concepts in catalysis and chemical reactors*, 2010, pp 163–188, Weinheim, Germany.
Shen J, Agblevor F: Modeling semi-simultaneous saccharification and fermentation of ethanol production from cellulose, *Biomass Bioenerg* 34:1098–1107, 2010.
Shen W, Zhang B, Cai W, Li Y, Xu Y: Hydrogen production by steam reforming of ethanol over an Ir/CeO2 catalyst: reaction mechanism and stability of the catalyst, *Int J Hydrogen Energy* 33:4377–4386, 2008.
Shieh JH, Fan LT: Estimation of energy (enthalpy) and exergy (availability) contents in structurally complicated materials, *Energ Sour* 6(1/2):1–46, 1982 Laramie.
Silva VMTM, Rodrigues AE: Novel process for diethylacetal synthesis, *Am Inst Chem Eng* 51(10):2752–2768, 2005.

Sjöström E, Alén R: *Analytical methods in wood chemistry, pulping, and papermaking*, Berlin, 2010, Springer-Verlag.

Sjöström E: *Wood Chemistry: Fundamentals and applications*, San Diego, California, 1993, Academic Press.

Sohel MI, Jack MW: Thermodynamic analysis of lignocellulosic biofuel production via a biochemical process: guiding technology selection and research focus, *Bioresour Technol* 102:2617, 2011.

Soimakallio S, Antikainen R, Thun R, et al: *VTT Tiedotteita* (2482), i, 1–220, A1–A33, B1–B5, C1–C3, 2009.

Song G, Shen L, Xiao J: Estimating specific chemical exergy of biomass from basic analysis data, *Ind Eng Chem Res* 50:9758, 2011a.

Song T, Pranovich A, Holmbom B: Effects of pH control with phthalate buffers on hot-water extraction of hemicelluloses from spruce wood, *Bioresource Technol* 102:10518–10523, 2011b.

Soudham VP, Alriksson B, Jönsson LJ: Reducing agents improve enzymatic hydrolysis of cellulosic substrates in the presence of pretreatment liquid, *J Biotechnol* 155:244–250, 2011.

Spaak G: Männen kring Carl Daniel Ekman och tillkomsten av världens första sulfitcellulosafabrik, Industrihistorisk skriftserie utg. Svenska cellulosa-och Trämassaföreningarna Nr 6, Stockholm, Sweden, 1957.

Stenius P: *Forest products chemistry, papermaking science and technology*, Jyväskylä, Finland, 2000, Gummerus Printing.

Stites RC, Hohman J: Patent WO 2010/068318 A1, June 17, 2010.

Stites RC, Tirmizi S, Hohman J, Deutch S: Patent US 2011/0124927, May 26, 2011.

Szargut J, Styrylska T: Angenäherte Bestimmung der Exergie von Brennstoffen, Brennst, *Wärme Kraft* 16:589, 1964 [in German].

Szwaja S, Naber JD: Combustion of n-butanol in a spark-ignition IC Engine, *Fuel* 89:1573–1582, 2010.

Takahara I, Saito M, Inaba M, Murata K: Dehydration of ethanol into ethylene over solid acid catalysts, *Catal Lett* 105:249–252, 2005.

Takahara I, Saito M, Matsuhashi H, Inaba M, Murata K: Increase in the number of acid sites of a H-ZSM5 zeolite during the dehydration of ethanol, *Catal Lett* 113(3–4):82–85, 2007.

Takei T, Iguchi N, Haruta M: Synthesis of acetoaldehyde, acetic acid, and others by the dehydrogenation and oxidation of ethanol, *Catal Surv Asia* 15:80–88, 2011.

Tanchoux N, Marcu I-C, Tichit D, Fajula F: Catalytic valorization of bioethanol over Cu-Mg-Al mixed oxide catalysts, *Catal Today* 147:231–238, 2009.

Teh K-Y, Lutz AE: Thermodynamic analysis of fermentation and anaerobic growth of baker's yeast for ethanol production, *J. Biotechnol.* 147:80, 2010.

Tokarev AV, Kirilin AV, Murzina EV, et al: The role of bio-ethanol in aqueous phase reforming to sustainable hydrogen, *Int J Hydrogen Energy* 35(22):12642–12649, 2010.

Trakarnpruk W, Arenamnart S: Ethanol conversion to ethylene using metal-mordenite catalysts, *Int J Catal Appl Sci Eng* 4(1):21–32, 2006.

Tsuchida T, Sakuma S, Takeguchi T, Ueda W: Direct synthesis of n-butanol from ethanol over nonstoichiometric hydroxyapatite, *Ind Eng Chem Res* 45:8634–8642, 2006.

Tsuchida T, Kubo J, Yoshioka T, Sakuma S, Takeguchi T, Ueda W: Reaction of ethanol over hydroxyapatite affected by Ca/P ratio of catalyst, *J Catal* 259:183–189, 2008.

Tyagi RD, Ghost TK: Studies on immobilized Saccharomyces cerevisiae. I. Analysis of continuous rapid ethanol fermentation in immobilized cell reactor, *Biotechnol Bioeng* 24:781–795, 1982.

Ueda W, Kuwabara T, Ohshida T, Morikawa Y: A low-pressure guerbet reaction over magnesium oxide catalyst, *J Chem Soc Chem Commun* (22):1558–1559, 1990.

Ullman F: *Ullman's encyclopedia of industrial chemistry*, ed 6, Weinheim, Germany, 2003, Wiley-VCH.

van Zyl WH, Lynd LR, den Haan R, McBride JE: Consolidated bioprocessing for bioethanol production using Saccharomyces cerevisiae, *Adv Biochem Eng Biotechnol* 108:205–235, 2007.

Vega JL, Clausen EC, Gaddy JL: Biofilm reactors for ethanol production, *Enzyme Microb Technol* 10:390, 1988.

Viikari L, Alapuranen M, Puranen T, Vehmaanperä J, Siika-aho M: Thermostable enzymes in lignocellulose hydrolysis, *Adv Biochem Eng Biotechnol* 108:121–145, 2007.

Wang D, Yang B, Zhai X, Zhou L: Synthesis of diethyl carbonate by catalytic alcoholysis of urea, *Fuel Process Technol* 88:807–812, 2007.

Wang L, Zhu W, Zheng D, et al: Direct transformation of ethanol to ethyl acetate on Cu/ZrO2 catalyst, *React Kinet Mech Catal* 101:365–375, 2010.

Waugh KC, Colley SW, Tabatabaei J, Wood MA: The detailed kinetics and mechanism of ethyl ethanoate synthesis over a Cu/Cr2O3 catalyst, *J Catal* 236:21–33, 2005.

Weinstein RD, Ferens AR, Orange RJ, Lemaire P: Oxidative dehydrogenation of ethanol to acetaldehyde and ethyl acetate by graphite nanofibers, *Carbon* 49:701–707, 2011.

Westereng B, Ishida T, Vaaje-Kolstad G, et al: The putative endoglucanase PcGH61D from Phanerochaete chrysosporium is a metal-dependent oxidative enzyme that cleaves cellulose, *PLoS One* 6(11):e27807, 2011.

Wertz J-L, Bédué O, Mercier JP: Cellulose science and technology, Boca Raton, Italy, 2010, CRC Press, Taylor and Francis Group, LCC.

Willför S, Sundberg A, Hemming J, Holmbom B: Polysaccharides in some industrially important softwood species, *Wood Sci Technol* 39:245–258, 2005a.

Willför S, Sundberg A, Pranovich A, Holmbom B: Polysaccharides in some industrially important softwood species, *Wood Sci Technol* 39:601–617, 2005b.

Willför S, Sundberg K, Tenkanen M, Holmbom B: Spruce-derived mannans – A potential raw material for hydrocolloids and novel advanced natural materials, *Carbohyd Polym* 72:197–210, 2008.

Willför S, Pranovich A, Tamminen T, et al: Carbohydrate analysis of plant materials with uronic acid-containing polysaccharides—a comparison between different hydrolysis and subsequent chromatographic analytical techniques, *Ind Crop Prod* 29:571–580, 2009.

Wu L, Shi X, Cui Q, Wang H, Huang H: Effects of the SAPO-11 synthetic process on dehydration of ethanol to ethylene, *Front Chem Sci Eng* 5(1):60–66, 2011.

Xiao Y, Li X, Yuan Z, Li J, Chen Y: Catalytic dehydration of ethanol to ethylene on TiO2/4A zeolite composite catalysts, *Catal Lett* 130:308–311, 2009.

Xu C, Willfor S: Spruce galactoglucomannan: isolation, characterization, and modification. In *Abstracts of papers, 243rd ACS national meeting & exposition, San Diego, CA, United States, March 25–29*, 2012.

Yang C, Meng Z: Bimolecular condensation of ethanol to 1-butanol catalyzed by alkali cation zeolites, *J Catal* 142:37–44, 1993.

Zaki TJ: Catalytic dehydration of ethanol using transition metal oxide catalysts, *Colloid Interface Sci* 284:606–613, 2005.

CHAPTER TWO

Biomass Pyrolysis

R.W. Nachenius*, F. Ronsse*, R.H. Venderbosch†, W. Prins*
*Department of Biosystems Engineering, Faculty of Bioscience Engineering, Ghent University, Ghent, Belgium
†BTG Biomass Technology Group BV, Enschede, The Netherlands

Contents

Advances in Chemical Engineering, Volume 42
ISSN 0065-2377
http://dx.doi.org/10.1016/B978-0-12-386505-2.00002-X

Abstract

Fast pyrolysis and slow pyrolysis (including torrefaction) are thermochemical processes involving the conversion of biomass into predominantly liquid or solid products respectively. Fast pyrolysis is generally employed to maximize the liquid bio-oil product yield, the benefit being that the bio-oil has a significant calorific value and the liquid may be handled with greater ease than conventional biomass. The bio-oil may be combusted directly or upgraded such that it may be used as a transportation fuel. The extraction of unique chemicals from the bio-oil is also a route for valorizing fast pyrolysis products. Slow pyrolysis processes are tailored to maximize the yield of the solid product. A process related to slow pyrolysis is torrefaction, where milder conditions are employed and the biomass is only partially pyrolyzed. The solids thus obtained differ from the original biomass in a number of ways, such as increased energy density, hydrophobicity, grindability, and reduced biodegradability. These characteristics allow for improved handling, transport, and utilization of the char and torrefied biomass within existing coal-based processes when compared to biomass. Char may also be used as a soil amendment in the form of biochar or processed further to form activated carbon.

1. INTRODUCTION

"Biomass" usually refers to all biological organic materials derived from living or recently living organisms. This excludes materials such as the fossil fuels (coal, oil, and natural gas). In terms of energy applications, "biomass" refers to the material derived directly from plants, but, strictly speaking, animal-derived materials also fall within its broad definition (Biomass Energy Centre, 2011). The various forms of biomass that are interesting in terms of energy applications include wood, energy crops, agricultural and forestry residues, and municipal, industrial, and food wastes.

The application of heat to biomass in an inert environment can drive various interesting thermochemical reactions. These reactions typically result in the production of a solid residue (char), condensable vapors (bio-oil or tar), and noncondensable gases. The condensed bio-oil may separate to form an aqueous phase and a water-insoluble phase consisting largely of lignin-derived molecules. The composition and the yields of the various products are dependent on the feed and the process parameters. In the existing literature, different terms are applied to the various biomass conversion techniques, differentiated by the biomass heating rate, the maximum temperature at which the conversion takes place, and the reaction medium.

Fast pyrolysis employs fast heating rates (generally above 100 °C/s) and temperatures between 450 and 550 °C, thereby maximizing the liquid (bio-oil) yield (typically 60–70 wt% on a dry-feed basis; Venderbosch and Prins, 2010). The dark brown or black bio-oil has a density of approximately 1200 kg/m^3. The viscosity ranges from 25 to 1000 cP (depending on the composition) and is generally acidic in nature (pH of 2–4) due to the presence of organic acids such as formic and acetic acid. In addition to handling and storing a liquid product rather than a solid, further benefits of bio-oil production through fast pyrolysis are listed below (Venderbosch and Prins, 2010):

- Decoupling liquid fuel production (scale, time, and location) from its utilization;
- Separating minerals on the site of liquid fuel production (to be recycled to the soil as nutrient);
- Increased biological stability (less rotting);
- Producing a renewable fuel for boilers, engines, turbines, power stations, and gasifiers;
- Secondary conversion to transportation fuels, additives, or specialty chemicals (biomass refinery);
- Primary separation of the sugar and lignin fractions with further subsequent utilization or upgrading (biomass refinery).

Slow pyrolysis is typified by a lower heating rate (below 80 °C/min) which favors the production of char. Typical temperatures are in the range of 350–750 °C. A variant of the slow pyrolysis process, torrefaction, is currently receiving significant academic and industrial attention. Torrefaction may be seen as a mild or incomplete form of slow pyrolysis. Compared to conventional slow pyrolysis processes, only partial thermochemical conversion and devolatilization take place as reduced temperatures are employed (200–300 °C). The benefits of torrefied biomass and char as solid energy carriers include increased energy density, hydrophobicity, and grindability, and reduced biodegradability in comparison with the unprocessed biomass feed. Through torrefaction and slow pyrolysis, the composition of the solid product approaches that of coal by becoming relatively more carbon rich at the expense of oxygen and hydrogen. This is indicated in the Van Krevelen diagram shown in Fig. 2.1. Torrefied biomass or char could therefore be incorporated within existing coal-based processes (as feed to gasifiers, as fuel for coal-fired boilers, or as a reducing agent in metallurgical smelting processes) as a "drop-in" substitute with greater ease than unprocessed biomass. Char can also be processed further to yield activated carbon which may be used as an industrial adsorbent. Another potential use of char is in its direct application as a soil amendment where it has

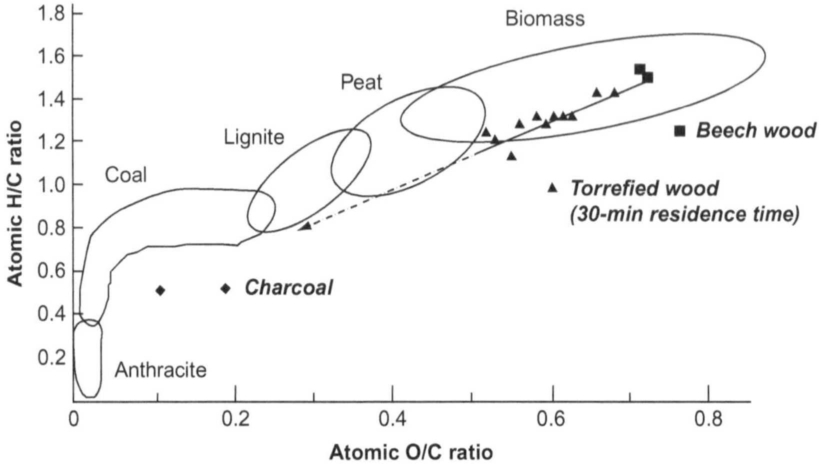

Figure 2.1 Van Krevelen diagram (Prins, 2005) as cited in van der Stelt et al. (2011).

been reported that increased fertility could be achieved in addition to the formation of stable carbon sinks. In such applications, the char is defined as "biochar" (Lehmann et al., 2006).

Woody biomass has been used extensively in pyrolysis applications as it is readily available and relatively homogeneous in comparison with other feedstocks. The utilization of other biomass sources such as agricultural residues, forestry residues, wastes, and energy crops could provide cheaper feed sources but also introduces further technical challenges.

The desired properties of char or torrefied biomass (calorific value, moisture content, particle size distribution, mineral content, surface area, pore size distribution) and bio-oil (calorific value, pH, moisture content, viscosity, composition, storage stability) are dependent on the intended product application. The definition of an optimal reactor (and optimal process) is inherently an economic concept which incorporates a large number of process factors that will affect the investment decision. These factors are discussed in this chapter. A critical review of the developed and proposed reactor technologies is also presented.

2. THERMOCHEMICAL CONVERSION OF BIOMASS

Numerous technologies are applicable to the upgrading of biomass through thermochemical conversion. These are highlighted in Table 2.1 together with the main differences between these various technologies. The processes of interest in this chapter will be limited to fast pyrolysis, slow pyrolysis, and torrefaction.

Table 2.1 Processes for the thermochemical conversion of biomass

Technology	Temperature	Heating rate	Pressure reaction	Reaction time	Medium	Target product
Torrefaction (van der Stelt et al., 2011)	200–300 °C	–	Atmospheric	<2 h	Oxygen-free atmosphere	Torrefied biomass
Slow pyrolysis (Williams and Besler, 1996)	>300 °C	<80 °C/min	Atmospheric	Hours–months	Oxygen-free atmosphere	Char
Fast pyrolysis (Bridgwater, 2012)	>300 °C	Up to 1000 °C/min	Atmospheric or vacuum	<2 s	Oxygen-free atmosphere	Bio-oil
Flash carbonization (Antal et al., 2003)	<800 °C	–	1 MPa	<30 min	Oxygen-free atmosphere/air	Char
Gasification (Bain and Broer, 2011)	600–1800 °C	–	0.1–8 MPa	–	Steam and oxygen or air	Gas
Hydrothermal carbonization (Funke and Ziegler, 2010)	180–220 °C	–	Water saturation pressure	Hours	Water	Char
Wet torrefaction (Yan et al., 2010)	200–260 °C	–	Water saturation pressure	5 min	Water	Char
Hydrothermal liquefaction (Dahmen et al., 2010)	300–350 °C	–	15–20 MPa	5–15 min	Water	Bio-oil
Supercritical water gasification (Antal et al., 2000)	>650 °C	–	>22 MPa	–	Supercritical water	Gas
Plasma pyrolysis (Zhao et al., 2001)	>3000 °C	–	–	–	Ionized gas	Gas

3. PYROLYSIS REACTIONS

Biomass consists largely of cellulose, hemicellulose, and lignin which are complex biopolymers that undergo various transformations as the reaction temperature is increased. This can be seen in the thermogravimetric analysis (TGA) results for the different biomass constituents as given in Fig. 2.2. The mass of the biomass decreases due to the evaporation of water during the initial heating stages, when the biomass temperature is increased to approximately 100 °C. As the temperature increases further to about 160 °C, bound water is removed from the biomass. The pyrolysis vapors at this stage exhibit a negligible heating value. At temperatures above 180 °C, the complex cellulose, hemicellulose, and lignin polymers break down and liberate a mixture of noncondensable gases and condensable vapors (Tumuluru et al., 2010). The condensed tar vapors containing water, acetic acid, methanol, acetone, phenols, and heavier tars are sometimes referred to as pyroligneous acid (Food and Agriculture Organization of the United Nations, 1987). Between 180 and 270 °C, the reactions are endothermic but have sometimes been seen to become exothermic at temperatures above 280 °C. It has been suggested that the exothermicity at this temperature range is due to the decomposition of vapors to coke and secondary vapors. There are contradictory results in the literature regarding the existence of this exothermic peak (Antal and Grønli, 2003). At temperatures above 400 °C, the less volatile components that still impregnate the

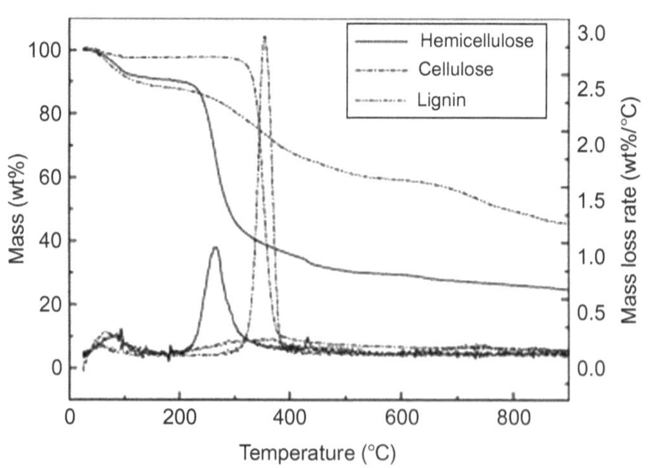

Figure 2.2 TGA curves for cellulose, hemicellulose, and lignin (Yang et al., 2007).

remaining solid char are gradually driven off from the residual, lignin-derived material, thereby yielding a solid product that has a higher fixed carbon content and a lower volatile carbon content. As the temperature is increased above 600 °C, the primary condensable components that are present in the gas phase undergo cracking and polymerization reactions, resulting in a decrease in the bio-oil yield (Dahmen et al., 2010; Tumuluru et al., 2010).

Fast pyrolysis, which aims at maximizing the bio-oil yield, would be promoted by rapid heating of the biomass to temperatures near 500 °C and the rapid removal of the condensable vapors (vapor residence time of a few seconds) to prevent further cracking and polymerization. To allow for the fast degradation of the biomass, small particles (<5 mm) are typically needed to allow for the rapid penetration of the heat into the particle (Venderbosch and Prins, 2010). Slow pyrolysis, aiming at a maximum solid product yield, requires lower heating rates or longer vapor residence times. Peak reaction temperature and solid reaction time have a significant influence on the solid product yield and quality. Although yields are important to overall process economics, care must also be given to the quality of the products. This is discussed in greater detail in Section 5.

A typical mass and energy balance for wood torrefaction is given in Fig. 2.3. It should be noted that yields and qualities can vary substantially with varying process conditions. The reaction equilibrium favors the presence of char, vapors, and gas over biomass. Research has indicated that the fixed carbon within the char is relatively independent of the reaction severity (Antal and Grønli, 2003) where a relative increase in severity is an increase in reaction temperature, an increase in residence time, or both.

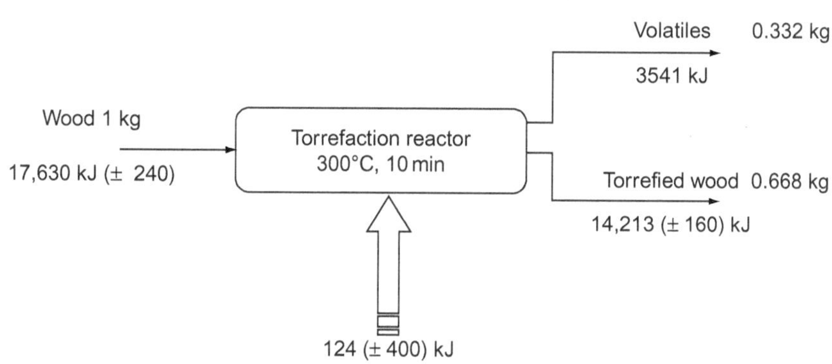

Figure 2.3 Mass and energy balance for torrefaction (Prins et al., 2006a).

3.1. Reaction kinetics

In pyrolysis, reaction kinetics for the rate of solid mass loss (as gas and vapor products are formed) are usually described as a function of the extent of reaction as shown in Eq. (2.1). Generally, first-order reaction kinetics are assumed ($n=1$). The validity of this assumption is questioned elsewhere in the literature (White et al., 2011).

$$\frac{d\alpha}{dt} = k(1-\alpha)^n \qquad [2.1]$$

where α is the extent of reaction, t is the reaction time, k is the reaction rate constant, and n is the reaction order.

The dimensionless extent of reaction, α, is defined by comparing the amount of solid biomass (on a dry, ash-free basis) that has been converted to the maximum amount of mass that is lost after extended periods of time when only char remains as solid product as shown in Eq. (2.2). According to this definition, the value of α will increase from 0 (unreacted biomass) to 1 (char).

$$\alpha = \frac{w_0 - w}{w_0 - w_f} \qquad [2.2]$$

where α is the extent of reaction, w_0 is the initial solid mass, w is the solid mass at a given time t, and w_f is the solid mass after no more mass loss is observable and this equates to the final mass of the char product.

The Arrhenius equation is used to describe the influence of the temperature on the reaction rate constant as shown in Eq. (2.3).

$$k = Ae^{(-E_a/RT)} \qquad [2.3]$$

where k is the reaction rate constant, A is the preexponential constant, E_a is the activation energy constant, R is the universal gas constant, and T is the reaction temperature.

It should be noted that there are some major reservations regarding the applicability of homogeneous reaction theory to heterogeneous processes like biomass pyrolysis. It has also been argued that, for processes where activation energy varies with experimental conditions, there is no real activated state and the Arrhenius equation is not applicable and/or the assumption that the reaction rate is proportional to the unreacted material does not hold. Nevertheless, much previous research has involved the evaluation of reaction rate coefficients by determining the preexponential constant (A) and the activation energy (E_a) according to the Arrhenius equations. To allow comparison between experimental results, this trend seems to be self-perpetuating (White et al., 2011).

The experimental quantification of the activation energy, the pre-exponential term, and the reaction order is subject to significant variation, even for an idealized biomass constituent such as microcrystalline cellulose. Variations in the observed biomass reaction kinetics are due to the natural variation in the composition of biomass but also inappropriately specified reaction model and measurement errors including incomplete mass balances (Wang et al., 2005). Inaccuracies in the quantification of the reaction kinetics and heats of reaction can also occur due to mass and heat transfer limitations, which is especially the case if large particles are used and high heating rates are applied (Grønli et al., 1999). Catalytic effects, which may be caused by the inorganic contaminants (ash) within the biomass, can also distort the analyses of reaction kinetics (Williams and Besler, 1996). These catalytic effects may be reduced by the application of hot water or dilute acid wash methods to remove these inorganic contaminants, thereby increasing the consistency of the results generated by the chosen thermal analysis method (Antal et al., 1998; Várhegyi et al., 1997).

3.2. Reaction models

Numerous reaction models have been proposed as a means of describing the solid- and gas-phase reactions taking place during the pyrolysis of biomass. The reason that so many models have been proposed is that biomass pyrolysis involves numerous complex chemical reactions (decarboxylation, condensation, cracking, and polymerization) without clearly defined reagents or products, thereby allowing the researcher many different routes of abstraction to simplify these complexities. Many of the reaction models were originally developed to describe the behavior of pure cellulose during pyrolysis as this is typically the most abundant constituent within woody biomass. Cellulose, hemicellulose, lignin, and total biomass pyrolysis models are discussed below. Only the reaction models that have found traction in the academic community are highlighted here. The numerous models are reviewed to a greater degree elsewhere in the academic literature (van der Stelt, 2010; White et al., 2011).

3.2.1 Global one-step model

A simple one-step model, shown in Fig. 2.4, can be used to describe the rate of mass loss from a solid substrate such as cellulose and has shown to give a reasonable approximation to the thermal mass loss behavior for cellulose

$$\text{Cellulose} \xrightarrow{k} \text{Char} + \text{Volatiles}$$

Figure 2.4 Global one-step model for cellulose pyrolysis (Repellin et al., 2010).

pyrolysis reactions (Antal et al., 1998). Here, only the reduction in the mass of the solid material is quantified and secondary reactions are ignored, which limits its applicability to situations where in-depth knowledge of the volatile formation is necessary (such as fast pyrolysis). This could be suitable for use when dealing with a simplified solid substrate such as cellulose, but such a reaction model does not have the flexibility to account for variation of the constituent content within more complex materials such as biomass.

3.2.2 Competitive reaction models

A more complicated reaction model is proposed by Broido and involves the irreversible formation of an intermediate anhydrocellulose or "active" form of cellulose which is subsequently converted in competitive, irreversible reactions that can form basic char or volatile tar (condensable vapors) as shown in Fig. 2.5. This char (C_A) can then undergo a series of further irreversible reactions, yielding successive chars (C_B, C_C) and volatiles (condensable vapors and noncondensable gases; Broido and Weinstein, 1972), as cited in Várhegyi et al. (1997) and White et al. (2011). The successive chars presumably display decreased volatile content. No further qualitative differences between these successive chars are given.

The Broido–Shafizadeh reaction model, shown in Fig. 2.6, is a simplified form of the Broido reaction model. It is also based on the irreversible formation of an intermediate product which is converted to either condensable volatiles (bio-oil) or chars and gases, but the subsequent conversion of the char product is no longer considered in this reaction model. From the reaction model, no selectivity is possible between char and gases and subequently no variations in the ratios of char and gas production should be observable. This model would be inappropriate at high temperatures (above 500 °C), where volatiles may be converted to char and gases. Reported

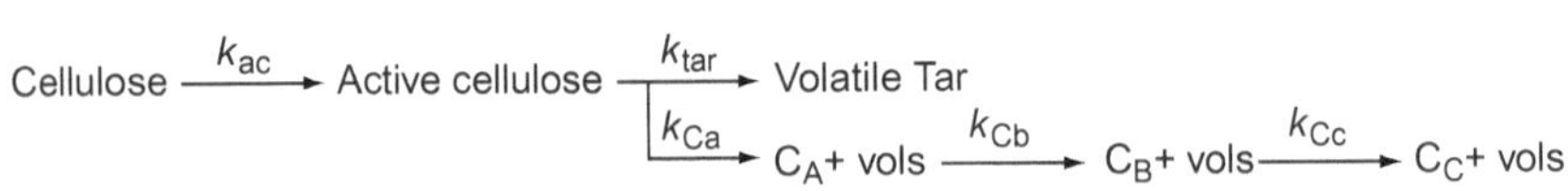

Figure 2.5 Broido model for cellulose pyrolysis (White et al., 2011).

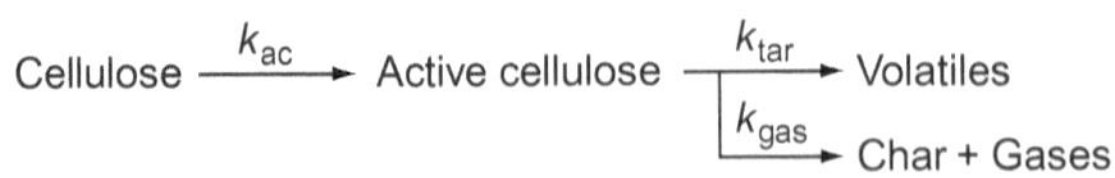

Figure 2.6 Broido–Shafizadeh model for cellulose pyrolysis (Mamleev et al., 2007; Várhegyi et al., 1997).

values for the preexponential terms and activation energies for the various reactions, using cellulose, are given in Table 2.2.

The Broido–Shafizadeh reaction model may be simplified further by the omission of the activation reaction step, since this step is not a rate-limiting step within the conversion of cellulose. This is shown in Fig. 2.7. Further insight into the reaction can be given by the estimation of stoichiometric coefficients to describe the distribution between char and gas products. Reported preexponential constants and activation energies are given in Table 2.3.

For the pyrolysis of hemicellulose (with xylan as a model reaction), a two-step reaction model was proposed by Di Blasi and Lanzetta as shown in Fig. 2.8. The model consists of two competitive reactions, one that yields volatile components (gases and condensable vapors) and one that yields intermediate products which are further converted to produce char and

Table 2.2 Arrhenius equation terms for the Broido–Shafizadeh model (Di Blasi, 1998)

Reaction	A (min^{-1})	E_a (kJ/mol)
k_{ac}	1.7×10^{21}	242.8
k_{tar}	1.9×10^{16}	197.7
k_{gas}	7.9×10^{11}	153.1

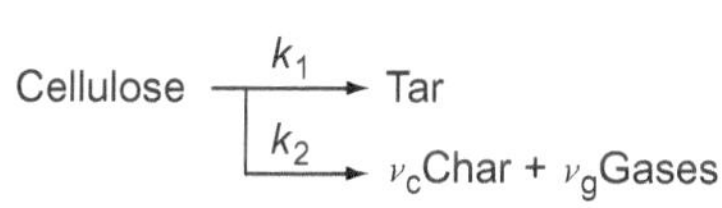

Figure 2.7 Modified Broido–Shafizadeh model for cellulose pyrolysis (Di Blasi, 1998).

Table 2.3 Arrhenius equation terms for the modified Broido–Shafizadeh model (Di Blasi, 1998)

Reaction	A (min^{-1})	E_a (kJ/mol)
k_1	4.0×10^{17}	217.5
k_2	1.6×10^{14}	179.0

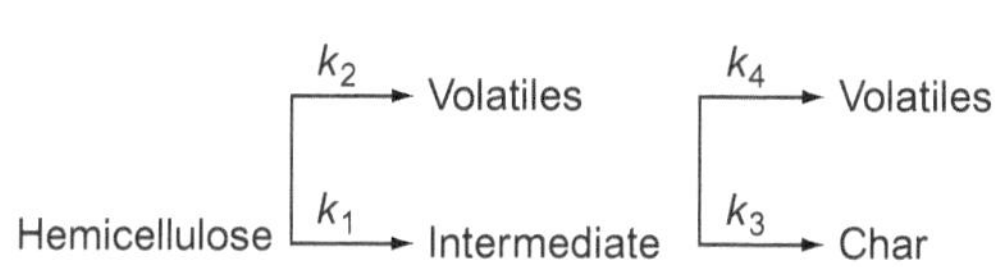

Figure 2.8 Di Blasi–Lanzetta model for hemicellulose pyrolysis (Di Blasi and Lanzetta, 1997).

additional volatiles (Di Blasi and Lanzetta, 1997). Reported preexponential terms and activation energies are given in Table 2.4.

A basic reaction model has been proposed in which biomass undergoes direct conversion into one of the three final products without intermediate products being formed as shown in Fig. 2.9. Each of the three independent, irreversible reactions is described by individual reaction rates, effectively affecting the reaction selectivity. Reported preexponential constants and activation energies are given in Table 2.5. One of the limitations of this form of reaction model is that biomass is treated as a homogeneous substance and the model does not allow for the variations that one would expect to find between various biomass feed materials unless one could obtain the specific reaction parameters of each type of biomass source individually.

A variation of the model takes into account the generation of water in combination with the gas-forming reactions and the cracking of condensable

Table 2.4 Arrhenius equation terms for the Di Blasi–Lanzetta model

Reaction	A (min^{-1})	E_a (kJ/mol)
k_1	1.04×10^{16}	66.19
k_2	1.99×10^{8}	91.48
k_3	2.06×10^{4}	56.36
k_4	3.52×10^{3}	52.59

Adapted from Di Blasi and Lanzetta (1997).

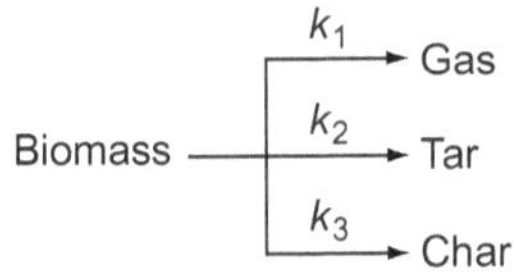

Figure 2.9 Three competitive reaction model for biomass pyrolysis (Shafizadeh and Chin, 1977), as cited in Di Blasi (1998).

Table 2.5 Arrhenius equation terms for the competitive reaction model (Shafizadeh and Chin, 1977), as cited in Di Blasi (1998)

Reaction	A (min^{-1})	E_a (kJ/mol)
k_1	7.80×10^{9}	140
k_2	1.20×10^{10}	133
k_3	6.48×10^{8}	121

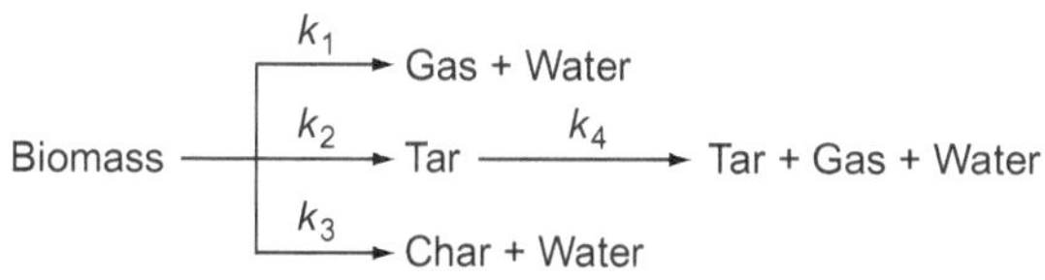

Figure 2.10 Modified three competitive reaction models for biomass pyrolysis (Chan et al., 1988), as cited in van der Stelt (2010).

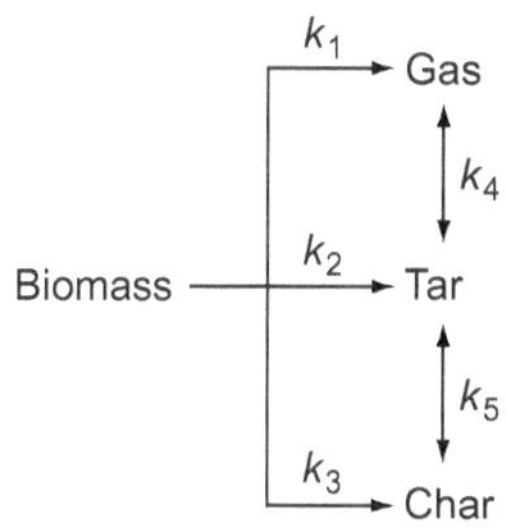

Figure 2.11 Modified three competitive reaction model 2 for biomass pyrolysis (Di Blasi et al., 1999), as cited in van der Stelt (2010).

vapors to form secondary tars, gases, and water as shown in Fig. 2.10. This reaction model accounts for the loss in condensable components at longer gas-phase residence times. These additional reactions obviously require more experimental data to quantify the preexponential constants and activation energies. In addition, char can also be formed from the secondary gas-phase reactions, leading to a slightly modified reaction model as shown in Fig. 2.11.

3.2.3 Pseudocomponent models

Reaction models in which biomass is differentiated into a number of "pseudocomponents" have been proposed as a method to account the differences between hardwood and softwood pyrolysis reactions (Gómez et al., 2005; Várhegyi et al., 1997) and to generate kinetic equations that are more generally applicable. The basis of such reaction models is that the conversion of individual pseudocomponents (e.g., cellulose, hemicellulose, and lignin) takes place at different rates without any interaction (synergistic or antagonistic effects) with other pseudocomponents. Particular attention must be given to how these pseudocomponents are quantified within the original biomass. Frequently, the pseudocomponents are categorized according to the constituents within the biomass, thereby allowing the reactions of cellulose, hemicellulose, and lignin to be modeled independently, for example, through a Broido–Shafizadeh type of reaction. Such an overall scheme is proposed by Koufopanos, where each individual reaction proceeds according to the Broido–Shafizadeh model with the exception that the formation of condensable vapors and noncondensable gases is linked (Koufopanos et al., 1989) as cited in Di Blasi (1998) and shown in Fig. 2.12. Reported preexponential constants and activation energies for various constituents, determined by the TGA of the individual constituents, are given in Table 2.6. It should be noted that the differentiation of biomass into cellulose, hemicellulose, and lignin fractions is subject to significant variability and is highly sensitive to the selection of analytical method.

A variation of this approach is the one in which cellulose reacts according to the Broido–Shafizadeh reaction model (Fig. 2.6), hemicellulose reacts according to the Di Blasi–Lanzetta (Fig. 2.8) model, and lignin reacts according to the global one-step model (Fig. 2.4; Rousset et al., 2006) as cited in Repellin et al. (2010).

Further categorization of pseudocomponents may be made by the differentiation of the various forms of lignin depending on whether the lignin contains relatively more hydrogen, oxygen, or carbon. This naturally increases the

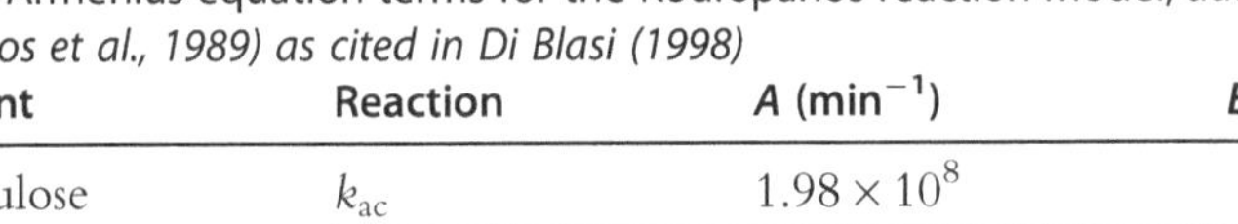

Figure 2.12 Koufopanos model for biomass pyrolysis (Di Blasi, 1998).

Table 2.6 Arrhenius equation terms for the Koufopanos reaction model, *adapted from (Koufopanos et al., 1989) as cited in Di Blasi (1998)*

Constituent	Reaction	A (min^{-1})	E_a (kJ/mol)
Hemicellulose	k_{ac}	1.98×10^{8}	72.4
	k_{tar}	6.60×10^{15}	174.1
	k_{gas}	1.50×10^{15}	172.0
Cellulose	k_{ac}	1.32×10^{16}	167.5
	k_{tar}	5.64×10^{17}	216.6
	k_{gas}	1.86×10^{15}	196.0
Lignin	k_{ac}	1.98×10^{14}	147.7
	k_{tar}	5.16×10^{10}	137.1
	k_{gas}	2.64×10^{9}	122.1

degree of complexity of the reaction model as more intermediate and final products are introduced, each requiring quantification of reaction rates. It is assumed that cellulose reacts to form either activated cellulose or char and water, while the activated cellulose degrades either to form levoglucosan or to a range of volatile vapors through competitive reactions. Hemicellulose is understood to convert to two different activated forms of hemicellulose according to fixed stoichiometry (and thus no selectivity). The first activated hemicellulose species is either converted to gases, light hydrocarbons, and char or to xylose. The second form of activated hemicellulose is converted to gases, vapors, and char. Finally, the various forms of lignin are subject to complex, competitive degradation reactions involving intermediate products and yielding gases, char, and condensable vapors (White et al., 2011). This reaction model is of interest for both fast and slow pyrolysis reactions but is not ideally suited to torrefaction reactions where predominantly hemicellulose degradation takes place.

3.3. Heat transfer and heat of reaction

Pyrolysis and torrefaction are thermally driven processes and therefore the application of sufficient heat to the biomass is crucial to the progress of these reactions. A significant amount of heat may be required for initial drying if the inherent moisture within the biomass feed is high. An indication of the relative heat of reaction is given in Fig. 2.13, which shows results generated through differential scanning calorimetry (DSC; Yang et al., 2007). The heat of reaction of biomass is determined by not only the composition of the

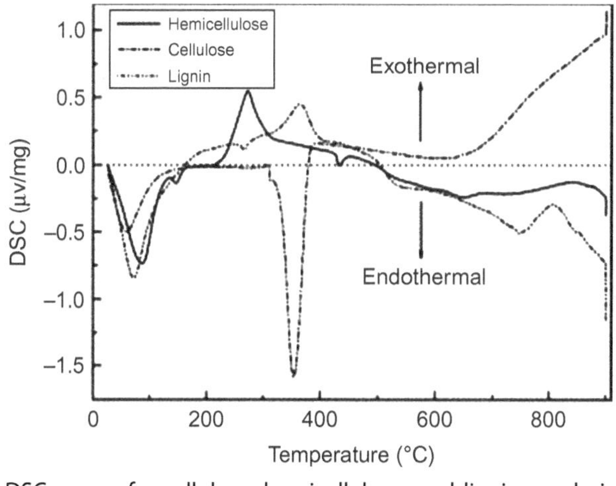

Figure 2.13 DSC curves for cellulose, hemicellulose, and lignin pyrolysis (Yang et al., 2007).

Table 2.7 Range of published heats of reaction for wood and biomass constituents

	Minimum $\Delta H_{Pyrolysis}$ (kJ/kg)	Maximum $\Delta H_{Pyrolysis}$ (kJ/kg)
Wood	−2300	418
Cellulose	−510	120
Hemicellulose	−363	42
Lignin	−455	79

Data from Rousset et al. (2006).

biomass (in terms of cellulose, hemicellulose, and lignin) but also the extent to which the degradation reactions have already taken place. Once cellulose, hemicellulose, and lignin have been broken down, further heat needs to be applied to complete devolatilization of the remaining char (Food and Agriculture Organization of the United Nations, 1987). The devolatilization process is inherently endothermic, whereas exothermic char-forming reactions may sometimes be observed (Yang et al., 2007). Typically, biomass pyrolysis is endothermic at temperatures below 280 °C and may be exothermic at temperatures above 280 °C (Tumuluru et al., 2010). It has also been shown that the reactions become more exothermic when taking place at increased pressures (Antal and Grønli, 2003). Data published in the literature regarding the heats of torrefaction of wood and biomass constituents can vary drastically for both wood and its constituents as shown in Table 2.7.

The transfer of heat to the biomass essentially consists of two interrelated, sequential processes: the external transfer of heat from the hot surface or heating medium to the biomass surface and the internal transfer of heat from the surface of the particle to the interior of the biomass particle. If the rate of external heat transfer limits the overall conversion of the biomass, the heat transfer regime is referred to as being "thermally thin." If internal heat transfer limits the overall heat transfer, it is referred to as being "thermally thick" (Di Blasi, 2000). The dimensionless Biot number, defined for a spherical particle as shown below in Eq. (2.4), is a typical measure of the ratio of external to internal heat transfer rates (Pyle and Zaror, 1984):

$$\mathrm{Bi} = \frac{hR}{K}, \qquad [2.4]$$

where Bi is the Biot number, h is the overall external heat transfer coefficient, R is the particle radius, and K is the thermal conductivity of the particle.

Biot numbers much larger than 1 indicate "thermally thick" heat transfer, and a Biot number significantly smaller than 1 would indicate operation in a "thermally thin" heat transfer regime (Di Blasi, 1996). Indicatively, a Biot number of 1 is obtained when a biomass particle of approximately 2 mm diameter is torrefied with a heat transfer coefficient of 100 W/m^2K (Prins, 2005).

To compare heat transfer rates with the intrinsic reaction rate, a "pyrolysis number" can be defined for a spherical particle according to Eqs. (2.5) and (2.6) (Pyle and Zaror, 1984):

$$\mathrm{Py} = \frac{K}{k\rho c_p R^2}; \quad \text{when Bi} \gg 1, \qquad [2.5]$$

$$\mathrm{Py} = \frac{h}{k\rho c_p R}; \quad \text{when Bi} \ll 1, \qquad [2.6]$$

where Py is the pyrolysis number, K is the thermal conductivity of the particle, h is the overall heat transfer coefficient, k is the apparent reaction rate constant, ρ is the density of the particle, c_p is the specific heat capacity of the particle, and R is the particle radius.

A first (rough) evaluation of the Biot and pyrolysis number indicates which set of simplified equations would be most suitable for modeling the overall process. For small particles such as sawdust (more typical in fast pyrolysis processes), the reaction is kinetically controlled, whereas for larger particles (larger than 50 mm diameter), heat transfer will completely govern the overall conversion of the biomass (Pyle and Zaror, 1984). However, when internal and external heat transfer rates are comparable and the overall heat transfer rate is close to the intrinsic rate of reaction (where the Biot number and pyrolysis number are close to 1), reaction kinetics and heat transfer processes need to be evaluated simultaneously.

A number of the properties that influence the rate of heat transfer (particle size, thermal conductivity, heat capacity, porosity) can change during pyrolysis and torrefaction. This is due to the change in the chemical and physical nature of the material in the reactor as the biomass is converted, the temperature sensitivity of the parameters with respect to temperature during conversion, and structural changes such as shrinkage and the formation of cracks that can take place within the char or torrefied biomass (Larfeldt et al., 2000).

3.4. Mass transfer

Vapors (including water) and gases released during pyrolysis and torrefaction move from the pores within the biomass and char particles and through the solid bulk volume. The flow of vapors and gases from inside a porous biomass particle can be modeled by applying Darcy's Law, thereby accounting for the pressure variations resulting from the generation of these vapors

(Di Blasi, 2000). Liquid droplets that are formed during pyrolysis may also be entrained with the vapors and gases and pass from the biomass bulk to the vapor space in the form of aerosols during fast pyrolysis. It has been shown that the char yield in fast pyrolysis increases with increasing particle size, and this might be due to an increased "cross-linking" between the droplets and the solid char if (gas-phase) residence times increase (Westerhof, 2011). In the case of torrefaction, where mostly solid products and gases are formed and it is unlikely to observe aerosol formation, mass transfer effects would be minimal. This is supported by results obtained from a sensitivity analysis based on a pyrolysis model which indicated that mass transfer was not important for the prediction of the char yield but that it did affect the secondary reactions that occur at higher temperatures (Grønli and Melaaen, 2000; Lanzetta et al., 1997).

3.5. Catalysis

Catalysts are generally not used in torrefaction and slow pyrolysis applications, although the inorganic material inherently present in the biomass has been shown to exhibit catalytic activity, particularly alkali and alkali-earth metals (Patwardhan et al., 2010). The presence of heterogeneous catalysts would have a limited effect on the initial devolatilization reactions due to the spatial separation of the solid biomass and catalytically active minerals. It has, however, been shown that an intermediate liquid compound (ILC) is produced via the initial reactions involved in cellulose fast pyrolysis (Lédé et al., 2002). This liquid may contact the inorganic catalysts to a greater extent than the solid, unreacted biomass, and this would explain how solid catalyst (whether inherently present in the biomass or actively added) may be effective. During fast pyrolysis, this ILC may be ejected from the particle as an aerosol by the flow of product gases. These aerosols and vaporized volatiles could then be brought into contact to an added catalyst.

The use of catalysts in fast pyrolysis processes aims at producing a bio-oil with increased chemical and physical stability (largely through a reduction in the oxygen content of the bio-oil), lowering pyrolysis temperatures, increasing yields of target components within the bio-oil, and improving miscibility for co-feeding with existing petrochemical refinery streams. The catalyst may be applied directly within the biomass pyrolysis reactor ("*in situ*" catalysis) or the condensable vapors may be handled separately in a dedicated, downstream reactor ("ex-bed" catalysis; Venderbosch and Prins, 2010). Solid, *in situ* catalysts can act as heat carriers for biomass pyrolysis applications, while ensuring that the reactive pyrolysis products are rapidly brought into contact with the catalyst. The catalyst would then be recovered with the char formed during pyrolysis. Burning off the char or coke formed during pyrolysis could then directly heat the catalyst particles which would be reintroduced to the pyrolysis reactor and allow for heat integration. Ex-bed catalysis has the benefit that the second, catalyst-containing reactor may operate at different temperatures, pressures, or flow regimes compared to the biomass pyrolysis reactor and this decoupling could be used to improve the efficacy of catalyst application.

In addition to the use of catalysts in biomass pyrolysis, they may be used for bio-oil upgrading (posttreatment). The composition of the bio-oil, containing numerous highly oxygenated compounds, may be controlled by using catalysts typically employed in the petrochemical industry. Fluidized catalytic cracking (FCC) catalysts, such as a ZSM-5 catalyst, may be used to crack the bio-oil into smaller molecules and liberating bound oxygen in the form of CO_2, CO, and H_2O. The hydrocarbon yields that have been observed from bio-oil cracking have typically been fairly low (20 wt% or 40% energetically) due to increased coke formation. Hydrotreating bio-oil is also possible at elevated temperatures and in hydrogen-rich atmospheres through the use of suitable hydrodeoxygenation catalysts. The majority of the hydrotreating work has been focused on the nonselective removal of oxygen from the bio-oil, although it may be argued that a selective removal of oxygen, whereby only certain oxygen-containing species are converted, may be more suitable. Research has shown that bio-oil produced during pyrolysis may be hydrotreated at temperatures between 300 and 400 °C, at residence times of longer than 1 h (Venderbosch and Prins, 2010). The hydrotreating processes typically operate at pressures of up to 20 MPa. The catalysts include NiMo-, CoMo-, and Ruthenium-based catalysts (Bridgwater, 2011).

4. FEED PROPERTIES RELEVANT TO REACTOR DESIGN

One of the major challenges presented by biomass is that it is subject to large variations in its chemical and physical properties. This is true, not only between the different plant species but even between the various parts of an individual plant. The handling of the biomass upstream of the reactor can also have a significant influence on the particle size distribution and moisture and ash content of the feed. Typical ultimate (elemental) analyses results for various biomass sources are given in Table 2.8; although the oxygen content reported for wheat straw seems to be a spurious result, its possible value could be 39 wt%.

Table 2.8 Ultimate analyses of typical biomass materials (McKendry, 2002)

Material	C (wt%)	H (wt%)	O (wt%)	N (wt%)	S (wt%)	Ash (wt%)
Cypress	55.0	6.5	38.1	–	–	0.4
Ash	49.7	6.9	43.0	–	–	0.3
Beech	51.6	6.3	41.4	–	–	–
Wood (average)	51.6	6.3	41.5	0	0.1	1
Miscanthus	48.1	5.4	42.2	0.5	<0.1	2.8
Wheat straw	48.5	5.4	3.9	0.5	0.4	4
Barley straw	45.7	6.1	38.3	0.4	0.1	6
Rice straw	41.4	5	39.9	0.7	0.1	–

Table 2.9 Cellulose, hemicellulose, and lignin content of various types of biomass (McKendry, 2002)

	Cellulose (wt%)	Hemicellulose (wt%)	Lignin (wt%)
Softwood	35–40	25–30	27–30
Hardwood	45–50	20–25	20–25
Wheat straw	33–40	20–25	15–20
Switchgrass	30–50	10–40	5–20

4.1. Biological constituent content

The various biomass constituents (e.g., the cellulose, hemicellulose, and lignin) all react at different temperatures. These are present in differing amounts for different biomass types as shown in Table 2.9. It should also be noted that significant variation may even be found between different tissues of the same type of plant.

The most abundant constituent of biomass is cellulose. Cellulose makes up the tough, water-insoluble material that is present in the protective cell walls of plants and thereby lends these properties to the portions of the plant where it is found (stems, stalks, trunks, and woody portions of plants; O'Sullivan, 1997). Cellulose is a polysaccharide homopolymer consisting of long, linear chains of up to 10,000 repeating units of D-glucopyranose molecules which are linked by β-1,4-glycosidic bonds (White et al., 2011). Due to the ordered structure of cellulose, semicrystalline behavior is observable, with hydrogen bonds stabilizing the cellulose molecules when aligned in parallel. Multiple cellulose fibers can be found together to form microfibrils which are grouped together to form cellulose fibers. These fibers are bound together by both hemicellulose and lignin. Cellulose undergoes thermal degradation at temperatures between 275 and 350 °C (Chen and Kuo, 2010).

Hemicellulose consists of shorter, more-branched polymers in comparison with cellulose and is found between cellulose fibers. Hemicellulose, as a heteropolymer, can consist of a variety of C5 sugars (pentoses such as xylose and arabinose) and C6 sugars (hexoses such as galactose, glucose, and mannose) as well as sugar acids (Saha, 2003). In deciduous wood (hardwood), hemicellulose consists largely of xylan (a pentose polysaccharide consisting of xylose), whereas coniferous wood (softwood) contains significantly more glucomannan (containing glucose and mannose) within the hemicellulose fraction (Ciolkosz and Wallace, 2011). Hemicellulose has a less ordered structure than cellulose. Due to the fact that hemicellulose is the least stable of the three major constituents of biomass, it starts degrading at lower temperatures via various dehydration, deacetylization, and depolymerization reactions at temperatures between 150 and 350 °C (Chen and Kuo, 2010), thereby contributing significantly to the overall conversion of biomass at the lower temperatures encountered during torrefaction.

Lignin is an amorphous, highly branched, aromatic polymer which binds cellulose and hemicellulose together. There are three basic types of phenylpropane-based monomers which make up lignin: sinapyl-, *p*-coumaryl-, and coniferyl-alcohol. Softwood contains lignin predominantly consisting of polymerized coniferyl-alcohol monomers in comparison with hardwood lignin which is largely based on coniferyl and sinapyl alcohol monomers. The lignin found in grassy biomass consists of all three types of monomers (Crocker and Andrews, 2010). Coniferous wood (softwood) usually contains slightly more lignin than deciduous (hardwood) species. Lignin undergoes thermal degradation between 250 and 900 °C, and the rate of these degradation reactions is known to be slower than those of cellulose and hemicellulose (Venderbosch and Prins, 2010). Lignin pyrolysis, in comparison with the pyrolysis of cellulose and hemicellulose, is also known to favor the production of char, thereby resulting in increased char yields that are observed during the pyrolysis of lignin-rich biomass (Antal and Grønli, 2003). The relative degradation of the three constituents over the pyrolysis temperature range is illustrated in the TGA curves given in Fig. 2.2.

In addition to cellulose, hemicellulose, and lignin, biomass also contains extractives such as fats, waxes, alkaloids, proteins, phenolics, simple sugars, pectins, mucilages, gums, resins, terpenes, starches, glycosides, saponins, and essential oils. Some of these compounds can be removed from the biomass through extraction with polar or nonpolar solvents (Mohan et al., 2006). The extractives present usually evaporate or are cracked as the biomass temperature increases during pyrolysis.

4.2. Moisture content

Water in the biomass feed has a major impact on the overall economics of the pyrolysis process in addition to the reactions themselves. The moisture within the biomass may be present in a number of forms. Water that is adsorbed onto the hydroxyl groups of cellulose and hemicellulose molecules through hydrogen bonds is typically referred to as "bound" water. Unbound, liquid water may also be found occupying the voids that are present within biomass if the moisture content (including bound water) exceeds the fiber saturation point (typically 30% of the oven-dry weight for most types of wood). Water vapor can also occupy the voids within biomass (Grønli and Melaaen, 2000). Moist biomass requires drying; thus, additional energy and increased processing capacity are needed. The heat of vaporization of the water must be accounted for by the combustion of feed, products, or another source of energy (Antal and Grønli, 2003). This consideration limits many high-moisture biomass sources unless sufficient drying can be achieved economically such as through sun drying in dry climates.

A benefit of performing any drying outside the reactor is that the pyrolysis vapors will have an increased calorific value as there will be less dilution with water vapor, although this could be at the expense of process complexity and cost. In the case of fast pyrolysis, water will be included in the final bio-oil product and this will reduce the calorific value of the bio-oil and affects the bio-oil viscosity which is known to range from 10 to 1000 cP. The organic material within the bio-oil is highly polar due to the large amount of oxygenated compounds. This results in the bio-oil being hydrophilic, and the presence of a distinct, aqueous phase is only observed if the water content of the bio-oil is above 30–45 wt%.

Water may also have an impact on reaction rates and product yields. Some catalytic effects have been claimed with an apparent reduction in the activation energy required to drive the wood pyrolysis reactions and resulting in faster reaction rates, although it has also been indicated that conversion in a steam atmosphere has practically no impact on either yields or kinetics (Antal and Grønli, 2003). One potential advantage of performing pyrolysis in steam atmosphere is the production of char with high surface areas (up to 820 m^2/g; Antal and Grønli, 2003) in comparison with typical values of below 500 m^2/g (Major et al., 2009). This high surface area could be created in a process similar to the steam activation process often employed in the production of activated carbon.

The conversion of biomass in hot, compressed, liquid water has also been investigated at laboratory scale and it was found that an energy-dense (higher heating value (HHV) of up to 26.5 MJ/kg) solid product is formed in addition to gases (mostly CO_2) and aqueous compounds (volatile acids and monosaccharides). No distinct liquid hydrocarbon phase was reported for wet torrefaction temperatures in the range of 200–260 °C. The reactions that are involved are reported to be mostly hemicellulose hydrolysis followed to a lesser degree by cellulose hydrolysis and monosaccharide degradation at higher temperatures. The heat of reaction of wet torrefaction was found to be of the same order of magnitude as that of conventional, dry torrefaction (Funke and Ziegler, 2010), with the advantage that feed material would not require drying prior to conversion. Subsequent drying of the solid product would be enhanced by its increased hydrophobicity as seen in its decreased equilibrium moisture content compared to unprocessed biomass when exposed to a humid atmosphere. Biomass that underwent wet torrefaction at 260 °C had a 5.3-wt% equilibrium moisture content compared to 15.6 wt% for the biomass (loblolly pine) in an atmosphere with a relative humidity of 83.6 wt% at 30 °C (Yan et al., 2009). A reduction in the mineral content is also achieved during wet torrefaction (Allen et al., 1996), similar to the leaching process described in Section 4.3, which could be beneficial to downstream gasification or combustion processes. Hydrothermal liquefaction aims at maximizing the liquid product by using higher temperatures (300–350 °C) than for wet torrefaction. The achievable bio-oil yields are approximately 50% by mass (Dahmen et al., 2010).

The mass flows of the streams being heated and cooled during pyrolysis are relatively larger when moist biomass is used due to the additional water within the process, and this would require highly efficient heat integration to ensure that minimal energy is lost from the process.

4.3. Ash content

Biomass feed sources contain varying levels of inorganic material (alkali metals, heavy metals, and chlorine) which are reported as the ash content through proximate analysis. The presence of inorganic minerals is also dependent on the type of biomass, its environment, the use of mineral fertilizers, and the amount of inorganic matter that is collected with the biomass during harvesting. Increased levels of alkali metals are generally seen in herbaceous biomass and agricultural and short-rotation energy crops in comparison with woody biomass (Yu et al., 2010). Alkali and alkali-earth metals, particularly sodium and potassium, and sulfur- and phosphorous-containing ammonia salts can have a catalytic effect during pyrolysis, favoring the production of char and gases at the expense of the condensable vapor product (Antal and Grønli, 2003; Patwardhan et al., 2010; Venderbosch and Prins, 2010). The mechanism whereby intermediate reaction products are brought into contact with the inherent inorganic material is discussed in Section 3.5. In addition to the effect on the bio-oil yield, a reduction in the mineral content of the biomass feed has also been shown to have a favorable impact on the heating value of the bio-oil product and to result in a reduction in the initial biomass decomposition temperature and an increase in the pyrolysis conversion rates (Raveendran et al., 1995). Through pyrolysis, the inorganic material will be concentrated in the solid char product and this can potentially limit char use in downstream conversion processes. If a relatively clean biomass source is used, the minerals in the char could be beneficial, as in the case of biochar, where these minerals are then returned to the soil and allow for nutrient recycling. The ash-related problems that have been observed in the downstream conversion of char include increased slagging, equipment fouling, and increased corrosion, particularly if chlorine and alkali metals are present (Dayton et al., 1999). Reducing the alkali metal content will only prevent the prevalence of slagging in downstream thermal processes if the fusion temperature of the ash within the char product is increased above the operating temperature of the downstream process (boiler, gasifiers, or reactor) and this is not always the case when leaching is applied. Reducing the chlorine content can be even more important as it minimizes volatilization of the alkali metals during char combustion or gasification, thereby keeping the alkali metals in the solid phase and away from sensitive process equipment such as heat exchanger tubes. Downstream equipment can thus be designed to operate at lower fouling rates (Turn et al., 1997).

The leaching of minerals from biomass commonly takes place in the sugar refining industry, where leaching removes the sucrose from the solid bagasse residue. During this water-based leaching process, significant removal of inorganic material is also achieved with the result that sugarcane bagasse has desirable fuel qualities (Olsson et al., 1997). A drawback relating to water leaching is the consumption and handling of water at an additional cost, not only in the leaching process but also in a subsequent water removal section. Inorganic material can also be removed from the product char. Through the leaching of char produced during pyrolysis at 550 °C, it was shown that almost all of the chlorine and approximately 90% of the potassium could be removed during a warm water (80 °C) leaching step, although this was not as efficient as leaching the minerals from the original biomass (Dayton et al., 1999). Design variables relating to leaching are the ratio of water (and the presence of acids) to biomass, the temperature of the water, its duration, and the application of mechanical means to dewater the biomass. These parameters need to be optimized. The addition of acid to the leaching water increases the efficiency of the alkali metal leaching. As a water wash is sufficient to reduce the release of alkali by 90% (Davidsson et al., 2002), an acid wash may not be warranted. The increased potassium content within the leachate stream might increase the value of the stream as a feed for the fertilizer industry (Dayton et al., 1999); however, the presence of heavy metals within the leachate could preclude this application.

4.4. Morphology

Another challenge in working with biomass is that there are large morphological variations within biomass feeds. Biomass varies in shape, size distribution, density, and strength. Woody biomass particles also display anisotropic behavior where physical properties depend on whether they are measured along, across, or tangential to the grain of the wood. Thermal conductivity is reduced by as much as 1/3 when measured across or tangential to the grain, while the gas permeability can be reduced by a factor of 10,000 (Di Blasi, 2000).

The particle size distributions of the feed material can affect the overall reaction progress. Large particles will experience reduced internal heating rates, and at some critical level, these limitations can impede the overall reaction progress. A narrow particle size distribution, with minimal difference between the larger and smaller particles, would ensure that the extent of pyrolysis of all the particles would be fairly uniform for a given reactor residence time.

The manner in which the biomass is fed to and through the reactor (in the case of continuous processes) can result in particle attrition. This can be beneficial as smaller particles may be pyrolyzed faster. Disadvantageously, large amounts of fine particles could not only pose additional technical problems in feeding and product recovery but also create a process safety risk.

5. PRODUCT SPECIFICATIONS RELEVANT TO REACTOR DESIGN

The products of thermochemical conversion can be categorized by their application. The specifications of the various products and their influence on the pyrolysis process are discussed below.

5.1. Noncondensable gases

The noncondensable or permanent gases that are formed during pyrolysis or torrefaction are, by definition, the gases that remain once the reaction vapors have passed through some form of condensation stage to separate the bio-oil from the permanent gases. In some applications, this separation may not be performed and a combined stream of reaction vapors and gases may be processed together, typically through combustion to generate the heat required to drive the process or to dry the biomass feed. In primitive slow pyrolysis processes such as early charcoal production processes, these gases are simply released to the atmosphere (Food and Agriculture Organization of the United Nations, 1987).

The composition of the noncondensable gases will be largely determined by the temperature at which the pyrolysis takes place and the temperature at which the vapors are condensed. At lower reaction temperatures (such as those encountered during torrefaction), these gases consist of carbon monoxide and carbon dioxide, whereas higher temperatures result in increased methane and hydrogen content. The variation in gas composition from the fast pyrolysis of 3 mm beech particles for varying temperatures is shown in Table 2.10 (Wang et al., 2005).

5.2. Bio-oil

Bio-oil (also commonly referred to as "tar") is used to describe the liquid produced from the condensation of the pyrolysis vapors. It is a brown to black liquid depending on the chemical composition and bio-oil has a heating value which is approximately equal to that of wood, although it has an increased density of up to 1200 kg/m^3. The viscosity of bio-oil can vary from 10 to 1000 cP, depending on the water content and the composition of the bio-oil (Venderbosch and Prins, 2010). The highly polar nature of the bio-oil results in the bio-oil being miscible with water and phase

Table 2.10 Variation in gas composition with changing fast pyrolysis temperature (Wang et al., 2005)

	Temperature (°C)				
	350	**400**	**450**	**550**	**800**
H_2 (vol%)	0.9	1.4	0.8	3.4	13.8
CO (vol%)	37.4	28.5	38.7	44.9	50.7
CO_2 (vol%)	60.5	68.6	58.5	37.2	13.5
CH_4 (vol%)	<0.1	<0.1	<0.1	10.9	13.9
C_2H_4 (vol%)	0.5	0.7	0.8	1.7	5.8
C_2H_6 (vol%)	0.4	0.5	0.7	0.9	1.1
C_3H_6 (vol%)	0.2	0.3	0.3	0.7	1.0
C_3H_8 (vol%)	0.1	0.2	0.2	0.2	0.1

separation typically occurs at water concentrations of above 30–45 wt%. The presence of organic acids in the bio-oil results in a pH between 2 and 4, and this impacts the downstream processing equipment. The factors affecting the properties of the bio-oil are the biomass feed material (particularly moisture content) and the reaction conditions.

By the fractional condensation of the vapors, various bio-oil fractions can be obtained, thereby steering the bio-oil properties to some extent (Westerhof et al., 2011). Bio-oil is not a stable product and the aging will tend to produce a bio-oil which is more viscous and could also allow for the formation of a distinct aqueous phase and a viscous organic phase. This aging process happens over a period of months at room temperature but is accelerated at higher temperatures where polymerization reactions are promoted (Venderbosch and Prins, 2010).

Bio-oil can be considered as the condensed fragments of the cellulose, hemicellulose, and lignin present in the biomass feed. The bio-oil typically contains a large number of different oxygenated compounds with molecular weights up to 10,000 g/mol. The larger molecules that are present in the bio-oil are not able to evaporate at the temperatures that are involved during typical pyrolysis; therefore, these molecules are entrained from the solid biomass as aerosols. Some molecules present in bio-oil are also the product of secondary degradation reactions such as oligomers of the primary volatile molecules. Analysis of these large molecules by gas chromatography is not possible as coking occurs when the bio-oil is vaporized after being brought onto the chromatographic column. Additionally, a large fraction will not be evaporated at all at temperatures below 300 °C (the usual injection temperature of a GC). More appropriate methods include solvent extraction methods combined with GCMS analysis. Detailed analyses results can be found elsewhere (Venderbosch and Prins, 2010). High-molecular-weight compounds can also be detected using liquid chromatography, including high-performance liquid chromatography and gel permeation chromatography.

5.2.1 Fuel oil

One of the most apparent uses for bio-oil is combustion for the generation of heat or power. Bio-oil typically has a lower heating value (LHV) of 14–18 MJ/kg, which is approximately 40–50% of that of a liquid hydrocarbon fuel on a mass basis (60% on a volumetric basis). The reduced LHV is due to both the presence of water and the high oxygen content of the bio-oils (Oasmaa and Czernik, 1999). The combustion characteristics of bio-oil are such that it may be used to replace heavy and light fuel oils in industrial boiler applications, but significant differences in ignition, viscosity, pH, and stability are reported. Bio-oil burns with low ash deposition rates, as inorganic material is concentrated in the char during pyrolysis (Venderbosch and Prins, 2010). Recently, co-combustion of pine-derived bio-oil with heavy fuel oil (HFO) was successfully proved in a 9-MW_{th} test boiler, using an optimized Y-jet steam-assisted atomizer. Bio-oil was fired at 2.6 MW_{th} and HFO at 4.7 MW_{th}. Combustion of the bio-oil gave a significant lower NO_x emission compared to HFO. The propensity of the ash from bio-oil to cause fouling appears to be about half that of HFO. Based on these results, Stork Technical Services claim to be capable of designing, producing, and installing burners suitable for combustion of pyrolysis oil on a commercial basis in a typical range of 5–100 MW_{th}. Existing boiler systems can be retrofitted to make them suitable for pyrolysis oil cofiring (Rinket and Toussaint, 2012).

5.2.2 Transportation fuel

Bio-oil may also be used directly as a fuel in diesel engines or gas turbines. Diesel engines are less sensitive to possible contaminants present in the bio-oil, particularly larger, slowly rotating diesel engines. One of the problems is that there is insufficient bio-oil available for large-scale testing and that major engine manufacturers are not showing sufficient interest into adapting their engines to accommodate bio-oil. Bio-oils are usually highly acidic and may contain abrasive particles and this can cause significant engine deterioration (wear and erosion of the injection needles); however, these effects may be mitigated by adapted engine design and selecting different metals for engine construction. The high viscosity and instability of the bio-oil can also cause problems to the fuel injection system of an engine, and carbon deposits may form in the combustion chamber and exhaust valves. Another problem associated with the use of bio-oil in diesel engines is that the cetane number is poor (typically below 10) and cetane-enhancing additives or the application of a dual fuel system are necessary (Venderbosch and Prins, 2010). A conventional one-cylinder 20-kW$_e$ diesel engine has been run on bio-oil. After modification of the fuel injection system, the engine could be fuelled with bio-oil for a "duration" experiment lasting 40 h without any notable effect on flue gas emissions and fuel consumption (Van de Beld et al., 2012).

The use of bio-oil within gas turbines has also been investigated, although this work has been limited. One of the hurdles in this type of research is again the limited availability of a standardized bio-oil product. Particularly, Orenda (Canada) is looking at ways to operate their GT 2500 gas turbine on bio-oil and has partnered with a bio-oil producer, Dynamotive (Canada), to promote development. The gas turbine requires modifications to the hot section and the combustion system in order to operate. Additionally, diesel is required to start the turbine after which the diesel is gradually substituted with bio-oil. The gas turbine demonstrated stable operation and was able to run at a maximum power output of 2.5 MW$_e$ (Oasmaa et al., 2005).

In addition to being fed to diesel engines or gas turbines, bio-oil (or slurries of char and bio-oil) may be fed to gasifiers for the production of synthesis gas or producer gas, after which the gaseous product may be converted to heat and power through a gas turbine. If synthesis gas is produced, it may also be used for the downstream production of Fischer–Tropsch fuels and chemicals. Some of the problems associated in feeding bio-oil to gasifiers are related to the bio-oil supply system which needs to be able to accommodate acidic material (Venderbosch and Prins, 2010).

Since bio-oil may not be suitable for direct application as a fuel in diesel engines (without modification), the bio-oil could be upgraded through cracking or hydrotreating, as discussed in Section 3.5. In addition to the catalytic upgrading of bio-oil (either during pyrolysis or in a decoupled process), the bio-oil may also be corefined with conventional petrochemical streams. The initial hydrotreating of bio-oil and subsequent corefining of bio-oil with conventional petrochemical feedstocks to an FCC process has also been shown to be possible at high efficiencies. The corefining of upgraded oils has only been proved at lab scale. Information regarding the catalytic upgrading of bio-oil has been given in Section 3.5.

5.2.3 Chemicals

Unfractionated bio-oil may be used as a substitute for phenols in the manufacture of particleboards. The ability of the bio-oil to promote cross-linking when mixed with a formaldehyde resin (thereby improving the strength of the particleboard) is attributed to the presence of lignin-derived compounds. Research has shown that up to 50% of the phenol may be replaced with bio-oil. Unfractionated bio-oil may also be reacted with ammonia, urea, or other amino compounds to form stable amines and amides which may be used as slow release fertilizers. Additionally, the lignin degradation products may also act as soil conditioners by regulating the soil acidity and the effects of excess metals in the soil. Dynamotive (Canada) has collaborated with fertilizer manufacturers, but no commercial products have reached the market. Bio-oil may be mixed with water to produce a product known as "liquid smoke." This product is commercially available and is sprayed onto meat products before cooking to impart a smoked flavor. Artificial flavorants based on bio-oil are produced by Red Arrow Products (USA) and ProFagus (Germany). The phenolic compounds in bio-oil such as phenol, cresol, eugenol, and xylenol and alkylated phenols are present in bio-oil and may be extracted at recoveries of up to 50%. The quantities of the lighter, more valuable phenolic compounds are limited due to incomplete cracking of the lignin. Once the phenolic compounds have been extracted, they could be used to replace creosote in wood-treatment applications. Numerous sugars are present in bio-oil, but only levoglucosan can be identified through gas chromatography. Levoglucosan is a product which has a high value but small market. It can be used as an intermediate for fermentation. Levoglucosan yields are increased through pretreatment steps such as demineralization and prehydrolysis of biomass. Levoglucosenone is also detectable within bio-oil, and it may be used for the production of antibiotics, pheromones, rare sugars, butenolide, immuno-suppressive agents, and other fine organic chemicals. Hydroxyacetaldehyde is also present in bio-oil. If extracted, it can be used in the browning of cheese and meat products. It may also be used as a precursor for glyoxal production. The carboxylic acids that are present within bio-oil, such as formic and acetic acid, can be used to form salts such as calcium formate and calcium acetate which may be used in the de-icing of roads, SO_2 removal during fossil fuel combustion, or as a catalyst during coal combustion. Furfural and furfuryl alcohol are also present in bio-oil (Venderbosch and Prins, 2010).

5.3. Char

Char is the generic term for the solid material that is produced through pyrolysis but may be referred to by different names based on the intended application. Torrefied biomass is a variant of char produced at lower temperatures. This is discussed below.

5.3.1 Charcoal

"Charcoal" is used to describe the char produced by slow pyrolysis when the intended use is as a fuel or as a reducing agent in some metallurgical smelting applications, where it is preferred over coal due to its lower ash content. The slow pyrolysis of hardwood biomass is one of the oldest chemical conversion processes known to man. Since prehistoric times, biomass has been pyrolyzed to form char which was subsequently used to fuel fires for light and heat and even for artwork (Antal and Grønli, 2003). The properties of various charcoals, obtained from various sources and produced at various temperatures, are given in Table 2.11, where the charcoals were prepared in an experimental unit at residence times of 2 h using powdered (177–250 μm) biomass (Cordero et al., 2001). Ultimate analysis refers to the quantification of the elements (typically, C, H, O, N, and S) present in biomass and char, whereas proximate analysis describes the charcoal in terms of fixed carbon, volatile carbon, and ash according to standardized analytical methods. The HHV indicates the suitability of the material as a fuel.

An increased volatile content of the charcoal promotes the ignition of the charcoal, but this typically results in more smoke during combustion and a reduction in the heating value of the charcoal. The volatile matter content of domestically-used charcoal is typically between 20 and 30 wt%. Industrial charcoal used in metallurgical applications has a typical volatile content of 15 wt%. The ash content can vary from about 0.5 to 5 wt% depending largely on the feed material selection. The LHV of charcoal ranges from 28 to 33 MJ/kg depending on feed material and pyrolysis conditions (Antal and Grønli, 2003). Increasing the pyrolysis temperature removes more volatile material and also tends to reduce the oxygen/carbon and hydrogen/carbon ratios, thereby yielding char with an increased fixed carbon, ash content and HHV. Lump charcoal, consisting of large particles, is a more desirable product for domestic fuel applications because it is easier to ignite and handle. Fine charcoal particles may be briquetted; however, this will have an impact on

Table 2.11 Proximate and ultimate analyses of lignocellulosic and carbonaceous materials

Biomass	$T_{Pyrolysis}$ (°C)	C (wt%)	H (wt%)	N (wt%)	O (wt%)	Fixed carbon (wt%)	Volatile carbon (wt%)	Ash (wt%)	HHV (MJ/kg)
Oak	–	50.3	6.1	0.2	41.1	14.1	83.6	2.3	19.725
	300	58.8	5.0	0.3	34.5	33.4	65.1	1.5	23.130
	350	75.7	3.3	0.6	19.1	55.3	43.4	1.3	27.345
	400	76.9	3.3	0.4	18.2	64.2	34.5	1.3	28.840
	450	81.2	3.0	0.4	13.6	76.4	21.8	2.8	30.650
	500	83.0	2.5	0.6	11.2	79.7	17.5	2.8	30.235
	550	87.1	2.4	0.5	6.9	82.2	14.7	3.1	32.720
	600	89.4	2.2	0.4	4.8	83.6	13.2	3.2	32.645
Pine	–	48.9	6.0	0.1	44.5	17.0	82.5	0.5	20.215
	300	57.8	5.0	0.2	36.5	31.3	68.1	0.6	22.840
	350	72.1	4.1	0.2	22.5	49.4	49.5	1.2	26.680
	400	74.7	3.6	0.2	20.3	62.2	36.5	1.3	28.245
	450	78.3	3.2	0.2	16.8	71.2	27.4	1.4	29.632
	500	81.8	3.0	0.2	13.4	78.1	20.2	1.7	30.738
	550	86.1	2.5	0.2	9.6	80.2	18.1	1.7	31.525
	600	87.4	2.2	0.3	8.5	84.9	13.4	1.7	31.913
Eucalyptus	–	47.3	6.0	0.1	46.5	17.9	82.0	0.1	20.080
	700	92.7	1.6	0.4	3.3	91.5	6.6	1.9	32.204
Wet straw	–	42.2	5.7	1.0	46.5	17.4	82.0	0.1	18.154
	600	65.2	2.2	1.5	6.1	63.5	11.8	24.7	24.541
Olive stones	–	49.0	6.1	0.8	42.0	19.5	78.3	2.2	20.230
	600	85.3	2.6	0.8	4.7	82.1	11.3	6.6	31.095
Almond shells	–	48.8	5.9	0.5	43.7	18.4	80.5	1.1	19.916
	600	86.8	2.6	0.8	6.1	86.5	9.8	3.7	32.390

Adapted from Cordero et al. (2001).

production costs due to additional binder requirements and will increase overall process complexity. Briquettes also have reduced heating values due to increased ash content if inorganic binders are used. Further information on pelletization and briquetting is presented in Section 6.8.

5.3.2 Torrefied biomass

Torrefied biomass refers to biomass which has undergone thermochemical conversion at significantly milder temperatures (200–300 °C) and at reduced reaction times compared to char. The benefit of such a conversion process is that the torrefied biomass product has superior properties in comparison with raw biomass when used within further thermochemical processes such as combustion, gasification, and co-combustion within coal-fired power plants. Torrefied biomass is produced with significantly higher energy efficiencies than charcoal. The improved properties include increased heating value, grindability, hydrophobicity, and a decrease in the susceptibility to biological degradation when compared to the unprocessed biomass (Ciolkosz and Wallace, 2011). The LHV may be increased from 17.7 to 20.4 MJ/kg through torrefaction. The bulk energy density will only be improved through a densification process such as pelletization or briquetting. In such an instance, the bulk energy density can be increased from 4.6 GJ/m^3 for torrefied biomass to between 14.9 and 18.4 GJ/m^3 for pellets of torrefied biomass. This can be compared to a bulk energy density of 5.8 MJ/m^3 for unprocessed wood (Kiel, 2007).

The purpose of torrefaction (including logistics, feed pretreatment, conversion, product handling, and the integration of heat within the process) is to produce a solid product that meets customers' requirement as economically and efficiently as possible. The use of torrefied biomass as a "drop-in" substitute for coal implies that the grindability and the combustion characteristics (heating value, reactivity, and density) are important product properties. The torrefaction process parameters that will have a significant bearing on the process feasibility include the nature of the biomass feed, the efficient transfer of heat to and within the biomass particles, and the integration of heat and energy flows within the torrefaction process.

The development of torrefaction at industrial scales (more information is given in Section 7.3) is relatively limited although it is gaining momentum, due to the increased interest in the use of biomass as a renewable source of energy and the fact that torrefied biomass can be integrated into existing coal-based technology. The increase in the energy density that can be achieved with torrefaction and pelletization also improves the logistics of utilizing remote biomass resources. It has been suggested that distributed torrefaction processes could be economically advantageous to fast pyrolysis processes as an intermediate preconversion process in the generation of synthesis gas and Fischer–Tropsch liquids (Magalhães et al., 2009; Uslu et al., 2008).

5.3.3 Activated carbon

Activated carbon, in granular or powdered forms, is typically used as an adsorbent for gaseous and waterborne impurities. To this end, the ideal activated carbon requires a large surface area to allow for sufficient adsorption to take place. The production of activated carbon requires a carbon-rich feed material which is low in inorganic material, such as char and coal, and polymers such as waste tires. Char that is used for activated carbon can be produced in slow pyrolysis methods similar to that of conventional charcoal and even by fast pyrolysis. The occurrence of secondary char-forming reactions would tend to reduce the surface area of the char through condensation in the pores within the activated carbon and should be avoided by minimizing vapor residence times. The adsorptive capacity of the char is increased through an activation step which involves treatment of the pyrolyzed char with an oxidizing agent or a strong acid or base. Important properties relevant for activated carbons are the Brunauer–Emmett–Teller (BET) surface area, porosity, and the hardness (when used in granular form). Carbon source, pyrolysis heating rate, peak pyrolysis temperature, and the activation process all affect the properties of the activated carbon to the extent that the BET surface area of the activated carbon can vary from 250 to 2410 m^2/g and the pore volume can vary from 0.022 to 91.4 cm^3/g (Ioannidou and Zabaniotou, 2007). Commercial grades of activated carbon have BET surface areas ranging from 500 to 1500 m^2/g. BET surface areas are generally less than 500 m^2/g if no activation is undertaken (Major et al., 2009).

5.3.4 Biochar

"Biochar" refers to char that is applied to soil in a deliberate manner as a soil amendment. Biochar improves the soil's properties such that the fertility is improved, and consequently crop yields are increased. Biochar reportedly acts as a habitat for microbes, adjusts the soil pH, improves soil water retention capacity, and increases the cation exchange capacity (CEC) which is important for the retention of nutrients within the soil. These benefits are believed to have resulted in the improved agricultural yields observable in the Terra Preta soils of South America which have significant anthropogenic carbon contents (Lehmann et al., 2006). The addition of char in combination with compost or manure to soil also has a significant history within Japanese agriculture (Ogawa and Okimori, 2010). The efficacy of using mineral fertilizers may be further increased by applying crushed, commercially available charcoal to the soil, as demonstrated by multiyear field trials in Brazil. Biochar has a low, inherent nutrient content and the combination of biochar with mineral fertilizer may have a synergistic benefit in terms of crop yield (Steiner et al., 2007). Not only char but also torrefied biomass can be used as a soil amendment where the microbial inoculation of the torrefied biomass has resulted in crop yield increases (Trifonova et al., 2009); however, the benefits would not last as long as for those of biochar as torrefied biomass consists of significantly more labile carbon.

A further benefit of biochar is that the carbon is less degradable than biomass and could therefore act as a carbon sink for CO_2 abatement strategies if its stability within the soil is sufficient. However, the effect of biochar on crop yields and its behavior as a stable carbon sink are subject to debate and the extent of these benefits is still being investigated (Lehmann et al., 2006).

The important physical properties of biochar include the bulk surface area, pore size distribution, particle size distribution, density, and packing. The surface area of the biochar affects soil aeration, water holding capacity, and nutrient retention as well as the microbial activity within the soil. The surface area of sandy soils is low and can therefore be increased through the application of biochar. Pore size distribution is important for the adsorption of small molecules and also provides a habitat for soil microbes. The surface areas of biochars are maximized at pyrolysis temperatures near 750 °C, although feed material selection would affect this. The particle size distribution is largely the result of the feed material selection and the occurrence of attrition during pyrolysis and product handling. The particle density and bulk density (which is also a function of particle size) are dependent on the properties of the initial biomass. Particle density tends to increase and bulk density tends to decrease with increasing pyrolysis temperatures (Downie et al., 2009).

The chemical heterogeneity of biochar has implications on the surface chemistry and the adsorption of aqueous species at the biochar surface. The CEC and anion exchange capacity of biochar strongly affect the retention of nutrients within the soil. The CEC tends to increase with increasing pyrolysis temperatures. The mineral content of the biochar, which is reported as ash, also affects soil behavior. Aromatic compounds, which are favored by high pyrolysis temperatures, tend to be more resistant to mineralization than aliphatic compounds and are therefore more stable once introduced into the soil (Chan and Xu, 2009).

6. PROCESS VARIABLES RELEVANT TO REACTOR DESIGN

6.1. Reactor temperature

The rates of all the involved pyrolysis reactions are increased at increased temperatures. By arbitrary definition, torrefaction is limited to conversion temperatures in the range from 200 to 300 °C (Prins et al., 2006b), and at these (relatively low) temperatures, reactions mostly involve the degradation of the less stable hemicellulose fraction. At higher temperatures, such as those employed during fast and slow pyrolysis, significant conversion of cellulose and lignin also occurs, yielding more condensable vapors and noncondensable gases at the expense of the char yield. Some of the molecules that are released from the biomass undergo cracking, polymerization, and condensation reactions which yield secondary char and lighter gases at the expense of the overall bio-oil yield. The increased devolatilization at such temperatures results in a greater fixed carbon content within the remaining char. The maximization of the bio-oil yield is typically achieved at temperatures close to 500 °C as this temperature is sufficient to allow for substantial devolatilization of the biomass while not being too high to promote substantial cracking of pyrolysis vapors into noncondensable gases or the polymerization of vapors into char. This is shown in Fig. 2.14. Further information has been given in Section 3.2.

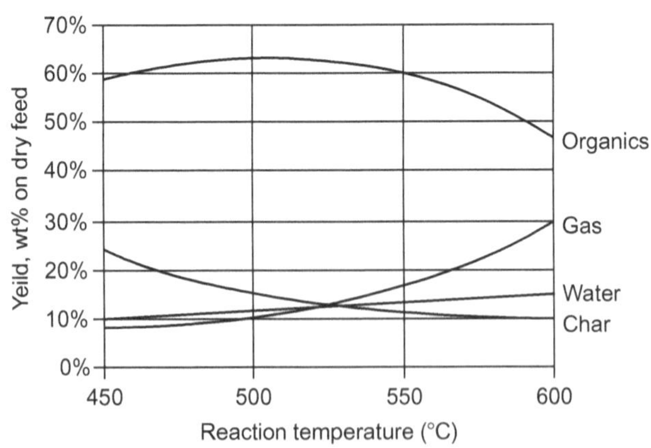

Figure 2.14 Changes in fast pyrolysis yields with increasing temperature (Bridgwater et al., 1999).

6.2. Reactor pressure

Pyrolysis is usually carried out at or near atmospheric pressure. Increasing reaction pressure favors the production of char due to increasing concentration of vapor-phase molecules, thereby increasing the reactions rates of the secondary char-forming reactions (Antal et al., 1996; Várhegyi et al., 1997). The effect of increasing pressure on the quality of torrefied biomass (expressed as product carbon content) in the pressure range of up to 4 MPa appears limited (Wannapeera and Worasuwannarak, 2012), as shown in Table 2.12.

A benefit of carrying out pyrolysis under reduced pressure is a reduction in the char yield, accompanied by a higher yield in condensable vapors, perhaps due to reduced effect on the secondary reactions. Existing vacuum pyrolysis processes typically employ larger particles and slower heating rates than fast pyrolysis, but the overall bio-oil yields are only slightly less than that of fast pyrolysis while more char is produced (Bridgwater et al., 1999)

Cellulose pyrolysis experiments have been carried out in sealed vessels while allowing the pressure to increase to between 3 and 14 MPa (Várhegyi et al., 1997). The char yield that was reported (36–40%) was significantly higher than what one would expect during atmospheric pyrolysis, though it should be noted that this could be either a pressure effect or the result of the extremely long residence time of the product vapors or both.

Experiments have also been performed at higher pressures of up to 1 MPa in the batch flash carbonization process, where oxygen is introduced into a pressurized reactor when the biomass within the batch reactor has reached a desired temperature. A flame front then moves through the reactor bed, thereby generating the heat required to drive the pyrolysis reactions once the supply of air has been stopped. The reactor in which the process

Table 2.12 Variation in carbon content (wt%, dry, ash-free) during torrefaction at different pressures

	200 °C	**225 °C**	**250 °C**
0.1 MPa	51.7	52.4	54.0
1 MPa	52.0	57.2	57.7
2 MPa	52.5	57.9	59.2
3 MPa	52.4	58.2	59.6
4 MPa	52.2	58.1	62.3

Adapted from Wannapeera and Worasuwannarak (2012).

is carried out may be viewed as a pressurized pyrolysis kiln. Kilns are discussed further in Section 7.2.1. It is reported that this process can achieve a char yield that approaches that of the thermodynamic equilibrium (when the additional oxygen is included in the equilibrium calculations) and that the reaction rates are generally higher than those reported for torrefaction and slow pyrolysis (Antal et al., 2003).

6.3. Heating rate

As the conversion of the biomass is temperature driven, particular emphasis must be on how heat is introduced to the biomass in the reactor. Higher heating rates and low vapor residence times are usually employed in fast pyrolysis where the yield of condensable vapors is maximized. In the case of slow pyrolysis, heating rates are generally reduced and are of lesser importance; however, heat transfer within the biomass particle and the biomass bulk should still be efficient as this influences the overall conversion rate in case where heat transfer is the limiting step. More importantly though, all particles should undergo a similar degree of reaction, and therefore heat should be supplied evenly throughout the biomass bulk. Nevertheless, particles will exhibit a greater extent of pyrolysis nearer to their surface compared to the interior of the particles. Torrefaction utilizes relatively lower temperatures, and therefore, the temperature driving force between the heat source and the reacting biomass could be limited, thereby exacerbating any heat transfer limitation.

6.4. Biomass residence time

Adequate time is necessary to allow sufficient heating to take place and for the reactions to proceed to the desired reaction extent. Primitive methods of slow pyrolysis for the production of charcoal from hardwood, such as the use of large pit and mound kilns, can take up to 20–30 days to achieve the necessary conversion depending on the size of the batch of biomass (up to 28 t of feed) being pyrolyzed (Food and Agriculture Organization of the United Nations, 1987). The pyrolysis of large pieces of biomass naturally hinders heat transfer and requires longer conversion times compared to the production of charcoal in the form of a powder. For fast pyrolysis, it has been shown that attention needs to be given to the residence time of the biomass (in addition to the residence time of the vapors) to ensure that sufficient devolatilization takes place. Suitable residence times range from 25 (for particles smaller than 2 mm) to 220 s (for particles less of 15 mm; Wang et al., 2005).

6.5. Vapor residence time

Secondary reactions involve the cracking and polymerization of non-condensable vapors, and these are promoted by increasing the vapor residence time and by increasing the interaction of pyrolysis vapors with char which can contain catalytically active minerals. Catalysis of gas-phase cracking and polymerization reactions by char appears almost absent if the chars are deprived of the minerals; thus, homogeneous vapor-phase cracking reactions are then dominant (Hoekstra, 2011; Ronsse et al., 2012). In low-temperature slow pyrolysis (as in the case of torrefaction), the amount of volatile material that is released and which can undergo secondary polymerization and cracking reactions is reduced while the lower temperatures also reduce the rates of these reactions. The influence of vapor residence time is expected to be less pronounced for torrefaction in comparison with fast or slow pyrolysis. In the case of fast pyrolysis, the secondary, vapor-phase reactions are undesirable as the bio-oil yield is reduced in favor of the production of gases and char and subsequently the vapor residence time should be kept below 2 s (Bridgwater, 2012).

6.6. Biomass conveying, mixing, and hydrodynamics

The degree of conversion that takes place within a particle is dependent on the time the particle is held at a given temperature. To control this residence time adequately, one would ideally need to achieve plug flow of the solid biomass in continuous reactor or use a batch reactor. In contrast to homogeneous reactions, the "composition" of the biomass (where composition refers to the degree of conversion of the individual particles) within the reactor can vary spatially. Solid-phase composition is regarded as an inconsequential parameter with respect to reaction kinetics (White et al., 2011). It is necessary to transfer heat to the solid phase and to transfer mass from the solid phase to the gas phase. By employing direct counter-current heating, it would be possible to achieve this more efficiently than direct cocurrent heating as heat and concentration profiles can be utilized more effectively (similar to the case of a counter-current heat exchanger).

6.7. Feed preparation

To some degree, feed preparation is necessary to allow the biomass feed to be successfully introduced into the process and to ensure that the final product meets specifications. Drying would be beneficial in the overall energy efficiency of the process, as discussed in Section 4.2. The leaching of mineral material can have an impact on the torrefaction and pyrolysis reactions and will also impact on the ash content of the final char product. This is discussed in Section 4.3. Manipulation of the particle size distribution of the biomass feed (chipping, grinding, or pelletization) may be required to allow for the feed material to be introduced efficiently into the reactor and to enable adequate heat transfer as discussed in Sections 3.3 and 4.4.

6.8. Product handling

For a given pyrolysis process, the solid product might not have the required particle size distribution or bulk (and therefore also calorific) density required for its intended application. Fine particles are susceptible to dust explosion, thereby posing a process safety risk. Furthermore, manipulation of the final product bulk density and particle size distribution could be optimized to minimize storage, handling, and transport costs. Densification processes such as compression, extrusion, and pelletization offer possibilities for altering the bulk density and the particle size distribution of the char or torrefied biomass product. These technologies are already prevalent in the production of wood pellets. Common compression processes include the briquetting roller press and extrusion processes, which can yield either larger briquettes (screw and ram extruders) or smaller pellets (pelletizers; Saidur et al., 2011). All densification processes require mechanical force and sometimes also heat to induce mechanical deformation and to increase the interparticular cohesion. The chemical and physical properties of the particles (predominantly plasticity) play a large role in determining the efficacy of the densification process. A more comprehensive discourse on these properties is given elsewhere (Grover and Mishra, 1996). Binders are additives that may be used to increase particle agglomeration. Organic binders are usually starch-based and inorganic binders, such as caustic soda, can also be used if they can be accommodated within the final use of the char or the torrefied biomass. Binders can be avoided for the briquetting of biomass at temperatures of between 250 and 300 °C under high pressure as thermal softening will promote binding (Grover and Mishra, 1996). It has been determined that, due to the minimal plasticity of a torrefied and pyrolyzed biomass such as rice husk, a binder is necessary to achieve any densification (Maiti et al., 2006). Pelletization would be possible without a binder if torrefaction conditions are sufficiently mild and lignin degradation within the biomass is negligible (Verhoeff et al., 2011). Interestingly, heat-softening materials such as biomass lignin and coal tar can also act as binders (Finney et al., 2009).

The densification of biomass through screw extrusion occurs at temperatures of up to 290 °C, and it has been noted that this results in a mechanically robust product with an outer layer that is partially carbonized (Grover and Mishra, 1996). It may therefore be possible to tailor the process to yield a product that maintains the mechanical strength of extruded biomass but which has been more extensively converted and therefore has the increased energy density and hydrophobicity which make the treated biomass attractive as an energy carrier. A more conventional approach would be to apply densification in series with a pyrolysis step.

The char particle size distribution can be reduced by grinding. Torrefied biomass and char have been shown to provide energy savings during grinding when compared to the biomass feed (Bergman and Kiel, 2005). The energy required to produce a powder with an average particle size of 0.2 mm is reduced by between 80% and 90% when biomass is torrefied at temperatures between 265 and 300 °C and for torrefaction times between 10 and 24 min, respectively (Bergman et al., 2005).

In fast pyrolysis processes, bio-oil is known to be thermally unstable as the bio-oil can continue to undergo secondary reactions while in the liquid phase (Bridgwater, 2012). For this reason, the bio-oil should not be stored at elevated temperatures (above 50 °C) so that these reactions are not accelerated. The filtering of the bio-oil prior to use is also necessary in certain applications such as firing in a gas turbine (Venderbosch and Prins, 2010). The downstream upgrading of bio-oil has been discussed in Section 3.5.

7. REACTOR TECHNOLOGY DEVELOPMENT

Several reactor alternatives have been used or are conceived as being applicable to pyrolysis or torrefaction. Generally, the slow pyrolysis is a more established technology, whereas fast pyrolysis and torrefaction are comparatively nascent technologies. Woody biomass is the generally preferred feedstock as it is more homogeneous and contains less ash than agricultural wastes and other biological residues, thereby simplifying process requirements. The utilization of cheaper feedstocks would, however, improve the economic feasibility of a particular process if it can be accommodated suitably.

7.1. Fast pyrolysis reactors

The aim of fast pyrolysis reactors is for the maximization of the bio-oil yield. Proposed reactor technologies are discussed below. A more exhaustive review may be found elsewhere (Bridgwater, 2012).

7.1.1 Entrained down-flow reactor

Entrained down-flow fast pyrolysis reactors have been developed by Georgia Tech Research Institute (USA), and further development has been undertaken by Egemin (Belgium). The process is based on small biomass particles (1–5 mm) being fed to a hot down-flow reactor. No solid heat carrier is used to promote heat transfer to the biomass and heat is transferred from the hot gas to the biomass. It was initially estimated that pyrolysis would be complete at residence times of less than 1 s; however, insufficient heat transfer was supposed to be responsible for the incomplete pyrolysis of the biomass feed at residence times of below 1 s, despite reactor temperatures of between 700 and 800 °C. Due to the incomplete pyrolysis, achievable bio-oil yields were less than 40 wt% on a dry-feed basis (Venderbosch and Prins, 2010). High gas flow rates are required to transfer heat to the biomass and this necessitates a large process. Additionally, the collection of the liquid product is difficult as the vapors are diluted by the hot gas reducing the partial pressure of the vapors. The plants at Georgia Tech and Egemin are no longer in operation, but research is ongoing in China (Bridgwater, 2012).

7.1.2 Ablative reactor

Ablative reactors are characterized by having the biomass pressed against a hot surface. This causes the biomass to "melt." The process may be further improved by increasing the pressure that the biomass exerts on the hot surface (by a centrifugal or mechanical force) and by mechanically moving the biomass over the hot surface. The molten biomass forms vapors at the interface between the biomass and the hot surface (approximately 600 °C) once the biomass is moved away mechanically. The removal of char from the biomass particle through abrasion and the direct transfer of heat from the hot surface mean that the particle size is less important to the overall heat transfer rates compared to the transfer of heat from a gas phase to the biomass. This does, however, produce fine char which is difficult to remove from the condensed bio-oil product. The vapors that are produced during ablative pyrolysis typically have a lower molecular weight due to vapor cracking occurring on the metal surface (Bridgwater, 1999). The process throughput is limited by the amount of heat that can be transferred to the hot surface from a source such as hot flue gas. The absence of an inert gas for heating the biomass results in smaller vapor streams leaving the reactor and more practical condensation of the bio-oil. The mechanical nature of the process also makes it more complex and costly (Bridgwater, 2012). The ablative pyrolysis of waste tire particles (1.3 mm) has been demonstrated by BBC (Canada) at a capacity of 10–25 kg/h with a liquid yield of 54 wt% for a residence time of 0.88 s. This process was subsequently scaled up to 25 t/d by Enervision (Canada) and operated until 1997 after which the plant was moved to Norway. The National Renewable Energy Laboratory (USA) developed a 20-kg/h vortex ablative reactor. A 30-kg/h reactor was also built in 1994 but was dismantled in 1997 (Bridgwater, 1999). Typical liquid yields were between 60% and 65% on a dry-feed basis (Bridgwater, 2012). Aston University (United Kingdom) developed a small, 3-kg/h reactor based on a heated plate which has been reported to achieve liquid yields of up to 80% on a dry-feed basis (Bridgwater, 1999; Peacocke and Bridgwater, 1994). A 6-t/d ablative pyrolysis plant has been built by Pytec (Germany) in 2006. This plant is undergoing testing, and a 50-t/d unit is being designed (Bridgwater, 2012).

7.1.3 Fluidized bed reactor

Fluidized bed reactors allow for the efficient transfer of heat to biomass particles, although a prerequisite is that the biomass particles should be fairly small (2–3 mm). Both bubbling fluidized beds and entrained, circulating fluidized beds (CFBs) may be employed for biomass fast pyrolysis. The design of fluidized beds is well understood; however, the scale-up of fluidized bed processes can be limited by overall heat transfer. The vapor residence time is controlled by adjusting the fluidization gas flow rates; however, in bubbling fluidized beds, the undesired entrainment of biomass particles can occur at high gas velocities. The reactor bed typically also contains small, inert sand particles to increase the rate of heat transfer to the biomass particles. Catalytically active solid materials may also be incorporated within the fluidized bed.

Bubbling fluidized beds require the char that is formed to be removed from the reactor due to catalytic ability of the char (particularly, if inorganic minerals are present) to promote undesirable, secondary vapor-phase reactions. This can be achieved through the removal of solids from the reactor bed. Fine char particles and other solids may also be removed from bed by entrainment. In such cases, cyclones are typically used to clean the vapor stream. The separated char particles can be burnt to generate heat to drive the pyrolysis process. Typical bio-oil yields are in the range of 70–75 wt% on a dry-feed basis. Bubbling fluidized beds were initially developed by the University of Waterloo (Canada) and have been developed further by numerous companies. Union Fenosa (Spain) built a 200-kg/h plant although this has now been dismantled. Dynamotive (Canada) has operated two pilot plants (75 and 400 kg/h) and has built two large-scale plants (100 and 200 t/d). These are not in operation at the moment. Fortum (Finland) built and operated a 500-kg/h unit which has been dismantled. Wellman (United Kingdom) has constructed a 250-kg/h plant but this has not been in operation either (Bridgwater, 2012).

At higher gas velocities, the solid material is completely entrained and flows upward within the reactor and exits together with the vapors. This means that the residence time of the biomass is approximately equal to that of the pyrolysis vapors. The higher gas velocities also increase the attrition of solid particles generating more fine particles which needs to be separated from the vapors. The char and sand that are separated from the vapors typically pass through a combustor where the char is burnt and the sand is heated. The hot sand is returned to the pyrolysis reactor to aid in heat transfer. This recirculation is the cornerstone of CFB reactor technology. Enel (Italy) operated a 650-kg/h plant built by Ensyn (Canada); however, this is not in operation any more. Red Arrow Products (USA) operates several Ensyn units with capacities of up to 1700 kg/h. Ensyn operates a 2000-kg/h plant and is planning to develop a 1000-t/d unit (Bridgwater, 2012).

7.1.4 Vacuum moving bed reactor

Vacuum pyrolysis is not really a form of fast pyrolysis, as lower heating rates are employed. It is, however, discussed in this section as, like fast pyrolysis, the aim is to increase the bio-oil yield. Operating under a vacuum reduces the residence time of the pyrolysis vapors, thereby limiting secondary, vapor-phase reactions. Due to the reduced heat transfer requirements, larger particle may be used when compared to conventional fast pyrolysis. An inert carrier gas is not needed. The vacuum pyrolysis process was developed by the University of Laval (Canada) and Pyrovac (Canada) resulting in a 3-t/h vacuum pyrolysis reactor. The process operates at an absolute pressure of approximately 20 kPa and temperatures of 450 °C. The heat is supplied indirectly to the reactor by the use of molten salts which are heated by the combustion of the noncondensable gases. A change of ownership in 2003 meant that work on the vacuum pyrolysis reactor was suspended; however, the unit has been acquired by NewEarth Renewable Energy (USA) for the production of torrefied biomass, charcoal, and bio-oil (Bridgwater, 2012; Venderbosch and Prins, 2010).

7.1.5 Screw reactor

Screw reactors (sometimes referred to as auger reactors) consist of one or more rotating helical screws which move the biomass together with a heat carrier (such as hot sand or metal beads) through a cylindrical shell. The heat may also be transferred across the reactor wall and no inert carrier gas is required. One of the limitations regarding screw reactors is that the residence time of the hot vapors is typically between 5 and 30 s, thereby promoting secondary vapor-phase reactions. Feed materials that are difficult to handle or have varying morphological characteristics can be accommodated in a screw reactor more practically than in fluid bed reactors (Bridgwater, 2012). ABRI-Tech (Canada) currently markets a 1-t/d and a 50-t/d process, and a 50-t/d unit is being constructed in the United States. Forschungszentrum Karlsruhe (Germany) has adapted a twin-screw reactor which has previously been used in the processing of shale oil, resulting in the construction of a 500-kg/h plant. Forschungszentrum Karlsruhe produces a slurry of char and bio-oil with the aim of feeding the slurry to an entrained-flow gasifier (Venderbosch and Prins, 2010).

7.1.6 Rotating cone reactor

The rotating cone reactor employs ablative pyrolysis with the addition of hot sand particles to enhance heat transfer. A mixture of biomass (with diameters of up to 10 mm) and hot sand enter near the bottom of an inverted cone which rotates at sufficient speed (up to 600 rpm) to carry the solids upward by the centrifugal forces generated during rotation. This is shown in Fig. 2.15. The sand and char that are produced within the reactor are transported to a combustion reactor where the char is burnt off and hot sand is returned to the pyrolysis reactor. The absence of a carrier gas simplifies

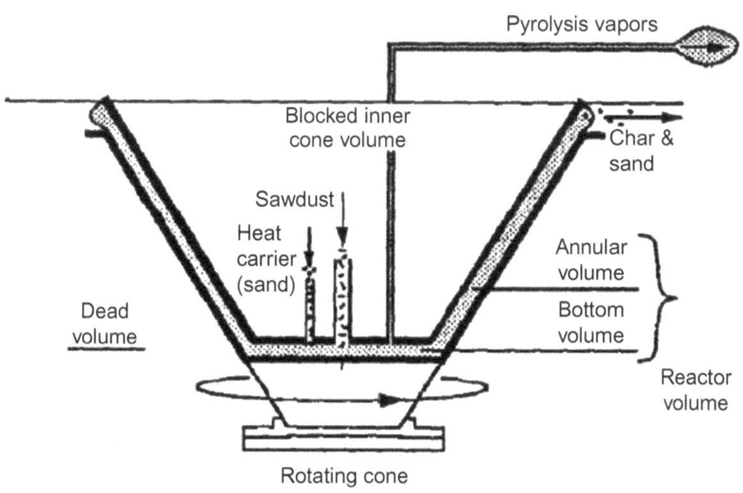

Figure 2.15 Rotating cone reactor (Saidur et al., 2011).

bio-oil condensation and bio-oil yields are approximately 70 wt% when using wood as a feed material. The rotating cone reactor was initially developed by Twente University (The Netherlands) at a scale of 20 kg/h (Wagenaar et al., 1994). Thereafter, the rotating cone reactor was developed further by Biomass Technology Group (The Netherlands), initially scaling up the reactor to 50 kg/h and then to 250 kg/h. The development culminated in a 50-t/d pyrolysis plant which was built for Genting Bio-oil (Malaysia) and operated on a feed of empty fruit bunches (a waste material from the palm oil industry). It has also been planned to construct another commercial scale unit (5 t/h) in Hengelo, The Netherlands (Venderbosch and Prins, 2010).

7.2. Slow pyrolysis reactors

The aim of slow pyrolysis reactors is to produce char at moderate temperatures and extended reaction times (in the order of a few hours or longer).

7.2.1 Kilns

Kilns are reactors where the heat required for pyrolysis is generated within the reactor and where the heat is transferred directly to the remaining biomass. Pit and mound kilns are the most simplistic and primitive forms of slow pyrolysis reactors used for the batchwise production of charcoal. This process involves the initial combustion of a fraction of the biomass in a pit or a mound to generate heat. Once sufficient heat has been generated, the pit or mound is sealed with soil and the remaining biomass is allowed to pyrolyze in the oxygen-free environment. A Casamance mound kiln, which incorporates a chimney to generate a draught, thereby directing the flow of hot vapors within the kiln and improving heat transfer, is depicted in Fig. 2.16.

The minimal infrastructure required to create such kilns allows the kilns to be constructed close to the biomass source at low capital costs. More advanced kilns are based on the construction of more permanent structures made from brick (beehive kilns) or reinforced concrete (Missouri kilns) to contain the pyrolysis reactions. Performing the slow pyrolysis reaction in fixed structures has the benefit of providing improved thermal insulation and reducing the possibility of air leakages into the biomass bed which is a typical problem when using pit and mound kilns.

The transfer of heat from the combusting biomass to the remaining biomass is inefficient, and it is not uncommon for significant quantities of the biomass feed to be either unreacted or completely burnt to ash. The extent of the pyrolysis is estimated by gauging the density and the color of the smoke that exits the kiln, and therefore, operator competence is vital in

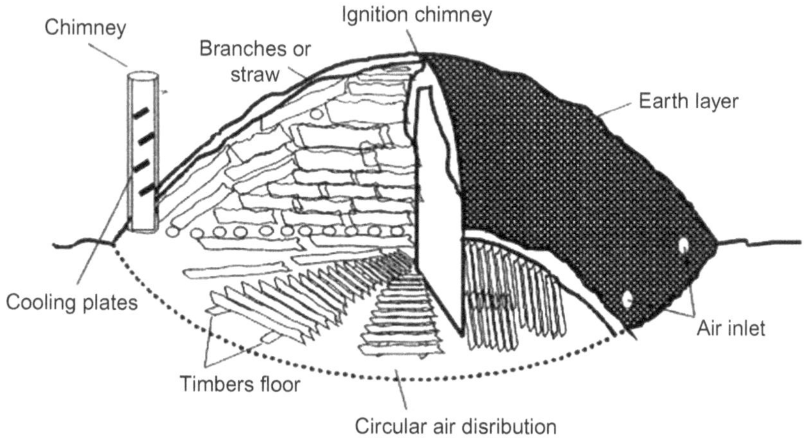

Figure 2.16 Casamance Mound Kiln (Schenkel et al., 1998).

ensuring good char yields. The nature of the smoke indicates the temperature within the kiln which in turn indicates when the flow of air into the kiln should be stopped. Thick white smoke leaves the kiln during the initial removal of moisture and is observable up to temperatures of 250 °C. This is followed by gray smoke (250–300 °C), minimal smoke (300–350 °C), and thin blue smoke (350–410 °C) (Kandpal and Maheshwari, 1993). Reaction times can vary from a few hours up to 30 days and batch sizes can be very large (containing up to 28 t of feed) due to cheap construction. The use of kilns is prevalent in the "third world" countries where biomass and labor are relatively cheaper in comparison with capital (Food and Agriculture Organization of the United Nations, 1987). The contamination of the charcoal with sand and mud is also possible depending on how the kilns are sealed and how the product charcoal is handled. Another disadvantage of kilns is that the release of the smoke directly to the atmosphere can violate emissions regulations and pose health risks to the workers operating the kilns, as has been the case for Missouri kilns in the United States. The batchwise operation of the kilns means that the composition of the flue gas varies with time, thereby hindering the capture of condensable vapors. One of the advantages of kilns is that they can handle large size ranges and this greatly reduces the amount of feed preparation that is required before pyrolysis. A typical charcoal yield for a Missouri kiln is between 20% and 30% (Antal and Grønli, 2003).

7.2.2 Retorts

Retorts operate differently to kilns in that the heat required to drive the pyrolysis reactions is generated outside of the pyrolysis vessel, and this heat is then transferred either directly (injecting the hot gas into the reactor) or indirectly (transferring heat across the reactor wall) to the biomass within the retort. Due to the improved utilization of heat and the fact that the remaining solid char does not come into contact with oxygen within retort systems, the char yields that are attainable are typically slightly higher (30–33%) than those of kilns (20–30%). Retorts were used within the wood distillation industry where the focus was on the recovery of the chemicals that are present within the vapors generated during pyrolysis. The separation of the pyrolysis gases from the remaining biomass bulk allows the extraction of chemicals such as methanol, acetic acid, and acetone (Food and Agriculture Organization of the United Nations, 1985). As these products could be produced within the petrochemical industry at lower costs, the wood distillation industry has declined since the 1930s (Demirbaş, 2001).

Early batch retorts consisted of sealed metal containers that were placed into wood or coal-fired furnaces. As the pyrolysis vapors that are generated near the completion of the slow pyrolysis reactions have a significant heating value, the batch process was improved by burning the pyrolysis vapors to provide the heat needed in integrated, multiple-batch retort processes. A simple, integrated system such as a Van Marion Retort, which consists of two batch reactors sharing a common combustion zone for the combustion of vapors, is able to achieve a charcoal yield of 30–32% (Antal and Grønli, 2003).

Other improvements, such as the Arkansas retort, included the use of steel wagons to move the biomass into and out of the retort as efficiently as possible, thereby increasing the throughput and efficiency of the process. Heat is transferred across the surface of the retort to the fixed bed of biomass in the wagon. Arkansas retorts were effective in pyrolyzing the required biomass but exposure to the corrosive environment and high temperatures meant that the maintenance of the wagons and the retorts became a major limitation. An advantage of the Arkansas retort was that the size of the biomass used can vary significantly (incorporating logs of up to 1.2 m in length and up to 12 cm in diameter), thereby reducing feed preparation requirements in comparison with the Reichert retort mentioned below. Additionally, the loss of charcoal through the production of fine particles is relatively small if the operating personnel is competent and total charcoal yields could be as high as 33% at feed rates of over 3000 t of dry wood per year (Food and Agriculture Organization of the United Nations, 1985).

The next step in the development of batchwise, retort-based, slow pyrolysis processes was the Reichert retort process developed by Evonik (previously Degussa, Germany) which is based on direct heating of the biomass to drive the pyrolysis reactions as shown in Fig. 2.17. The indirect transfer

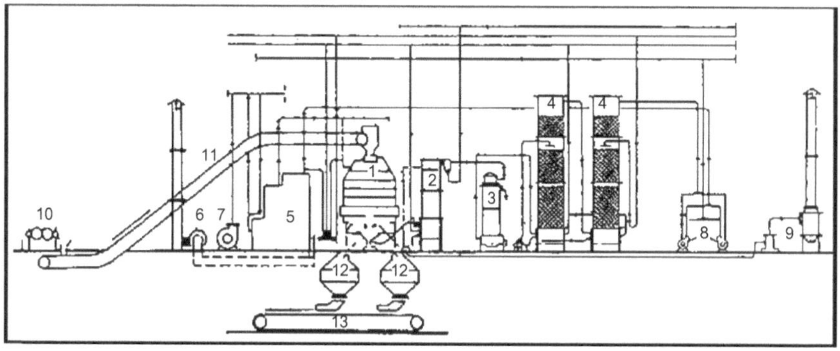

Figure 2.17 Reichert retort process (Food and Agriculture Organization of the United Nations, 1985). 1. Large space retort, 2. tar stripper, 3. water cooler, 4. scrubber for residual gas, 5. combustion chamber and heat exchanger, 6. off-gas fan, 7. combustion air fan, 8. forced gas circulation fan, 9. dust collector, 10. wood preparation, 11. charging conveyor, 12. charcoal cooler, and 13. charcoal conveyor.

of heat was the source of the mechanical problems experienced by Arkansas retorts and direct heating was seen as a method of avoiding these increased maintenance costs. The direct heating is achieved by passing combusted pyrolysis vapors (450–550 °C) vertically through a fixed bed of biomass consisting of wood logs that are approximately 30 cm long and 10 cm thick. Size reduction would be necessary to achieve such dimensions and to ensure thereby that the lumps of char can be removed from the retort by gravity after the 12–18 h needed to complete the pyrolysis reactions. Small pieces of wood should be avoided so that the pressure drop over the biomass bed does not restrict the flow of the hot gas. The combined pyrolysis vapors are cooled to remove condensable products, whereas the permanent gases are combusted to act as a source of heat for pyrolysis reactions. Multiple retorts are used so that the combusted offgas from one retort can be used to heat the biomass within the next retort. Charcoal plants based on the Reichert retort are still in operation and run at capacities of up to 30,000 t of charcoal per year. The capital costs required for the installation of a Reichert process are known to be high but are still being operated by ProFagus (Germany) (Dahmen et al., 2010; Food and Agriculture Organization of the United Nations, 1985).

The Lambiotte process is a continuous process based on the concept of a moving bed, which uses gravity to transfer the wood logs (30 cm in length, 8 cm in diameter) through the reactor, as shown in Fig. 2.18. At the bottom of the retort, cold noncondensable pyrolysis gases (40 °C), which have not passed through the heater, absorb heat from the descending charcoal and transfer this heat upward toward unreacted biomass, thereby cooling the

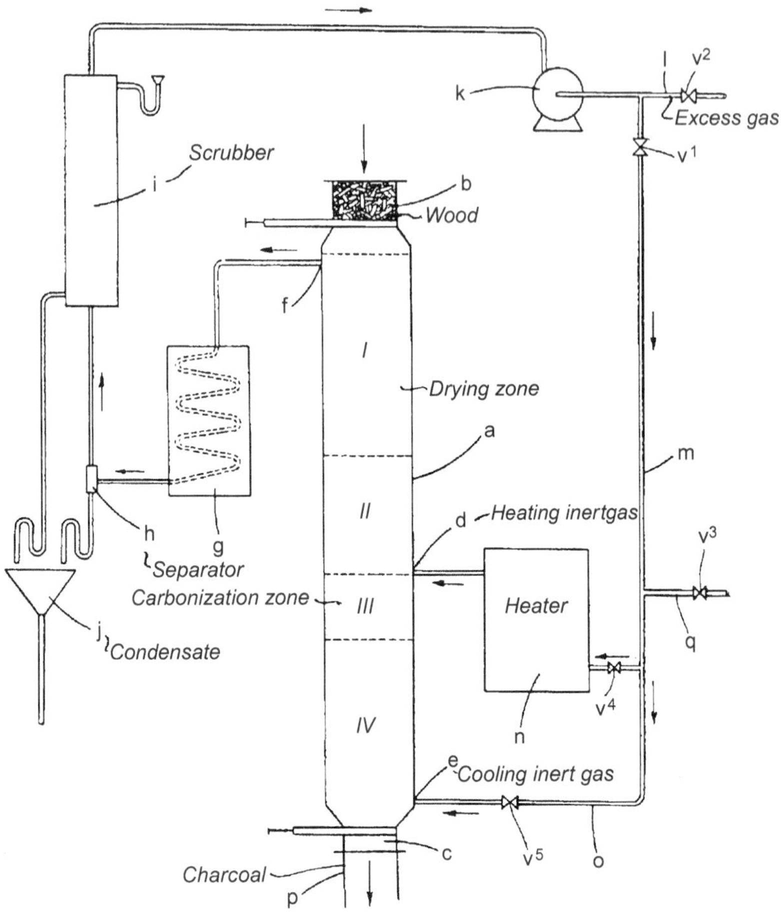

Figure 2.18 Lambiotte Process (Lambiotte, 1942). (a) Retort, (b) inlet chamber, (c) charcoal outlet, (d) hot inert gas inlet, (e) cold inert gas inlet, (f) gaseous product outlet, (g) cooler, (h) separator, (i) scrubber, (j) condensate collector, (k) recycle fan, (l) purge line, (m) combined recycle line, (n) heater, (o) cold recycle line, and (v^{1-5}) control valves.

char product. Additional heat is introduced by feeding combusted pyrolysis gases (300–600 °C) just above the section where the carbonization reactions become exothermic in nature. The combined, hot gas flow is sufficient to drive the pyrolysis reactions. By manipulating the flow of the hot and cold gas streams, it is possible to achieve adequate control of the biomass bed while ensuring that the residence time of the pyrolysis vapors is not excessive, thereby preventing unwanted decomposition reactions. The combined gas that leaves the top of the retort is cooled from approximately 100 °C to recover condensable material while the remaining gases can be recycled to the retort (Lambiotte, 1942). The recovery of heat from the charcoal product, in addition to the continuous operation of the Lambiotte process, ensures that thermal efficiencies are higher than that of other retort processes (Food and Agriculture Organization of the United Nations, 1985). The moisture content of the biomass feed material should be reduced to below 30% to ensure that the process does not require energy from external fuel sources. The increased degree of automation that is required for the continuous process tends to increase capital requirements, whereas the labor requirements per ton of charcoal produced are reduced. Reported capacities of Lambiotte retorts are up to 12,000 t of charcoal per year (Dahmen et al., 2010).

Lurgi (Germany) has also developed a continuous retort and this is in operation in Australia; however, specific details of the process were not available at the time of writing this book chapter.

7.3. Torrefaction reactors

Torrefaction attempts to produce a solid product with favorable properties in comparison with biomass at higher energetic efficiencies than slow pyrolysis processes. It should be noted that some of the information regarding the development of torrefaction technology was divergent and peer-reviewed references were not always available. Generally, technologies that have been patented but have not progressed to some level of commercialization have been excluded.

7.3.1 Screw reactors

Screw reactors (sometimes referred to as auger reactors) are based on the principle of a rotating, helical screw which is used to continuously feed biomass through a heated tubular shell. The background behind this concept is given for fast pyrolysis reactors in Section 7.1.5. Generally, in the case of torrefaction, the heat transfer rates do not need to be as high as for fast pyrolysis. Although the heat transfer to the biomass bulk can be improved by radial mixing, it could be that the heat transfer within the biomass particles controls the overall heat transfer rate (Dahmen et al., 2010), thereby necessitating small particles for efficient heat transfer or larger reactors. Small particles are also necessary to ensure the smooth passage of the biomass material through the screw reactor. The flow of the biomass and char particles through the reactor is controlled by manipulating the rotational frequency of the screw and the movement of biomass through the reactor may be approximated as being plug flow. It has been noted that a degree of filling of about 45% of the screw volume can result in a defined mass flow that is almost independent of the rheological properties of the granular biomass; however, particle mixing (and therefore also the transfer of heat from the reactor wall) remains poor under these conditions (Roedig and Klose, 2008). The fact that mechanical force is used to move the solid material is an advantage if the particle dimensions and interparticular cohesion are such that the particles could cause blockages in gravity fed processes. Feed preparation would be an additional cost if the feed material is not available with a suitable particle size distribution. The heating of the reactor may also be done in stages with different sections of the screw reactor being raised to different temperatures and achieving a controlled temperature profile, although this is more prevalent in small, lab-scale reactors that utilize electrical heating.

The capacity of individual screw reactors is limited as increasing screw reactor diameters reduce the ratio of heat transfer surface area to reactor volume when indirect heating is applied. Therefore, the benefits of economies of scale are poor in comparison with other pyrolysis and torrefaction reactor technologies (Bergman and Kiel, 2005). Large-scale processes would require multiple screw reactors operating in parallel. It has been estimated that the maximum production capacity for a single torrefaction line would be between 50,000 and 60,000 t/y (Uslu et al., 2008). Smaller reactors can be incorporated into small processes which could be designed to be easily moved to the locations where biomass is available. This is a benefit when the biomass is not readily available at one, centralized location. An additional benefit of screw reactors is the ability to accommodate feeds with a wide range of morphological characteristics (from fine powder to lumpy, sticky, and fibrous materials), although screw reactors are not recommended for materials which have a tendency to foul the heat transfer surface (Waje et al., 2006).

Particle size reduction might be necessary in order to feed the screw effectively. The small size of the char particles that are produced (either from the small particles in the feed or through the attrition of larger particles) could also necessitate briquetting or pelletization, depending on end-user requirements. Pelletization has also been shown to increase the volumetric energy density of torrefied biomass, and this would have a beneficial effect on the cost of downstream logistics (Uslu et al., 2008). Pechiney (France) operated a torrefaction process based on a screw reactor with a production capacity of 12,000 t of torrefied biomass per year using wood as feed material. The plant was decommissioned in the 1990s due to economic reasons involving energy losses during the cutting and sieving of the biomass feed (Bergman and Kiel, 2005). It has also been reported that screw reactors are currently being developed by BTG (The Netherlands), Biolake (The Netherlands), FoxCoal (The Netherlands), Agri-tech Producers (USA), and ETPC (Sweden) (Chew and Doshi, 2011; Kleinschmidt, 2010, 2011).

7.3.2 Multiple/rotary hearth furnaces

Rotary hearth furnaces, such as Wyssmont, Buell, or Herreshoff furnaces, are reactors which were developed for mineral processing and can accommodate small feed particles with a range of morphological characteristics. These reactors consist of multiple of circular hearths (heated surfaces) which are stacked and rotate within an insulated outer shell, as shown in Fig. 2.19. The hearths rotate and the biomass, which enters the furnace at the top, moves over the top hearth and drops down to the subsequent hearth. Ploughs and rabble arms within the furnace are used to move and agitate the biomass as it moves downward from one hearth to the next. Similar to screw reactors, particle size reduction could be required depending on the feedstock. The fine charcoal product might also require pelletization or briquetting depending on end-user requirements. Multiple hearth furnaces are currently being developed by CMI-NESA (Belgium) and Wyssmont (USA; Chew and Doshi, 2011; Kleinschmidt, 2010, 2011). A patent application has been made in the United States with Andritz (Germany) as assignee. This patent is based on the pressurized torrefaction of wood in a rotary hearth furnace where the increased pressure allows for a greater mass flow of hot gas into the reactor and increase heat transfer to the wood (Rawls et al., 2012).

7.3.3 Belt conveyor furnaces

Belt conveyors can move biomass through a heated reaction zone. Multiple belts can be used where the biomass drops down from one belt to the next, thereby allowing for some mixing to take place. Other possibilities involve the use of oscillating conveyor belts where the oscillation allows for mixing while maintaining plug flow of the biomass through the reactor. The technology of Stramproy is based on the oscillating conveyor belt and is reportedly used in their installation at the site in Steenwijk (The Netherlands). This facility is to provide 90,000 t/y of torrefied biomass to the Amercentrale power plant operated by Essent in the Netherlands. Although the agreement has been signed, the latest indication is that Stramproy has not yet started delivering torrefied biomass on a continuous basis. One of the benefits of such a reactor is that there should be little particle attrition and this would

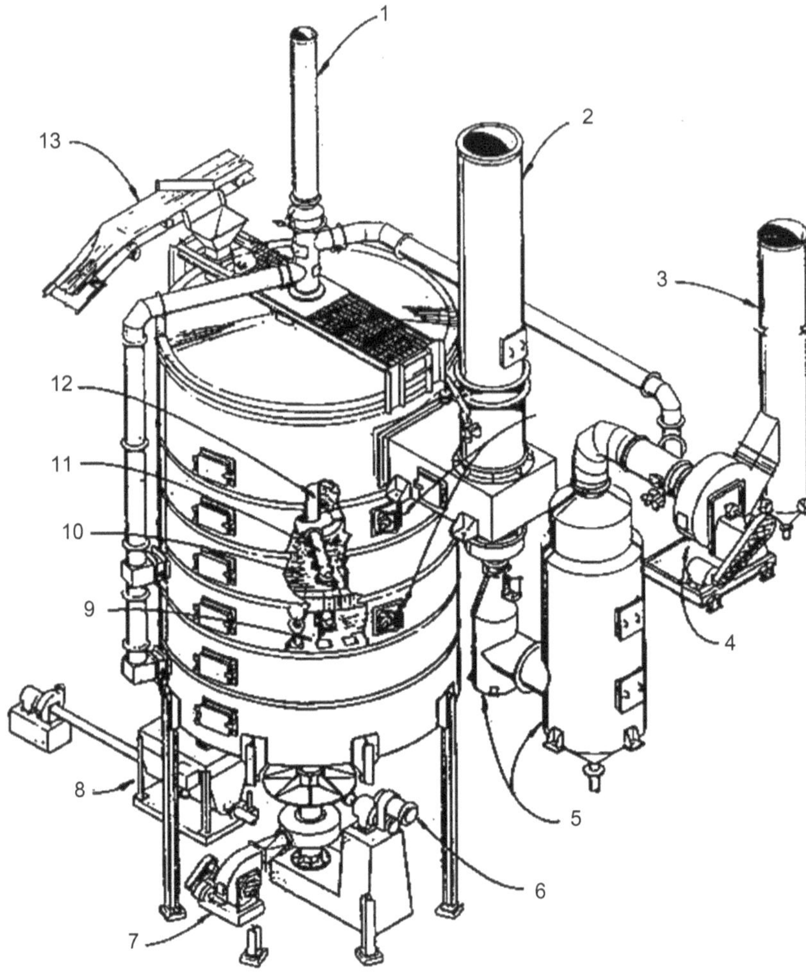

Figure 2.19 Herreshoff rotary hearth furnace (Food and Agriculture Organization of the United Nations, 1985). 1. Air stack to cool central staff, 2. emergency by-pass stack, 3. main exhaust stack, 4. induced draft fan, 5. air pollution controls, 6. center shaft drive, 7. cooling fan for central shaft, 8. charcoal cooler, 9. out hearth, 10. in hearth, 11. rabble arm and teeth, 12. center shaft, and 13. feed conveyor.

tend to reduce the production of fine material. This would make product handling easier and reduce explosion risks and product losses. As there is limited contact between the gas phase and solid phase, it may be expected that the rates of heat and mass transfer from the bulk solid material passing on the conveyor belt would be reduced. If large biomass particles are available, it may be possible to overcome this by using belt conveyors that have some small holes to allow the passage of the hot gas vertically through the bed of biomass moving horizontally through the reactor with care to ensure that the biomass particles are not blown off the conveyor belt. Since conveyor belts are commercially available, the use of a conveyor belt furnace could have cost advantages in comparison with more complicated reactors systems, but there might be limitations to the scaling up of the conveyor belt furnace and multiple parallel reactors could be required for high production rates. This technology is being developed by Stramproy Green Investments (The Netherlands) and Agri-Tech Producers (Chew and Doshi, 2011; Kleinschmidt, 2010, 2011).

7.3.4 Rotary drum reactors

Rotary drums are used for the heating of bulk solids. In the case of a rotary kiln, the heat that is required for the process is produced within the drum through the controlled combustion of a fuel within the rotary drum. In contrast, rotary retorts operate based on heat being transferred (either directly or indirectly) to the drum from an external source. The biomass passes through the inclined, horizontal, rotating drum under the force of gravity on a continuous basis. This type of reactor could have an advantage over a screw reactor in that the heat transfer is not limited by the surface area of the rotary drum if direct heating is applied. This type of reactor would also have an advantage over moving bed reactors in that small particles could be employed without the risk of excessive pressure drop occurring when combusted gases are circulated. Possible drawback of such a process is that some backmixing within the largely plug flow operation could occur. Excessive attrition between char particles could also lead to the production of excessive char fines that would either require briquetting or could be entrained with the gases leaving the rotary drum, thereby creating a process safety risk. An example of a comparable industrial application is the use of a rotary retort for the pyrolysis of particulate oil shale in the Galoter process, although this process is more focused on the pyrolysis vapors than the solid residue (coke). The maximum production capacity of this reactor concept is estimated to be 5 t/h (van der Stelt, 2010). It has been stated that the application of a "rotating drum" reactor for torrefaction is being developed by TorrCoal (The Netherlands) which can operate at a capacity of 35,000 t of torrefied biomass per year (Brouwers, 2011). Other groups developing rotary drum reactors are EBES (Austria), 4Energy Invest (Belgium), and BioEndev/ETPC (Sweden) (Chew and Doshi, 2011; Kleinschmidt, 2010, 2011).

7.3.5 Toroidal fluidized bed reactors

A toroidal, dilute-phase, fluidized bed reactor has been developed by Torftech (United Kingdom) and the application of this technology on the torrefaction of biomass is being undertaken by Topell Energy (The Netherlands) (Chew and Doshi, 2011; van der Stelt, 2010). The first torrefied biomass should have been produced and delivered to the Essent power plant in Geertruidenberg (The Netherlands) during the beginning of 2011 (Topell Energy, 2010), but no confirmation of this could be found. In the toroidal fluidized bed reactor, heat is directly transferred to the biomass by hot combustion gas which enters the cylindrical reactor at the bottom and which passes through angled blades. The velocity of the gas increases on passing through the blades and is directed in such a way that a toroidal vortex is formed which fluidizes and mixes the biomass particles within the reactor. The high gas flow rates and the thorough mixing allow for efficient heat transfer to the biomass particles which enter the reactor from above. Large particles leave the reactor by passing radially out of the fluidized region where the particles drop to the discharge point. Fine particles are entrained with the gas flow and leave the top of the reactor to be collected by a gas cyclone (Shu et al., 2000). This reactor has the benefit of having high heat and mass transfer rates but this will not be advantageous for large particles where the overall reaction progress is limited by heat transfer within the particle rather than from the gas to the biomass particle. The configuration of a toroidal fluidized bed reactor is given in Fig. 2.20.

7.3.6 Moving bed reactors

ECN (The Netherlands) is developing a moving bed reactor for torrefaction application within their "BO_2" process which also incorporates pelletization as a means to increase the energy density of the final torrefied biomass. Biomass enters the top of a vertical reactor and the bed descends as torrefied biomass is continuously removed from the bottom of the reactor. The biomass is directly heated by recirculating the gases and vapors that are produced during torrefaction. It is foreseeable that excessive pressure drop could be encountered if the size of the particles in the reactor is too small. Excess gases and vapors are combusted. In the process proposed by ECN, the combusted gases and vapors are then used to indirectly heat a recirculating gas stream and also to dry the biomass feed. The biomass does not therefore come into contact with combusted gases. The justification for the moving bed design is based on the benefits of low cost and high throughput. It is envisaged that the moving bed reactor technology will be used as the basis for a plant producing

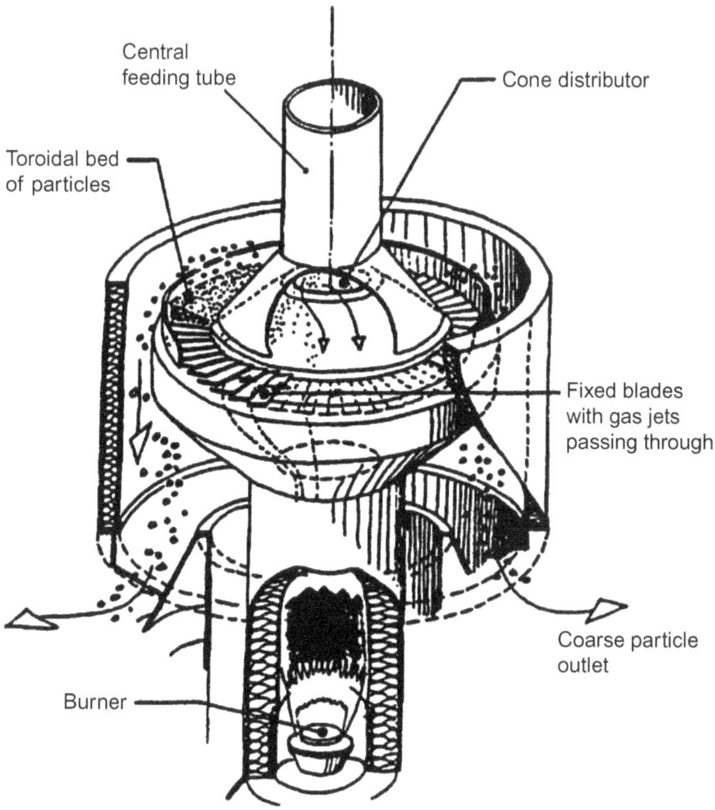

Figure 2.20 Toroidal fluidized bed reactor (Shu et al., 2000).

70,000 t of "BO_2" pellets per year, although it is estimated that designs based on this type of reactor technology could range from 60,000 to 100,000 t of pellets per year (Kiel et al., 2008).

In addition to ECN, a moving bed reactor (TORSPYD reactor) is also being developed by Thermya (France) for the production of torrefied biomass (Bio-Coal). This process requires the biomass to be crushed before torrefaction at temperatures that are typically below 250 °C. The hot gas, which drives the torrefaction reactions, passes upward through the descending bed, thereby achieving counter-current flow. The modeling of a 2.8-t/h TORSPYD reactor has indicated that, at the low temperatures used, biomass degradation reactions only take place in the bottom 7% of the reactor volume, whereas the rest of the reactor volume is necessary for the heating and drying processes. The biomass is assumed to have 20 wt% initial moisture content (Ratte et al., 2011). An industrial scale process, with a production capacity of 20,000 t/y, has been constructed by Thermya for LMK Energy in Mazingarbe (France). At the time of writing, this process was reported to be starting commissioning and the possibility of increasing the capacity to 60,000 t/y was under consideration (Chew and Doshi, 2011; Kleinschmidt, 2010, 2011; Thermya—Energy, 2011).

7.3.7 Microwave reactors

A process has been developed whereby microwave radiation provides the required energy for torrefaction. This process is said to be under development by Rotawave (United Kingdom); however, the indications of the state of development and the details surrounding the process are limited (Rotawave Ltd., 2011). Information regarding microwave torrefaction is available through issued patents which report that the torrefaction should be performed through a combination of conventional thermal heating to reduce the moisture to 3% after which microwave radiation (3–8 GHz) can be used to heat the biomass to achieve a necessary torrefaction temperature of 280 °C. The benefit of using microwaves is that the heating can be controlled more accurately and the biomass bed can be heated uniformly, thereby avoiding temperatures where possible exothermic pyrolysis reactions are initiated. The limitations in the thermal diffusivity of biomass material can be overcome through the use of microwaves. Pelletization is also incorporated to improve the energy density and material-handling properties of the torrefied biomass (Grassi, 2011). The perceived benefit in improved heating efficiency would need to surpass the additional costs of such a system before this type of reactor technology becomes preferable over simpler alternatives.

8. CONCLUSIONS

Biomass is a sustainable source of solid fuel. Pyrolysis and torrefaction of biomass may be employed to yield varying amounts of gaseous, liquid, and solid products that are of interest for downstream applications. In the case of fast pyrolysis, liquid bio-oil is the desired product and its yield is maximized by conversion at temperatures close to 500 °C and short vapor residence times. The composition of the bio-oil may be manipulated to a certain degree with the appropriate selection of process conditions, the application of catalysts and methods such as fractional condensation. The ease of handling liquids and the increased volumetric energy density of bio-oil in relation to biomass are some of the reasons that bio-oil could be seen as a more useful energy carrier than biomass. The bio-oil may be utilized as a fuel or be processed further to produce chemicals where its use as a food flavorant is established. A commodity bio-oil does not yet exist as potential customers have limited material for testing and this in turn results in producers not being able to scale up their processes to the level where bulk bio-oil can become commercially available as a commodity fuel.

The aim of both slow pyrolysis and torrefaction is to maximize the yield of the solid product. The improved properties of the torrefied biomass and char, produced by torrefaction and slow pyrolysis, respectively, include increased energy density, hydrophobicity, grindability, and reduced biodegradability when compared to the initial biomass feed, thereby allowing for cheaper transport, handling and processing of the torrefied biomass or char. Other applications of char include use as a soil amendment or as an adsorbent (activated carbon) after undergoing an activation step. Slow pyrolysis is an established technology and the range of suitable reactors varies from very simplistic batch kilns to automated, continuous processes. Plentiful, cheap biomass (or a reduced need for high product yields) and relatively cheap labor costs (in comparison with capital costs) tend to favor the simplistic reactors whereas the opposite is true for the complex automated processes such as the Reichert retort and the Lambiotte process. Torrefaction is a process where significant recent development has been taking place. Technology developers are in the process of commercializing their torrefaction processes. Some contracts with industrial electricity providers have been signed although, at the time of writing this chapter no evidence of continuous operation of any of the torrefaction plants was found. The development of a variety of torrefaction reactors (screw reactors, multiple hearth furnaces, conveyor belt furnaces, rotary drum reactors, toroidal fluidized bed reactors, moving bed reactors, and microwave reactors) indicates that there is no certainty in which technology has the advantage.

The majority of the pyrolysis reactors are focused on the conversion of wood as a feedstock which is simplified by the homogeneity that may be associated with woody biomass. The incorporation of other feed streams such as energy crops, agricultural residues, food waste, and industrial waste may be attractive due to reduced costs but will tend to complicate the torrefaction process. The economic feasibility of a commercial pyrolysis or torrefaction process is influenced by the suitability of the chosen reactor.

The appropriate design of a reactor depends on the highly variable properties of the biomass feed (including the biological composition, moisture content, ash content, and biomass particle morphology), the required reaction conditions (temperature, pressure, and reaction time), and the product requirements from the end user (particle size distribution, ash content, and moisture content). From the indications of the results available in the literature, the critical considerations for reactor design are the reaction kinetics (of which numerous models are presented) and the heat transfer to and within the particles. Through the incorporation of additional processes such as grinding, drying, and mineral leaching of the biomass feed and pelletization or briquetting of the product, the gap between biomass feedstock and a usable product may be bridged, although these additional steps all have cost implications.

Many facets of chemical engineering are of importance to pyrolysis and torrefaction processes. In the case of fast pyrolysis, the efficient transfer of heat is necessary to ensure that the biomass can achieve the reaction temperature rapidly and that the vapors spend a minimal amount of time at these conditions. The reaction models will always be a gross simplification of the true chemical reactions that are involved but these models still need refinement to account for the effects of catalysis and to further understand the secondary vapor-phase reactions. An accurate, combined reaction kinetic and heat transfer model would reduce the technical risks of scaling up of bench-scale fast pyrolysis units. A refinement of the bio-oil analysis techniques would also be of great value, since almost half the compounds present in bio-oil cannot be identified by current gas chromatography. In the cases of slow pyrolysis and torrefaction, an intimate understanding of the transfer of heat to the surface of the biomass particle and then through the particles themselves is needed so that it can be ensured that the required extent of pyrolysis is achieved. The processing of the char or torrefied biomass by means of grinding, briquetting, and pelletization is understood on a largely empirical level and any fundamental insight into these processes would allow for greater confidence in their design. Over and above the importance of chemical engineering to the pyrolysis and torrefaction processes, attention should also be given to the application of the products. New and innovative niche applications would assist in the commercialization of these technologies, thereby creating a footing from which the production of bulk energy carriers would be a step closer.

REFERENCES

Allen SG, Kam LC, Zemann AJ, Antal MJ: Fractionation of sugar cane with hot, compressed, liquid water, *Ind Eng Chem Res* 35(8):2709–2715, 1996.

Antal MJ, Grønli M: The art, science, and technology of charcoal production, *Ind Eng Chem Res* 42(8):1619–1640, 2003.

Antal MJ, Croiset E, Dai X, et al: High-yield biomass charcoal, *Energy Fuel* 10(3):652–658, 1996.
Antal MJ, Várhegyi Gá, Jakab E: Cellulose pyrolysis kinetics: revisited, *Ind Eng Chem Res* 37 (4):1267–1275, 1998.
Antal MJ, Allen SG, Schulman D, Xu X, Divilio RJ: Biomass gasification in supercritical water, *Ind Eng Chem Res* 39(11):4040–4053, 2000.
Antal MJ, Mochidzuki K, Paredes LS: Flash carbonization of biomass, *Ind Eng Chem Res* 42 (16):3690–3699, 2003.
Bain R, Broer K: Gasification. In Brown R, editor: *Thermochemical processing of biomass—conversion into fuels, chemicals and power,* 2011, John Wiley and Sons, Ltd., pp 47–77.
Bergman P, Kiel J: Torrefaction for biomass upgrading. In *Proceedings of the 14th European biomass conference and exhibition*, 2005.
Bergman P, Boersma A, Zwart R, Kiel J: *Torrefaction for biomass co-firing in existing coal-fired power stations "Biocoal",* Energy Research Centre of the Netherlands, 2005.
Biomass Energy Centre: *What is biomass,* Website, 2011.
Bridgwater A: Principles and practice of biomass fast pyrolysis processes for liquids, *J Anal Appl Pyrolysis* 51(1–2):3–22, 1999.
Bridgwater AV: Review of fast pyrolysis of biomass and product upgrading, *Biomass Bioenergy* 38:68–94, 2012.
Bridgwater AV: Thermochemical processing of biomass: conversion into fuels, chemicals and power. In Brown RC, editor: *Chapter Upgrading Fast Pyrolysis Liquids,* 2011, John Wiley and Sons, Ltd., pp 157–199.
Bridgwater A, Meier D, Radlein D: An overview of fast pyrolysis of biomass, *Org Geochem* 30 (12):1479–1493, 1999.
Broido A, Weinstein M: Kinetics of solid-phase cellulose pyrolysis. In Wiedemann HG, editor: *Proceedings of the third international conference on thermal analysis*, Davos, Switzerland, August 1971, Basel, 1972, Birkhäuser Verlag, pp 285–296.
Brouwers J: Commercialisation Torr-Coal torrefaction technology. In *2011 central European biomass conference: torrefaction workshop*, 2011.
Chan K, Xu Z: Biochar: nutrient properties and their enhancement. In Lehmann J, Joseph S, editors: *Biochar for environmental management: science and technology*, London, Washington, 2009, Earthscan, pp 67–84.
Chan WCR, Kelbon M, Krieger-Brockett B: Single-particle biomass pyrolysis: correlations of reaction products with process conditions, *Ind Eng Chem Res* 27(12):2261–2275, 1988.
Chen W-H, Kuo P-C: A study on torrefaction of various biomass materials and its impact on lignocellulosic structure simulated by a thermogravimetry, *Energy* 35(6):2580–2586, 2010.
Chew J, Doshi V: Recent advances in biomass pretreatment—torrefaction fundamentals and technology, *Renew Sustain Energy Rev* 15(8):4212–4222, 2011.
Ciolkosz D, Wallace R: A review of torrefaction for bioenergy feedstock production, *Biofuels Bioproducts Biorefining* 5(3):317–329, 2011.
Cordero T, Marquez F, Rodriguez-Mirasol J, Rodriguez J: Predicting heating values of lignocellulosics and carbonaceous materials from proximate analysis, *Fuel* 80 (11):1567–1571, 2001.
Crocker M, Andrews R: The rationale for biofuels. In Crocker M, editor: *Thermochemical conversion of biomass to liquid fuels and chemicals*, Cambridge, UK, 2010, Royal Society of Chemistry, pp 1–28.
Dahmen N, Henrich E, Kruse A, Raffelt K: Biomass liquefaction and gasification. In Vertes A, Qureshi N, Blaschek H, Yukawa H,editors: *Biomass to biofuels: strategies for global industries*, Chichester, UK, 2010, John Wiley and Sons Ltd., pp 91–122.
Davidsson K, Korsgren J, Pettersson J, Jäglid U: The effects of fuel washing techniques on alkali release from biomass, *Fuel* 81(2):137–142, 2002.

Dayton DC, Jenkins BM, Turn SQ, et al: Release of inorganic constituents from leached biomass during thermal conversion, *Energy Fuel* 13(4):860–870, 1999.
Demirbaş A: Biomass resource facilities and biomass conversion processing for fuels and chemicals, *Energy Conversion Manage* 42(11):1357–1378, 2001.
Di Blasi C: Kinetic and heat transfer control in the slow and flash pyrolysis of solids, *Ind Eng Chem Res* 35(1):37–46, 1996.
Di Blasi C: Comparison of semi-global mechanisms for primary pyrolysis of lignocellulosic fuels, *J Anal Appl Pyrolysis* 47(1):43–64, 1998.
Di Blasi C: The state of the art of transport models for charring solid degradation. In *Polymer international*, vol. 49(10), 2000, John Wiley & Sons, Ltd., pp 1133–1146.
Di Blasi C, Lanzetta M: Intrinsic kinetics of isothermal xylan degradation in inert atmosphere, *J Anal Appl Pyrolysis* 40–41:287–303, 1997.
Di Blasi C, Signorelli G, Di Russo C, Rea G: Product distribution from pyrolysis of wood and agricultural residues, *Ind Eng Chem Res* 38(6):2216–2224, 1999.
Downie A, Crosky A, Munroe P: Physical properties of biochar. In Lehmann J, Joseph S, editors: *Biochar for environmental management: science and technology*, London, Washington, 2009, Earthscan, pp 13–32.
Finney KN, Sharifi VN, Swithenbank J: Fuel pelletization with a binder: part I—identification of a suitable binder for spent mushroom compost-coal tailing pellets, *Energy Fuel* 23(6):3195–3202, 2009.
Food & Agriculture Organization of the United Nations: Industrial charcoal making, *Paper 63*, 1985.
Food & Agriculture Organization of the United Nations: Simple technologies for charcoal making, *Paper 41*, 1987.
Funke A, Ziegler F: Hydrothermal carbonization of biomass: a summary and discussion of chemical mechanisms for process engineering, *Biofuels Bioproducts Biorefining* 4 (2):160–177, 2010.
Gómez CJ, Várhegyi Gá, Puigjaner L: Slow pyrolysis of woody residues and an herbaceous biomass crop: a kinetic study, *Ind Eng Chem Res* 44(17):6650–6660, 2005.
Grassi G: *Torrefaction of lingo-cellulosic biomasses and mixtures*, 2011 (0179701–A1).
Grønli M, Antal MJ, Várhegyi G: A round-robin study of cellulose pyrolysis kinetics by thermogravimetry, *Ind Eng Chem Res* 38(6):2238–2244, 1999.
Grønli MG, Melaaen MC: Mathematical model for wood pyrolysis - comparison of experimental measurements with model predictions, *Energy Fuel* 14(4):791–800, 2000.
Grover P, Mishra S: *Biomass briquetting: technology and practices*, Food and Agriculture Organization of the United Nations, 1996.
Hoekstra E, *Fast pyrolysis of biomass*, Universiteit Twente, 2011.
Ioannidou O, Zabaniotou A: Agricultural residues as precursors for activated carbon production—a review, *Renew Sustain Energy Rev* 11(9):1966–2005, 2007.
Kandpal J, Maheshwari R: A decentralized approach for biocoal production in a mud kiln, *Bioresour Technol* 43(2):99–102, 1993.
Kiel J: Torrefaction for biomass upgrading into commodity fuels. In *IEA Bioenergy Task 32 workshop: "fuel storage, handling and preparation and system analysis for biomass combustion technologies"*, 2007.
Kiel J, Verhoeff F, Gerhauser H, Meuleman B: BO_2-technology for biomass upgrading into solid fuel—pilot-scale testing and market implementation. In *16th European biomass conference and exhibition*, 2008.
Kleinschmidt C: *Statusoverzicht en impactanalyse van torrefactie in Nederland*, KEMA, 2010.
Kleinschmidt C: Overview of international developments in torrefaction. In *2011 central European biomass conference: torrefaction Workshop, KEMA*, 2011.
Koufopanos CA, Lucchesi A, Maschio G: Kinetic modelling of the pyrolysis of biomass and biomass components, *Can J Chem Eng* 67(1):75–84, 1989.

Lambiotte A: *Process of continuous carbonization of cellulosic materials*, 1942 (2289917).
Lanzetta M, Di Blasi C, Buonanno F: An experimental investigation of heat-transfer limitations in the flash pyrolysis of cellulose, *Ind Eng Chem Res* 36(3):542–552, 1997.
Larfeldt J, Leckner B, Melaaen MC: Modelling and measurements of heat transfer in charcoal from pyrolysis of large wood particles, *Biomass Bioenergy* 18(6):507–514, 2000.
Lédé J, Blanchard F, Boutin O: Radiant flash pyrolysis of cellulose pellets: products and mechanisms involved in transient and steady state conditions, *Fuel* 81(10):1269–1279, 2002.
Lehmann J, Gaunt J, Rondon M: Bio-char sequestration in terrestrial ecosystems—a review. In *Mitigation and adaptation strategies for global change*, vol. 11, Netherlands, 2006, Springer, pp 395–419.
Magalhães AI, Petrović D, Rodriguez AL, Putra ZA, Thielemans G: Techno-economic assessment of biomass pre-conversion processes as a part of biomass-to-liquids line-up, *Biofuels Bioproducts Biorefining* 3(6):584–600, 2009, John Wiley and Sons, Ltd.
Maiti S, Dey S, Purakayastha S, Ghosh B: Physical and thermochemical characterization of rice husk char as a potential biomass energy source, *Bioresour Technol* 97(16):2065–2070, 2006.
Major J, Steiner C, Downie A, Lehmann J: Biochar effects on nutrient leaching. In Lehmann J, Joseph S,editors: *Biochar for environmental management: science and technology*, London, Washington, 2009, Earthscan, pp 271–287.
Mamleev V, Bourbigot S, Yvon J: Kinetic analysis of the thermal decomposition of cellulose: the main step of mass loss, *J Anal Appl Pyrolysis* 80(1):151–165, 2007.
McKendry P: Energy production from biomass (part 1): overview of biomass, *Bioresour Technol* 83(1):37–46, 2002.
Mohan D, Pittman CU, Steele PH: Pyrolysis of wood/biomass for bio-oil: a critical review, *Energy Fuel* 20(3):848–889, 2006.
O'Sullivan A: Cellulose: the structure slowly unravels. In *Cellulose*, vol. 4, Netherlands, 1997, Springer, pp 173–207.
Oasmaa A, Czernik S: Fuel oil quality of biomass pyrolysis oils—state of the art for the end users, *Energy Fuel* 13(4):914–921, 1999.
Oasmaa A, Peacocke C, Gust S, Meier D, McLellan R: Norms and standards for pyrolysis liquids. End-user requirements and specifications, *Energy Fuel* 19(5):2155–2163, 2005.
Ogawa M, Okimori Y: Pioneering works in biochar research, Japan, *Soil Res* 48(7):489–500, 2010.
Olsson JG, Jäglid U, Pettersson JBC, Hald P: Alkali metal emission during pyrolysis of biomass, *Energy Fuel* 11(4):779–784, 1997.
Patwardhan PR, Satrio JA, Brown RC, Shanks BH: Influence of inorganic salts on the primary pyrolysis products of cellulose, *Bioresour Technol* 101(12):4646–4655, 2010.
Peacocke G, Bridgwater A: Ablative plate pyrolysis of biomass for liquids, *Biomass Bioenergy* 7 (1–6):147–154, 1994.
Prins MJ: *Thermodynamic analysis of biomass gasification and torrefaction*, Universiteit Eindhoven, 2005.
Prins MJ, Ptasinski KJ, Janssen FJ: More efficient biomass gasification via torrefaction, *Energy* 31(15):3458–3470, 2006a.
Prins MJ, Ptasinski KJ, Janssen FJ: Torrefaction of wood: part 1. Weight loss kinetics, *J Anal Appl Pyrolysis* 77(1):28–34, 2006b.
Pyle D, Zaror C: Heat transfer and kinetics in the low temperature pyrolysis of solids, *Chem Eng Sci* 39(1):147–158, 1984.
Ratte J, Fardet E, Mateos D, Héry J-S: Mathematical modelling of a continuous biomass torrefaction reactor: TORSPYD™ column, *Biomass Bioenergy* 35(8):3481–3495, 2011.
Raveendran K, Ganesh A, Khilar KC: Influence of mineral matter on biomass pyrolysis characteristics, *Fuel* 74(12):1812–1822, 1995.

Rawls J, Stromberg B, Weston J, Jiang X, Hunt T: *Method and system for the torrefaction of lignocellulosic material*, 2012 (20120042567).
Repellin V, Govin A, Rolland M, Guyonnet R: Modelling anhydrous weight loss of wood chips during torrefaction in a pilot kiln, *Biomass Bioenergy* 34(5):602–609, 2010.
Rinket M, Toussaint A: Experience with firing pyrolysis oil on industrial scale, *IEA Bioenergy Agreement Task 34 Newsl* 31:3–4, 2012.
Roedig M, Klose W: Modelling of coal pyrolysis using a twin screw reactor. In *Proceedings of the 4th Ulcos seminar*, 2008.
Ronsse F, Bai X, Prins W, Brown RC: Secondary reactions of levoglucosan and char in the fast pyrolysis of cellulose, *Environ Prog Sustain Energy* 31(2):256–260, 2012.
Rotawave Ltd: *Rotawave applications: targeted intelligent energy system*, Website, 2011.
Rousset P, Turner I, Donnot A, Perré P: Choix dun modèle de pyrolyse ménagée du bois à léchelle de la microparticule en vue de la modélisation macroscopique, *Annu For Sci* 63 (2):213–229, 2006.
Saha B: Hemicellulose bioconversion, *J Ind Microbiol Biotechnol* 30:279–291, 2003.
Saidur R, Abdelaziz E, Demirbas A, Hossain M, Mekhilef S: A review on biomass as a fuel for boilers, *Renew Sustain Energy Rev* 15(5):2262–2289, 2011.
Schenkel Y, Bertaux P, Vanwijnbserghe S, Carre J: An evaluation of the mound kiln carbonization technique, *Biomass Bioenergy* 14(5–6):505–516, 1998.
Shafizadeh F, Chin PP: Thermal deterioration of wood. In Goldstein IS, editor: *Wood technology: chemical aspects*, Washington, 1977, American Chemical Society, pp 57–81.
Shu J, Lakshmanan V, Dodson C: Hydrodynamic study of a toroidal fluidized bed reactor, *Chem Eng Process Process Intensification* 39(6):499–506, 2000.
Steiner C, Teixeira W, Lehmann J, et al: Long term effects of manure, charcoal and mineral fertilization on crop production and fertility on a highly weathered Central Amazonian upland soil, *Plant Soil* 291:275–290, 2007.
Thermya—Energy, *Thermya news—TORSPYD construction on schedule*, Website, 2011.
Topell Energy, *First factory for energy from biomass in Duiven*, 2010.
Trifonova R, Postma J, Schilder M, van Elsas J: Microbial enrichment of a novel growing substrate and its effect on plant growth. In *Microbial ecology*, vol. 58, New York, 2009, Springer, pp 632–641.
Tumuluru J, Sokhansanj S, Wright C, Boardman R: *Biomass torrefaction process review and moving bed torrefaction system model development*, Idaho National Laboratory, US Department of, Energy, 2010 (INL/EXT-10–19569).
Turn SQ, Kinoshita CM, Ishimura DM: Removal of inorganic constituents of biomass feedstocks by mechanical dewatering and leaching, *Biomass Bioenergy* 12(4):241–252, 1997.
Uslu A, Faaij AP, Bergman P: Pre-treatment technologies, and their effect on international bioenergy supply chain logistics. Techno-economic evaluation of torrefaction, fast pyrolysis and pelletisation, *Energy* 33(8):1206–1223, 2008.
Van de Beld B, Holle E, Florijn J: The use of pyrolysis oil and pyrolysis oil derived fuels in diesel engines for CHP applications, *Appl Energy*, 2012.
van der Stelt M: *Chemistry and reaction kinetics of biowaste torrefaction*, Technische Univesiteit Eindhoven, 2010.
van der Stelt M, Gerhauser H, Kiel J, Ptasinski K: Biomass upgrading by torrefaction for the production of biofuels: a review, *Biomass Bioenergy* 35(9):3748–3762, 2011.
Várhegyi G, Antal MJ, Jakab E, Szabó P: Kinetic modeling of biomass pyrolysis, *J Anal Appl Pyrolysis* 42(1):73–87, 1997.
Venderbosch R, Prins W: Fast pyrolysis technology development, *Biofuels Bioproducts Biorefining* 4(2):178–208, 2010.
Verhoeff F, Pels J, Boersma A, Zwart R, Kiel J: ECN torrefaction technology heading for demonstration. In *Proceedings of the 19th European biomass conference and exhibition*, 2011.

Wagenaar B, Prins W, van Swaaij W: Pyrolysis of biomass in the rotating cone reactor: modelling and experimental justification, *Chem Eng Sci* 49(24, Part 2):5109–5126, 1994.
Waje SS, Thorat BN, Mujumdar AS: An experimental study of the thermal performance of a screw conveyor dryer, *Drying Technol* 24(3):293–301, 2006.
Wang X, Kersten SRA, Prins W, van Swaaij WPM: Biomass pyrolysis in a fluidized bed reactor. Part 2: experimental validation of model results, *Ind Eng Chem Res* 44 (23):8786–8795, 2005.
Wannapeera J, Worasuwannarak N: Upgrading of woody biomass by torrefaction under pressure, *J Anal Appl Pyrolysis* 96:173–180, 2012.
Westerhof R: *Refining fast pyrolysis of biomass*, Universiteit Twente, 2011.
Westerhof RJM, Brilman DWF, Garcia-Perez M, et al: Fractional condensation of biomass pyrolysis vapors, *Energy Fuel* 25(4):1817–1829, 2011.
White JE, Catallo WJ, Legendre BL: Biomass pyrolysis kinetics: a comparative critical review with relevant agricultural residue case studies, *J Anal Appl Pyrolysis* 91(1):1–33, 2011.
Williams PT, Besler S: The influence of temperature and heating rate on the slow pyrolysis of biomass, *Renew Energy* 7(3):233–250, 1996.
Yan W, Acharjee TC, Coronella CJ, Vásquez VR: Thermal pretreatment of lignocellulosic biomass, *Environ Progress Sustain Energy* 28(3):435–440, 2009.
Yan W, Hastings JT, Acharjee TC, Coronella CJ, Vásquez VR: Mass and energy balances of wet torrefaction of lignocellulosic biomass, *Energy Fuel* 24(9):4738–4742, 2010.
Yang H, Yan R, Chen H, Lee DH, Zheng C: Characteristics of hemicellulose, cellulose and lignin pyrolysis, *Fuel* 86(12–13):1781–1788, 2007.
Yu C, Zheng Y, Cheng Y-S, Jenkins BM, Zhang R, VanderGheynst JS: Solid–liquid extraction of alkali metals and organic compounds by leaching of food industry residues, *Bioresour Technol* 101(12):4331–4336, 2010.
Zhao Z, Huang H, Wu C, Li H, Chen Y: Biomass pyrolysis in an argon/hydrogen plasma reactor, *Eng Life Sci* 1(5):197–199, 2001.

CHAPTER THREE

Hydrotreating of Triglyceride-Based Feedstocks in Refineries

David Kubička*, Vratislav Tukač†

*Department of Renewable and Environmental Technologies (UniCRE-RENTECH), Research Institute of Inorganic Chemistry (VUANCH), Chempark Litvínov, Litvínov, Czech Republic

†Department of Organic Technology, Institute of Chemical Technology Prague, Technická 5, Praha, Czech Republic

Contents

Abstract

This chapter deals with some important aspects of deoxygenation of triglycerides that has become a new refining technology. After the introduction of basics of deoxygenation chemistry, the thermodynamic aspects of deoxygenation are discussed. Then hydrodynamics in conventional hydrotreaters is compared with the deoxygenation system, and predictions on the vegetable oils deoxygenation are made. Kinetics of triglyceride deoxygenation reactions including kinetics of the individual reaction steps

Advances in Chemical Engineering, Volume 42
ISSN 0065-2377
http://dx.doi.org/10.1016/B978-0-12-386505-2.00003-1

involved are discussed for the system CoMo on alumina. Next factors influencing catalyst deactivation, particularly those related to feedstock, are analyzed. Finally, commercial, available deoxygenation technologies are overviewed. The attention is focused on both stand-alone and coprocessing technologies.

LIST OF SYMBOLS

A. LATIN LETTERS

A pre-exponential factor of reaction, dimension depends on the reaction
Bo Bodenstein number
c concentration, mol/m^3
c_p specific heat capacity, J/kg/K
C_{H_2} concentration of hydrogen, kmol/m^3
C_S concentration of liquid substrate, kmol/m^3
C concentration, wt%
d_p diameter of the catalyst particles, mm
d_{pi} diameter of inert fines, mm
D diameter of the reactor, m
D_j substance diffusivity, m^2/s
E activation energy of reaction, J/mol
ΔH_r heat of reaction, kJ/mol
k kinetic constant of reaction, dimension depends on the reaction
kc mass transfer coefficient, m/s
L length of the bed, m
n reaction order to substrate, 1
P pressure, MPa
Q volumetric flow rate, m^3/s
Q_m mass flow rate, kg/h
r reaction rate, mol/s/m^3
R universal gas constant, 8.31434 J/mol/K
t time, s
T temperature, K
U linear velocity of the fluid, m/s
V volume, m^3
W mass of catalyst, kg
X conversion level, 1

B. GREEK LETTERS

ε bed voidage, 1
λ heat conductivity, J/m/K/s
μ dynamic viscosity, Pa s
ν kinematic viscosity, mm^2/s
ρ density, kg/m^3
σ surface tension, mN/m

C. SUBSCRIPTS AND SUPERSCRIPTS

calc calculation
cat catalyst
exp experimental
G gas
i index of reaction
j index of substance
L liquid
s solid phase
wt weight

ABBREVIATIONS

AGO atmospheric gas oil
LGO lite gas oil
RO rapeseed oil

1. INTRODUCTION

Several reasons can be identified for the recent interest in automotive fuels derived from renewable energy resources. The major ones include (i) the limited reserves of petroleum that could be replaced, at least partially, by renewable feedstocks; (ii) the national security, as the main petroleum reserves are found in politically unstable regions; (iii) the increased awareness of climate changes that have been attributed to the increasing atmospheric concentrations of CO_2 due to the ever-increasing consumption of fossil fuels; and (iv) the support of rural areas in Western countries aimed at supporting employment and cultural landscape in the areas (Kubičková and Kubička, 2010; Schaub and Vetter, 2008).

Consequently, a fast development and commercialization of two large-scale biofuel production technologies, namely, bioethanol and fatty acid methyl esters (FAME) production, have followed to satisfy the current demand for biofuels, that is, automotive fuels derived from renewable energy resources. Bioethanol is produced typically from sugar cane (Brazil), corn (United States), and wheat and sugar beet (Europe) and is consumed in gasoline-driven cars. There are two common alternatives: (i) gasoline–bioethanol blends with up to 10 vol.% of ethanol that can be used in the standard spark-ignition engines and (ii) ethanol-based fuels containing typically 70–90 vol.% of ethanol, which require specially modified engines (the so-called flexi fuel vehicles; Kubičková and Kubička, 2011).

FAME (biodiesel) is produced by transesterification of vegetable oils (VOs) with methanol as an alternative fuel for compression-ignition engines. It is used either as a blending component of the standard diesel (in volumetric concentrations up to 7% or 10%) or as a stand-alone fuel that requires that a car is equipped with a modified compression-ignition engine. The recognition that transesterification of VOs yields a diesel fuel having significantly better fuel properties than neat VOs led to the spread of this technology. Biodiesel provides some benefits in comparison with conventional diesel fuel. These include its biodegradability and better lubricity that allows reducing the amount of added diesel lubricity additives. On the other hand, there are some disadvantages connected to the use of biodiesel (Kubičková and Kubička, 2011).

From the refining perspective, both technologies are rather agriculturally based technologies and the refiners have to purchase these products to blend them as it is required by the legislation in many European countries. Moreover, neither bioethanol nor biodiesel is fully compatible with the current hydrocarbon fuels and existing engine technologies. It seems, thus, only logical that there is a demand for alternative refining technologies. As a result, renewable (green) diesel has become, in the recent years, a viable alternative to the conventional biodiesel. The renewable diesel concept is based on a fundamentally different approach than production of biodiesel even though it uses the same feedstock. In contrast to biodiesel production, which is based on mere reduction of the molecular weight of triglycerides due to their transesterification with methanol and leads to modification of physical properties while keeping chemical properties intact, renewable diesel production is based on modification of chemical as well as physical properties. Consequently, the renewable diesel is much closer in terms of chemical composition as well as fuel properties to the conventional diesel fuel than biodiesel.

The relevant issues pertaining, in particular, to the accomplishment of catalytic transformation of triglycerides into a mixture of hydrocarbons being suitable as diesel fuel blending components have been addressed by several authors over the past 5 years (Kubička, 2008; Kubičková and Kubička, 2010, 2011; Lestari et al., 2009; Murzin et al., 2012). While the fundamental aspects of deoxygenation chemistry have been covered in these reviews in detail, a view focused on chemical engineering aspects of deoxygenation including an overview of commercial applications is still missing. Thus, the aim of this review is to address the most important issues relevant to commercialization of deoxygenation technologies. After a brief summary of the main aspects of deoxygenation chemistry, the next chapters will be devoted to thermodynamics, hydrodynamics, and kinetics of triglyceride deoxygenation. Then deactivation aspects will be discussed, and the final part will be focused on the existing commercial technologies and their most salient features. In particular, the advantages and disadvantages of the two basic concepts, namely, stand-alone processing of triglycerides and coprocessing of triglycerides with petroleum fractions in existing or modified hydrotreating units, will be analyzed.

2. BASICS OF DEOXYGENATION CHEMISTRY

The production of renewable diesel is based on the conversion of triglycerides into hydrocarbons, that is, on the complete deoxygenation. Reflecting the nature of carbon–oxygen bonds in triglycerides, deoxygenation can proceed via several different reaction routes which take place either in the presence or in the absence of hydrogen. The method utilizing hydrogen is, in fact, more common. It secures a stable catalytic activity, and it is hence used in the commercialized deoxygenation technologies. Several deoxygenation pathways have been reported in the presence of hydrogen (Huber et al., 2007; Kubička and Kaluža, 2010; Kubičková and Kubička, 2010; Laurent and Delmon, 1994a). They have been classified as hydrodeoxygenation (HDO), (hydro)decarboxylation, and decarbonylation.

HDO can be described as a total hydrogenation that yields hydrocarbons and water as the only reaction products. All oxygen atoms are eliminated during HDO in the form of water. As all naturally occurring fatty acid moieties of triglycerides have an even number of carbon atoms, the resulting hydrocarbons, except propane that originates from the glycerol backbone of triglycerides, have an even carbon number too (Kubička and Kaluža, 2010). In contrast, in (hydro)decarboxylation (HDC), any oxygen atom in triglycerides is eliminated as carbon dioxide. Consequently, only hydrocarbons with odd carbon atom numbers are formed from the fatty acid moieties. The prefix hydro- is sometimes used to stress that hydrogen is involved in the reaction. It has been proposed that hydrogen is needed to break the fatty acid moiety loose from the triglyceride. The released fatty acids undergo subsequent decarboxylation to yield hydrocarbon and CO_2 (Kubička and Kaluža, 2010). Both reaction pathways are schematically depicted in Fig. 3.1.

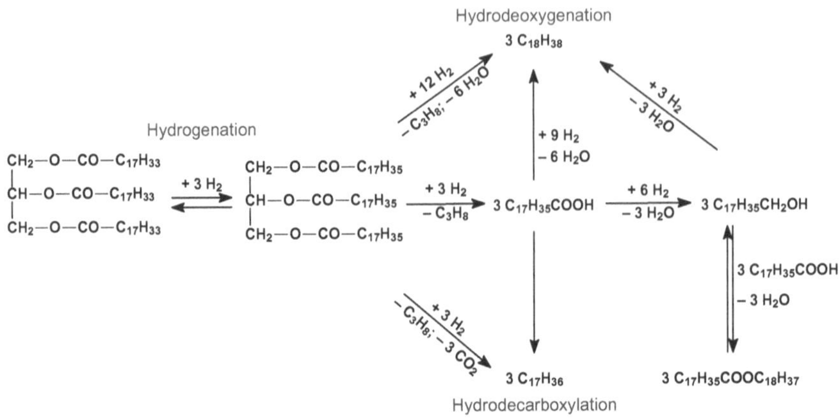

Figure 3.1 A schematic view on the transformation of triglycerides into hydrocarbons under hydrotreating conditions. *Based on Kubička (2008).* (For color version of this figure, the reader is referred to the online version of this chapter.)

In addition, decarbonylation has been proposed as a plausible reaction pathway as well (Huber et al., 2007). Unfortunately, the complex reaction network involving hydrogen, CO, CO_2, water, and methane under typical deoxygenation reaction conditions makes it difficult, if not impossible, to provide an unequivocal evidence whether carbon oxides found in reaction products originate from decarboxylation and subsequent hydrogenation of carbon dioxide or decarbonylation or whether both decarboxylation and decarbonylation take place in parallel.

The understanding of the deoxygenation reaction pathway is of immense practical importance as it determines the theoretical hydrogen consumption which is extremely important for any refining process. The reaction scheme (Fig. 3.1) indicates that hydrodecarboxylation should be the preferred course of deoxygenation if hydrogen consumption was the only criterion. However, this would be true only if the secondary reactions, in particular, carbon dioxide hydrogenation to carbon monoxide and even methane, could be avoided. It can be easily seen that if all carbon dioxide formed was to be hydrogenated to afford methane the total hydrogen consumption of the HDC route would be larger than the total hydrogen consumption of the HDO path. Moreover, both reaction pathways differ in the carbon atom efficiency. While in HDO theoretically, only three carbon atoms per each triglyceride molecule are not recovered as liquid hydrocarbons, in HDC it is six carbon atoms.

The occurrence of both reaction pathways is affected by catalytic system used as well as by the reaction conditions. For example, palladium on carbon catalysts was shown to catalyze virtually and exclusively the hydro(decarboxylation) pathway (Immer et al., 2010; Kubičková et al., 2005; Maier et al., 1982; Snåre et al., 2006), while typical hydrotreating catalysts (sulfided NiMo, CoMo) provide both HDO and HDC products, that is, *n*-octadecane and *n*-heptadecane, respectively (Bezergianni et al., 2010a,b; Donnis et al., 2009; Kubička, 2008; Kubička and Kaluža, 2010; Kubička et al., 2009, 2010; Mikulec et al., 2009; Priecel et al., 2011). The extent to which each of the pathways contributes to the overall deoxygenation depends on reaction conditions and catalyst composition. HDO was shown to be preferred over HDC at higher hydrogen pressures and lower reaction temperatures (Kubička, 2008). The origin of this selectivity dependence will be discussed in Section 3. The role of the sulfided catalyst is an interesting one, too. It has been recently observed by Kubička and Kaluža (2010) that sulfided nickel catalyst provided virtually and exclusively the HDC products, while sulfided molybdenum catalyst yielded almost solely HDO products. However, both sulfided catalysts were significantly less active than the sulfided NiMo catalyst that afforded both HDC and HDO products.

The experimental studies have demonstrated that milder reactions that typically used in hydrotreating of petroleum middle distillate fractions are sufficient to achieve complete deoxygenation of triglycerides. Over fresh sulfided catalysts, total conversion of triglycerides was obtained at as low temperatures as 280 °C at moderate hydrogen pressures (1–5 MPa; Donnis et al., 2009; Kubička and Kaluža, 2010; Kubička et al., 2010; Priecel et al., 2011). Similar reaction conditions were applied successfully also in the case of palladium catalysts (Immer et al., 2010; Kubičková et al., 2005; Simakova et al., 2009; Snåre et al., 2006, 2008). As the aim of this chapter is to describe upgrading of triglycerides in a refinery, the discussion will be focused mostly on sulfided hydrotreating catalysts that are the most likely ones to be used commercially. In fact, the available data on the already commercialized technologies indicate the use of hydrotreating catalysts.

3. THERMODYNAMIC ASPECTS OF DEOXYGENATION

Thermodynamic assessment of any reaction system is essential prior to any industrial application. While, nowadays, there are many different databases containing relevant thermodynamic data for a wide variety of reaction systems, the relevant data for deoxygenation of triglycerides have been until recently missing. Smejkal et al. (2009) have applied Joback's contribution method for the assessment of the relevant thermodynamic data of tristearate, which was chosen as a model compound representing triglycerides. Considering that the majority of triglycerides in VOs are composed of C_{18} fatty acid moieties and that hydrogenation of their double bonds is very facile at deoxygenation reaction conditions, tristearate seems to be an ideal model component to describe triglycerides. The applicability of Joback's contribution method for estimating thermodynamic properties of tristearate was verified on butyl stearate for which there are available data in commercial databases. Based on the very good agreement between these data and data estimated by Joback's contribution method, it was argued that the method is suitable for the assessment of thermodynamic data of tristearate (Smejkal et al., 2009).

The thermodynamic prediction for 7 MPa indicated that in the temperature range of 270–350 °C, which is relevant for total deoxygenation of triglycerides, *n*-octadecane is the thermodynamically preferred product. Its concentration was predicted to be in the range of 70–80%. This means that, at an elevated pressure of 7 MPa and high temperature, the HDO reaction is preferred over hydrodecarboxylation resulting in the formation of hydrocarbons having the same number of carbon atoms in their molecules as there were in the original fatty acid moieties (in case of tristearate it means *n*-octadecane). On the other hand, the relative importance of hydrodecarboxylation increases with the decreasing reaction pressure. For example, the decrease in reaction pressure from 7 to 0.7 MPa at a constant temperature of 310 °C results in a change in the predicted composition (based on thermodynamic data) from ca. 80/20 (C_{18}/C_{17}) to about 55/45 (C_{18}/C_{17}; Smejkal et al., 2009). This can be attributed to the mole changes in the reaction system—during hydrodecarboxylation, the total number of moles in the system increases from 4 to 7, while during HDO, the total number of moles in the system decreases from 13 to 10 (transformation of 1 mol of tristearate is considered). These results are in line with experimental observations (Kubička, 2008).

The thermodynamic predictions showed that, in excess of hydrogen, *n*-octadecane should be the exclusive hydrocarbon product, which was not found experimentally. In order to account for the observed formation of *n*-heptadecane, the H_2/oil ratio was varied until reasonable agreement with the experimental data was found, which was at the value of H_2/oil ratio equal to 7. It was thus suggested that under the reaction conditions, the reaction was limited by hydrogen diffusion to the active sites, that is, by hydrogen mass transfer through the stagnant liquid film formed by reactants and/or products on the outer surface of catalyst particles (Smejkal et al., 2009). Apart from the mass transfer limitations itself, the availability of hydrogen is negatively affected by the low solubility of hydrogen in VOs (Piqueras et al., 2008; Santacesaria et al., 1994; Schmidt and Schomacker, 2007). A calculation confirming these previous reports is given below.

From the practical point of view, heat of reaction is key information obtainable from thermodynamic data. Heat effects of chemical reactions influence both reaction rates in the catalytic bed and operation conditions and engineering process design. Industrial high-capacity trickle-bed reactors which are preferentially used for hydrodesulphurization (HDS) treatment of crude oil distillates, as well as promising HDO of VOs, are considered to be adiabatic. This is due to the bad radial heat conductivity by contact points among particles of catalyst. Overheating of the catalytic bed that would follow could result in catalyst deactivation by excessive coking, undesired cracking of VO, and vaporization of reaction mixture. The most important parameter that characterizes the heat effect during the reactor operation is the adiabatic temperature rise defined by the following equation:

$$\Delta T_{\mathrm{ad}} = c_0(-\Delta H_r)/(\rho c_p) \qquad [3.1]$$

The practical meaning of this parameter is that it makes possible the assessment of the increase in temperature of the reaction mixture caused by chemical reaction carried out with 100% conversion. These conditions are common for both HDS and HDO processes. The most important parameter of Eq. (3.1) is the value of reaction enthalpy at reaction temperature.

There is a brief list of the main reactions relevant for deoxygenation and HDS, and their respective heats of reaction are given in Table 3.1. Heat of reaction was computed here at 280 °C using the process simulator Aspen Plus (AspenTech, 2011).

From the comparison of the reaction heats of some reactions which can occur in HDO and HDS processes, presented in Table 3.1, it follows that all important reactions, except decarboxylation, produce thermal energy, that is, exothermic. Magnitudes of reaction heats are comparable for both HDO and HDS, but very high concentration of triglycerides in VOs in conjunction with high heat of possible triglyceride hydrolysis poses an insuperable problem to operate HDO of pure VO in industrial scale trickle-bed reactors. The only practicable arrangement is to dilute VO to get proper adiabatic temperature rise, and/or to coprocess VO together with crude oil fractions by HDS process.

Table 3.1 Heat of selected reactions characteristic for deoxygenation and hydrodesulphurization at 280 °C

Reaction ID	Reaction path	$(\Delta H_r)_{280\,°C}$ (kJ/mol)
Hydrolysis	Glyceryl oleate → oleic acid and glycerol	−176.3
Saturation	Oleic acid → stearic acid	−53.8
HDO	Stearic acid → *n*-octadecanol and water	−65.8
HDO	Stearic acid → *n*-octadecane and water	−178.5
Decarboxylation	Stearic acid → *n*-heptadecane and carbon dioxide	+18.8
HDS	Thiophene → *n*-butane and hydrogen sulfide	−262.3
HDS	Benzothiophene → ethylbenzene and hydrogen sulfide	−159.1
HDS	Dibenzothiophene → biphenyl and hydrogen sulfide	−124.8
HDN	Quinoline → *n*-propylbenzene and ammonia	−245.5

4. HYDRODYNAMICS ASPECTS

Catalytic deoxygenation of VOs in refineries resembles much catalytic hydroprocessing of middle distillates obtained by crude oil fractionation, namely, atmospheric gas oil (AGO) HDS. The principal issue that needs to be addressed is focused on scale-up of laboratory kinetic data to full industrial scale process. The crucial task in evaluation of reaction kinetics parameters is to eliminate the influence of both internal and external mass and heat transfer on kinetic results. On the other hand, productivity and selectivity of industrial hydroprocessing reactors are frequently determined by hydrodynamics of flowing fluids. It is well known (Levenspiel, 1999) that reactor behavior is governed by four interrelated quantities. In process design, it is necessary to specify the input—known kinetics and flow pattern, and based on these data, it is possible to predict the output results. Mass transfer limitations caused by both internal diffusion in catalyst particles and external transport of hydrogen to surface of the catalyst can also be the reaction rate constraining features, that is, limiting the reactor productivity.

To evaluate hydrodynamics and mass transfer conditions in a trickle-bed reactor, information about physical properties of used fluids needs to be available in order to calculate flow and mass transfer criteria. Also the type of flow regime, such as trickling, pulsing, or spray flow, determines deviations from the plug flow caused by axial dispersion of fluids in fixed catalytic bed. Moreover, maldistribution of liquid velocities in the cross-section of the catalytic bed, catalyst bypassing as a result of wall flow, and uneven wetting of the outer surface of catalyst particles are important for trickle-bed reactor behavior. All these phenomena have a profound influence on the overall effectiveness of the catalytic reactor. Therefore, in Section 4.1, the key hydrodynamic parameters of VOs and middle distillates will be compared in order to allow an assessment of VO deoxygenation in conventional hydrotreating reactors using hydrodesulfurization reactors as a benchmark.

4.1. Flow characteristic of VOs and middle distillates

A comparison of key physical properties (viscosity, density, and surface tension) of VO (rapeseed oil) with petroleum middle distillates (AGO) is given in Table 3.2. It can be clearly seen that the main difference is in viscosity, which is 4–10 times higher for rapeseed oil than for AGO. The other parameters are also higher in the case of rapeseed oil, but only by 10% for density and 20–30% for surface tension. The presented values are at temperatures where they can be obtained experimentally, which are significantly lower than the typical temperatures used during HDS and deoxygenation. Hence, the usual (Reid et al., 2001) physical approximations of temperature dependence of physical properties were used to estimate these important physical properties under relevant reaction temperatures. Due to the difference between common temperatures needed for deoxygenation (temperatures less than 300 °C are generally sufficient to achieve complete conversion) and for HDS process, which is operated at around 350 °C,

Table 3.2 Comparison of important properties of middle distillate (AGO) and rapeseed oil (RSO)

Physical property	Unit	Middle distillate	Rapeseed oil
Density, ρ(15 °C)	kg/m^3	844	920
Density, ρ(70 °C)	kg/m^3	806	883
Kin. viscosity, ν(25 °C)	mm^2/s	6.84	62.14
Kin. viscosity, ν(150 °C)	mm^2/s	0.95	3.98
Surface tension, σ(40 °C)	mN/m	27	33
Surface tension, σ(85 °C)	mN/m	22.5	29

the physical properties of middle distillate and rapeseed oil were extrapolated up to temperature of 300 °C. The approximations and the resulting estimates of physical properties calculated here for both feedstocks are presented in Table 3.2.

The temperature dependence of kinematic viscosity can be approximated by Eq. (3.2), using the values of the coefficients of the equation presented in Table 3.3 that were determined here by fitting measured viscosities in the range of 25–150 °C (Fig. 3.2).

$$\ln(v) = b_1 + \frac{b_2}{T} + \frac{b_3}{T^2} \qquad [3.2]$$

Table 3.3 The temperature dependence of kinematic viscosity: coefficients of Eq. (3.2)

Liquid	b_1	b_2	b_3
AGO	1.1526	−2299.4	7.5524×10^5
Rapeseed oil	3.7084	−3678	1.1363×10^6

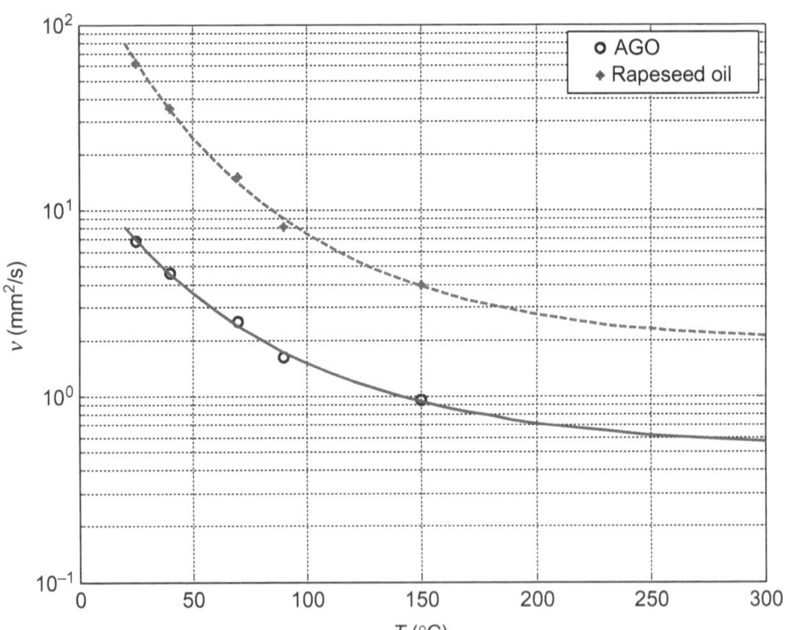

Figure 3.2 Temperature dependence of kinematic viscosity of rapeseed oil (RSO) and middle distillate (AGO). (For color version of this figure, the reader is referred to the online version of this chapter.)

A comparison of the solution of Eq. (3.2) with experimental data is presented in Fig. 3.2. Rapeseed oil has almost an order of magnitude higher viscosity than middle distillate in the experimental region which is also predicted for the extrapolated region. As a consequence, it can be expected that the flowing features of rapeseed oil will be worse in comparison with AGO.

The temperature dependence of density can also be described by a polynomial expression given by Eq. (3.3). The coefficients of this equation are presented in Table 3.4 and were estimated by fitting the experimentally measured values in the range of 15–70 °C. The graphical comparison of experimental data and prediction based on Eq. (3.3) for rapeseed oil and AGO is shown in Fig. 3.3. The higher density of rapeseed oil by approximately 10 % in

Table 3.4 The temperature dependence of density: coefficients of Eq. (3.3)

Liquid	a_1 (g/dm^3)	a_2 (g/dm^3 K)	a_3 (g/dm^3 K^2)
AGO	0.85442	−0.00068484	0
Rapeseed oil	0.93052	−0.00069805	2.3377×10^{-7}

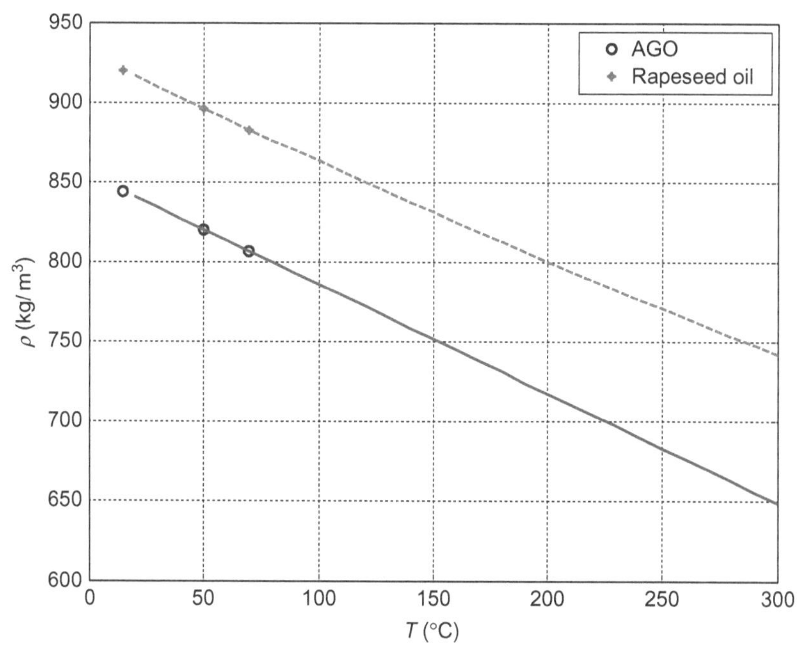

Figure 3.3 Temperature dependence of density of rapeseed oil and middle distillate (AGO). (For color version of this figure, the reader is referred to the online version of this chapter.)

comparison with AGO, which was found at temperatures below 100 °C was also predicted by the Eq. (3.3) for 300 °C. Once again this will have an impact on the hydrodynamics of the hydrotreating system.

$$\rho = a_1 + a_2 T + a_3 T^2 \qquad [3.3]$$

The surface tension data were adopted from the literature (Lüft and Spicher, 2007) and subsequently were approximated using Eq. (3.4) proposed by Macleod (1923). This equation can be simplified into Eq. (3.5) with parameter $k' = 2.7354$ for AGO and 2.6586 for rapeseed oil. The graphical representation of the surface tension temperature dependence is then given in Fig. 3.4. It demonstrates that similar to the other investigated physical properties, also estimated surface tension of rapeseed oil at 300 °C is higher than the surface tension of AGO.

$$\text{const.} = \frac{M\sigma^{1/4}}{\rho^L - \rho^G} \qquad [3.4]$$

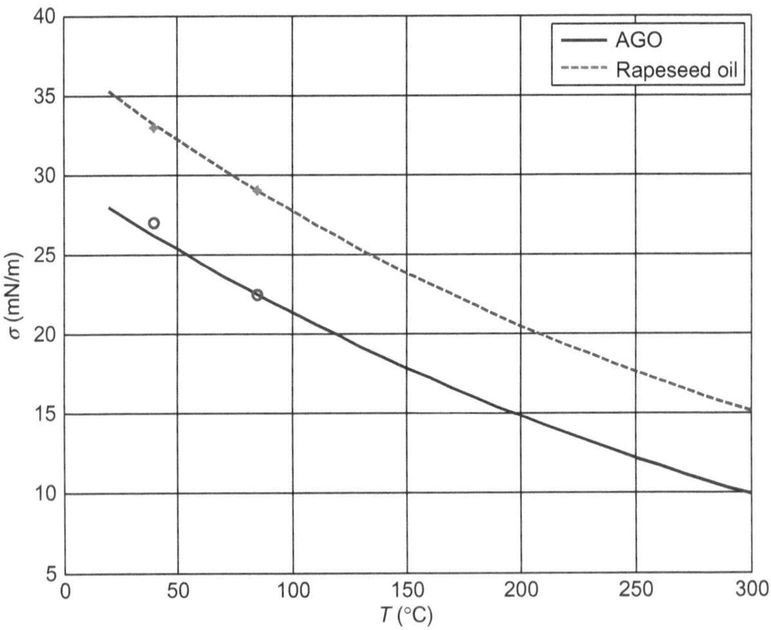

Figure 3.4 Temperature dependence of surface tension of rapeseed oil and middle distillate (AGO). (For color version of this figure, the reader is referred to the online version of this chapter.)

$$k' = \frac{\sigma^{1/4}}{\rho^L} \qquad [3.5]$$

In summary, the flow characteristics determining properties of rapeseed oil such as viscosity, density, and surface tension are less convenient in comparison with those of middle distillate and hence more energy needs to be spent to transport VOs into the reactor. Furthermore, higher pressure drop along the trickle-bed reactor can also be expected for VOs deoxygenation in comparison with AGO hydrotreating.

The driving force of the hydrogen liquid–solid mass transfer as well as of the reaction rate in porous catalysts is the concentration of hydrogen dissolved in the liquid phase reactants, here AGO and VO, respectively. Some data for hydrogen solubility in triglycerides could be found in the literature. For example, hydrogen solubility in soybean oil was reported by Fillion to follow Eq. (3.6) with parameters $H_0 = 8.84$ MPa m^3/kmol and $\Delta E = -5000$ kJ/kmol (Fillion and Morsi, 2000):

$$\mathrm{He}_{H_2} = H_0 \exp\left(\frac{-\Delta E}{RT}\right) \qquad [3.6]$$

The equilibrium hydrogen concentration (kmol/m^3) in rapeseed oil was given by Bern et al. (1975) and can be calculated according to Eq. (3.7) with P(MPa), T(K). Hydrogen solubility in light vacuum gas oil was reported in the literature (Cai et al., 2001) and allows to estimate hydrogen solubility in a petroleum fraction to be hydrotreated:

$$C^* = 0.203 \exp\left(\frac{-5900}{RT}\right) P \qquad [3.7]$$

Using the available data, solubility of hydrogen in both VOs and a petroleum fraction as a function of temperature was calculated here. The results are summarized in Fig. 3.5. It can be observed that while similar hydrogen solubility is predicted for petroleum distillate and rapeseed oil, lower hydrogen solubility is predicted for soybean oil at 300 °C, which is the temperature used here to represent reaction conditions.

Apart from hydrogen solubility which is comparable for VOs and gas oil fractions, hydrogen diffusivity in those liquids can play an important role as well. Hydrogen diffusivity in soybean oil was taken from the literature (Fillion and Morsi, 2000), while diffusivity in diesel was calculated using Wilke Chang equation for benzene as a model compound. The results are reported as a function of temperature in Fig. 3.6. Moreover, comparison

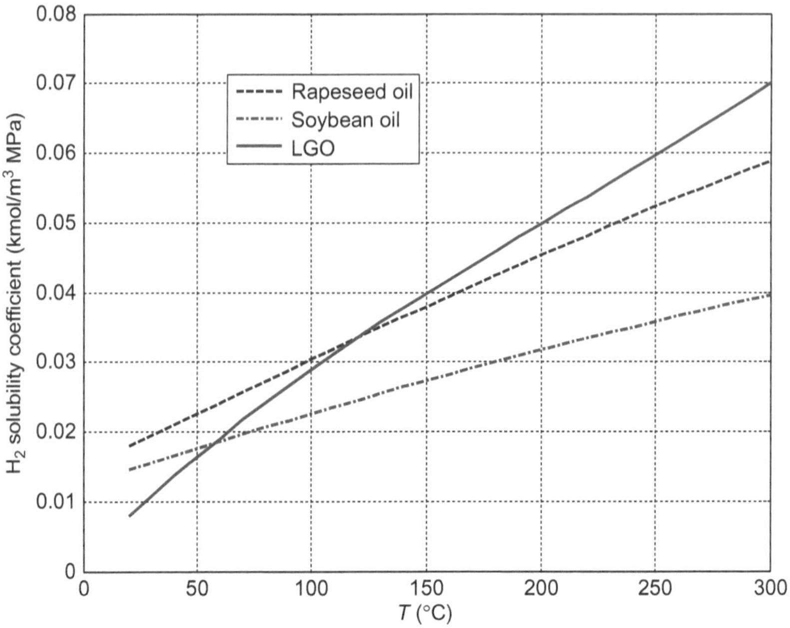

Figure 3.5 Temperature dependence of hydrogen solubility in rapeseed and soybean oil and light vacuum gas oil. (For color version of this figure, the reader is referred to the online version of this chapter.)

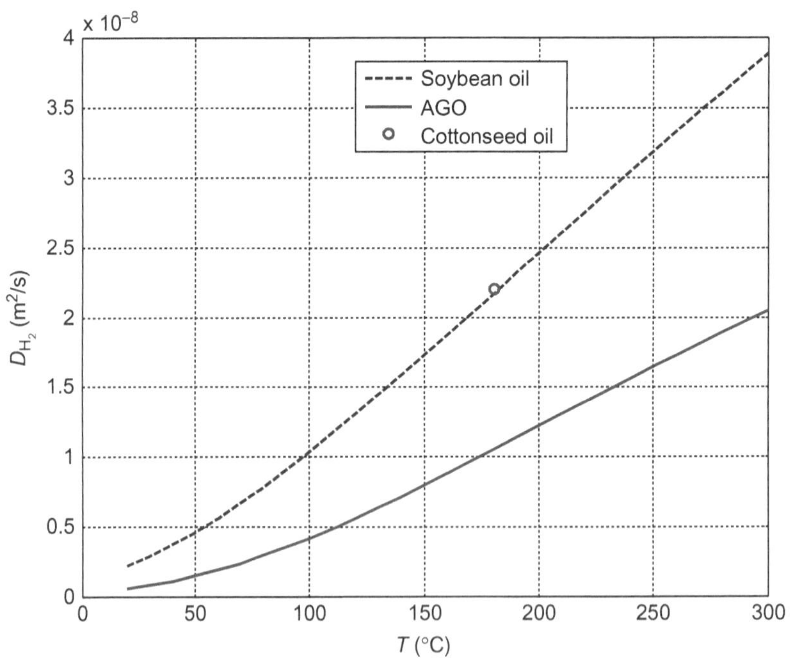

Figure 3.6 Temperature dependence of hydrogen diffusivity in soybean oil and middle distillate (AGO). (For color version of this figure, the reader is referred to the online version of this chapter.)

of triglyceride and hydrogen diffusivity in VO is also available in the literature (Andersson et al., 1974). Hydrogen diffusivity in cottonseed oil at 180 °C was 2.2×10^{-8} m^2/s, whereas triglyceride diffusivity in the same oil was 100 times slower (1.8×10^{-10} to 4.2×10^{-10} m^2/s at 131–192 °C, respectively). The higher hydrogen diffusivity in VOs as compared with middle distillate (Fig. 3.6) compensates its lower solubility in the same liquid, and the transport phenomena in VO and AGO can be estimated to be approximately the same.

4.2. Hydrodynamics and mass transfer in HDO and HDS reactors

Continuous catalytic deoxygenation of VOs in trickle-bed reactors to the mixture of hydrocarbons (alkanes) is still performed and investigated more in laboratory and pilot scale than in industrial scale, which is contrary to HDS process with capacities exceeding 100 tons of fuel per hour in one reactor. This was the reason to compare transport properties of laboratory HDO trickle-bed reactor with an industry scale HDS reactor.

Parameters of the industrial trickle-bed reactor in Tables 3.5 and 3.6 were adopted from the literature (Mary et al., 2009). Parameters and conditions of the laboratory HDO reactor were calculated here and correspond to the laboratory reactor conditions at 280 °C, used for HDO kinetics evaluation.

A comparison of hydrodynamic conditions in a laboratory HDO reactor and an industry scale HDS trickle-bed reactor is presented in Table 3.6. The elimination of both external and internal mass transfer diffusion resistance is the key assumption to measure intrinsic kinetics in a laboratory reactor. Among the most important hydrodynamic phenomena, which strongly depend on the operating conditions, are catalyst wetting, liquid maldistribution, axial dispersion, and gaseous reactant transfer.

Incomplete catalyst utilization in a pilot reactor may occur especially in the trickle flow regime and can have two main causes. First is due to reactor-scale liquid maldistribution that may leave certain portions of the bed poorly irrigated by channeling and wall flow. A proper design of liquid distributors and/or sufficient layer of packing to reach equilibrium liquid distribution solve this obstacle. The other cause of incomplete catalyst utilization is particle-scale incomplete external wetting. This occurs at sufficiently low liquid mass velocity when the available liquid flow is insufficient to cover all the external surface of the catalyst particles with a continuous liquid film. Solution is to fill void space among catalyst particles by fine inert material to

Table 3.5 Operation parameters of trickle-bed reactors

	Reactor	
Parameter	**Laboratory**	**Industrial (Mary et al., 2009)**
L, length of the bed (m)	0.34	5
D, diameter of the reactor (m)	0.017	1.5
d_p, diameter of the catalyst particles (mm)	0.25–0.5	1.5
d_{pi}, diameter of inert fines (mm)	0.1	–
Dilution of catalyst by inert ratio (mL_{in}/mL_{cat})	1.5	0
U_L, velocity of the liquid (m/s)	4×10^{-5}	1×10^{-3}
U_G, velocity of the gas (m/s)	5.5×10^{-2}	5×10^{-2}
Reaction type	HDO	HDS
Flow regime	Trickling downflow	Trickling/pulsing flow
Type of catalyst	$NiMo/\gamma Al_2O_3$	$NiMo/\gamma Al_2O_3$
Inert	Silicon carbide	–
ε, bed voidage	0.3	0.4
T, temperature (°C)	280	327
P, pressure (MPa)	3.5	3.5
n, reaction order to substrate	1	1
X, conversion level	0.999	0.999
C_{H_2}, concentration of hydrogen (mol/m^3)	197	230
C_S, concentration of liquid substrate (kmol/m^3)	2.68	0.235
ρ_L, density of the liquid (kg/m^3)	753	750
ρ_G, density of the hydrogen (kg/m^3)	5	5
η_L, dynamic viscosity of the liquid (Pa/s)	16×10^{-4}	5×10^{-4}
η_G, dynamic viscosity of gas (Pa/s)	1×10^{-4}	1×10^{-4}
E, activation energy (kJ/mol)	75	100
ΔH_r, heat of reaction (kJ/mol)	–	250
D_{H_2}, molar diffusivity in oil (m^2/s)	3.6×10^{-8}	1×10^{-8}
D_{TG}, molar diffusivity in oil (m^2/s)	3.6×10^{-10}	–

Table 3.6 Comparison of hydrodynamic conditions of HDO and HDS (Mary et al., 2009)

Conditions	Criteria	HDO	HDS
Catalyst-bed dilution	V_{inert}/V_{cat}	1	0
Liquid maldistribution	Wall effects $\frac{D}{d_p} \geq 25$	170–85	500–25
	Channeling (mm) $d_{pi} \geq \frac{d_p}{10}$	0.04	0–0.3
Axial dispersion	$\frac{L}{d_p} \geq \frac{20n}{Bo_L} \ln\left(\frac{C_0}{C_f}\right)$	1771	1667–789
Wetting efficiency	$\frac{\eta_L U_L}{\rho_L d_p^2 g} \geq 5 \times 10^{-6}$	7.07×10^{-5}	7.55×10^{-6}–5×10^{-6}
Limiting reactant gas	$\frac{D_{eB} C_{Bi}}{b D_{eA} C_A^*} \gg 1$	0.14	0.014
Mass transfer	Internal $\frac{r(1-\varepsilon)\left(\frac{d_p}{2}\right)^3}{D_c C} \leq 1$	6×10^{-5}–1×10^{-5}	1–0.315
	External $k_{ext} \geq \left(\frac{10 d_p}{C}\right) r(1-\varepsilon)$	4×10^{-3}	1×10^{-3}–4.2×10^{-5}

decrease the bed porosity and increase liquid holdup by pseudostatic liquid lenses at particles contact points.

The most reaction systems could be classified (Wu et al., 1996) as being liquid- or gas reactant limited. The value of the ratio of the liquid reactant flux to the catalyst particle to the gas reactant flux to the catalyst particle delineates two categories. For value $\gg 1$, the reaction can be considered as gas reactant rate limited, while values $\ll 1$ mean that it is the liquid reactant that is rate limiting. For liquid-limited reactions, an upflow reactor should be preferred as it provides for complete catalyst wetting and for the fastest transport of the liquid reactant to the catalyst. For gas-limited reactions, a downflow reactor, especially at partially wetted conditions, is to be preferred as it facilitates the direct transport of the gaseous reactant to the catalyst by unwetted surface of catalyst particles.

Magnitudes of parameters presented in Table 3.6 give evidence on fulfilling of criteria allowing to neglect external effects in the HDO kinetic results, such as wall effects and channeling, axial dispersion and catalyst wetting, and both external and internal mass transfer. Due to the low triglyceride diffusivity in VOs and/or low sulphur compound concentration in middle distillates, both reactions HDO and HDS are liquid reactant limited. This means that complete wetting of catalyst surface is key presumption of full catalyst utilization. To guarantee it, dilution of catalyst by fine inert or high linear liquid flow velocity has to be used.

5. KINETICS ASPECTS

Deoxygenation of triglycerides involves a very complex set of reactions carried out in liquid phase on heterogeneous catalysts by gaseous hydrogen. In contrast to HDS, the concentration of liquid substrate is very high, due to the fact that the feed liquid consists almost exclusively of triglycerides. Consequently, also the necessary amount of hydrogen exceeds the amount needed for HDS. At high capacity operations in trickle-bed reactors, kinetic models like Langmuir–Hinshelwood and Hougen–Watson are overshadowed by uncertainties in hydrodynamics, mass transfer effects and frequent catalyst deactivation. As a result, simple power law kinetics is often more useful to predict reactor behavior (Levenspiel, 1999). Moreover, the underlying reaction network is developed on the basis of presumed reaction pathways.

To evaluate a set of chemical reactions which represent only important changes in chemical composition, experiments with different intermediates were carried out. The above is the reason for choosing the first-order kinetic model to evaluate reaction rates of triglycerides deoxygenation and also to use oleyl alcohol, oleic acid, and oleyl oleate as original substrates in HDO experiments. Detailed description of catalytic activity measurements is presented elsewhere (Kubička and Kaluža, 2010).

Catalytic tests were performed in a bench-scale trickle-bed reactor of 17 mm inner diameter. Food grade rapeseed oil with the following concentration of important fatty acid moieties, oleic acid 62%; linoleic acid 20%; linolenic acid 9%; and stearic acid 3%, was used as a liquid feed, and refinery hydrogen gas (>99% purity) was used as a gas feed. The reactor was loaded prior to the experiments with catalyst diluted by an inert (SiC) to ensure sufficient catalyst-bed length and to improve the reaction-heat transfer, liquid distribution, and catalyst wetting. The particle sizes of the catalyst and SiC were 0.25–0.5 mm and <0.1 mm, respectively. In the vertical axis of the reactor, a thermo-well of the outer diameter equal to 5 mm was located to ensure axial temperature profile measurement. The reaction temperature was set during the experiments in the range of 260–280 °C. The contact time of the liquid feed with the catalyst (V/F) was varied in the range of 0.2–5 h by varying the catalyst mass (*W*) and rapeseed oil feed rate (Q_m). The hydrogen-to-rapeseed oil molar ratio and hydrogen pressure were constant at 50 and 3.5 MPa, respectively. The liquid products (collected and separated at 80 °C) were withdrawn after stabilization of reaction conditions (6 h) in 2-h intervals and analyzed by off-line gas chromatography after separation of the water phase. The gas-phase products were collected once at every reaction condition and analyzed off-line using a standard three-column GC setup with flame ionization and thermal conductivity detectors enabling the detection of both permanent gases and hydrocarbons at the same time.

5.1. Mechanism of triglycerides deoxygenation

General lumped reaction pathway (Kubička and Kaluža, 2010) incorporates hydrogenation of olefinic bond to saturated triglycerides and parallel hydrogenolysis and decarboxylation to alkanes and to stearic acid. Stearic acid also reacts to yield C_{17} and C_{18} alkanes and stearyl alcohol and also by consecutive way to stearyl stearate. The following part of chapter on kinetics of rapeseed oil transformations and HDO of intermediates is adapted from the published data (Tukač et al., 2012). In Fig. 3.7, the reaction scheme of triglyceride deoxygenation (both unsaturated and saturated) to C_{18} and C_{17} alkanes, acids, and esters is presented.

Chemical compounds involved in HDO reactions are specified in more detail in Table 3.7, together with their IUPAC name and CAS number.

The activity of the catalysts in deoxygenation can be described by considering at the conversion of triglycerides, that is, evaluating the rate of disappearance of triglycerides. The dependencies have been obtained by using the pseudo-first-order kinetics to fit the relevant experimental data for catalysts at three reaction temperatures. The results are depicted in Fig. 3.8, where it is seen that the experimental data followed the empirical pseudo-first-order-kinetics rather well.

Furthermore, it can be seen that the conversion increases with increasing reaction temperature and that complete conversion of triglycerides can be

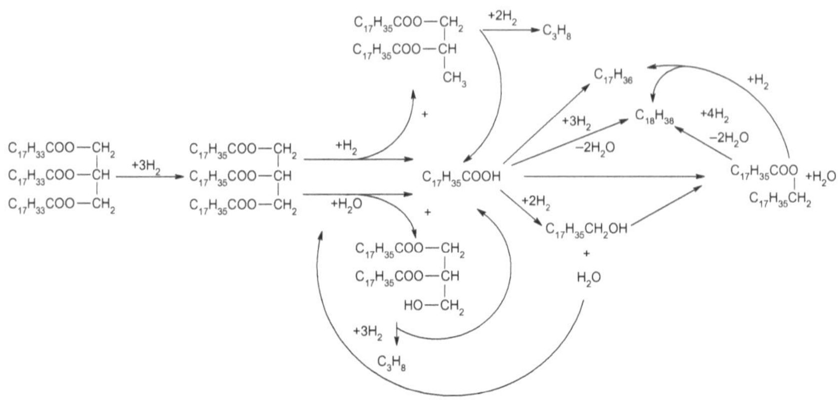

Figure 3.7 The reaction pathways involved in conversion of triglycerides into hydrocarbons.

Table 3.7 Chemical compounds important in vegetable oil deoxygenation

Name	IUPAC name	Formula	CAS No.
Glyceryl trioleate	2,3-Bis[[(*Z*)-octadec-9-enoyl]oxy]propyl (*Z*)-octadec-9-enoate	$(C_{17}H_{33}COO)_3C_3H_5$	122-32-7
Glyceryl tristearate	1,3-Di(octadecanoyloxy) propan-2-yl octadecanoate	$(C_{17}H_{35}COO)_3C_3H_5$	555-43-1
Glyceryl distearate	(2-Hydroxy-3-octadecanoyloxypropyl) octadecanoate	$(C_{17}H_{35}COO)_2C_3H_5OH$	504-40-5
Propanediol distearate	1,3-Propanediol distearate	$(C_{17}H_{35}COO)_2C_3H_6$	17367-44-1
Oleyl oleate	[(*Z*)-Octadec-9-enyl] (*Z*)-octadec-8-enoate	$C_{17}H_{33}COOCH_2C_{17}H_{33}$	3687-45-4
Oleyl stearate	[(*Z*)-Octadec-9-enyl] octadecanoate	$C_{17}H_{35}COOCH_2C_{17}H_{33}$	17673-50-6
Stearyl stearate	Octadecyl octadecanoate	$C_{17}H_{35}COOCH_2C_{17}H_{35}$	2778-96-3
Oleic acid	(*Z*)-Octadec-9-enoic acid	$C_{17}H_{33}COOH$	112-80-1
Stearic acid	Octadecanoic acid	$C_{17}H_{35}COOH$	57-11-4

Table 3.7 Chemical compounds important in vegetable oil deoxygenation—cont'd

Name	IUPAC name	Formula	CAS No.
Oleyl alcohol	(*Z*)-Octadec-9-en-1-ol	$C_{17}H_{33}CH_2OH$	143-28-2
Stearyl alcohol	Octadecan-1-ol	$C_{17}H_{35}CH_2OH$	112-92-5
Glycerol	Propane-1,2,3-triol	$C_3H_5(OH)_3$	56-81-5
Heptadecene	Heptadec-1-ene	$C_{17}H_{34}$	6765-39-5
Octadecene	Octadec-1-ene	$C_{18}H_{36}$	112-88-9
Heptadecane	Heptadecane	$C_{17}H_{36}$	629-78-7
Octadecane	Octadecane	$C_{18}H_{38}$	593-45-3
Propane	Propane	C_3H_8	74-98-6

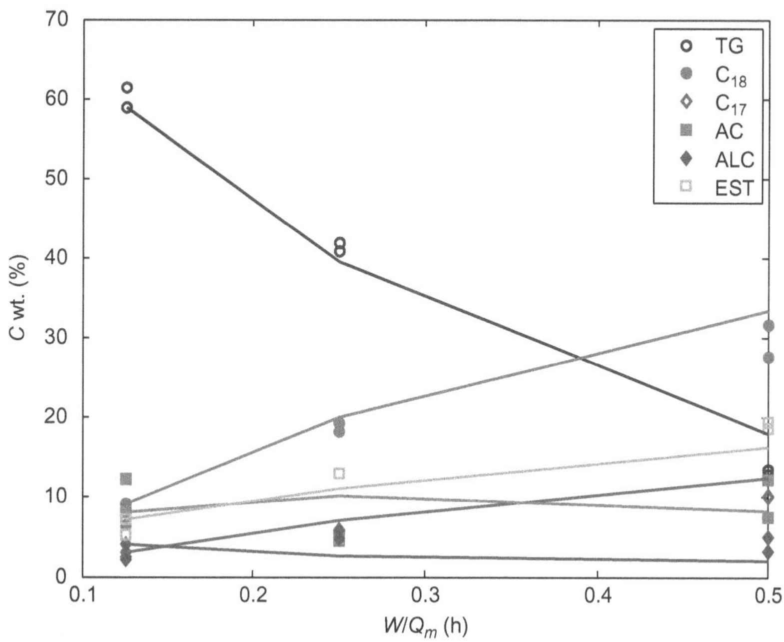

Figure 3.8 Concentrations of representative group of compounds for rapeseed oil deoxygenation. The solid lines represent the solution of the first-order pseudo-homogeneous plug flow isothermal model of the reactor and the points are the experimentally measured data. CoMo catalyst, 300 °C and 3.5 MPa. (TG, triglycerides; C_{17} C_{18}, alkanes; AC, fatty acids; ALC, fatty alcohols; EST, esters). (For color version of this figure, the reader is referred to the online version of this chapter.)

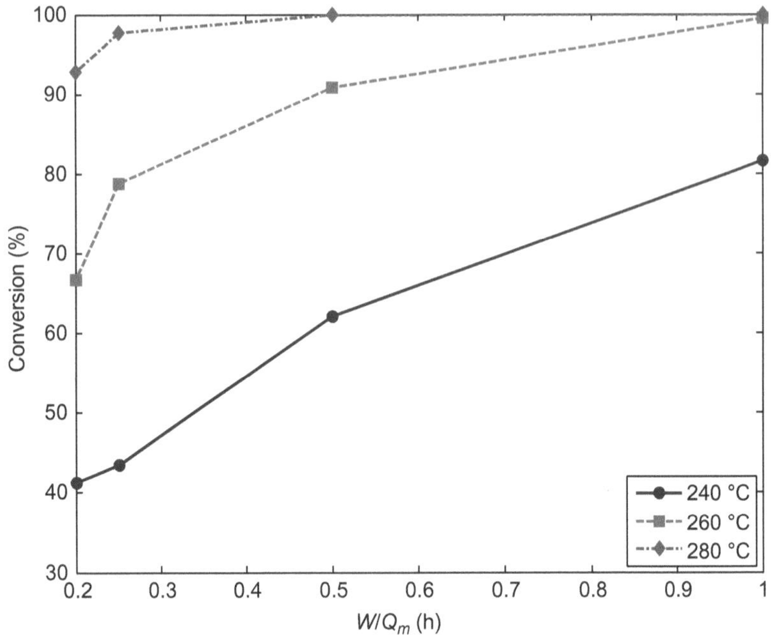

Figure 3.9 Conversion of triglycerides over the CoMo catalyst at different temperature as a function of contact time. (For color version of this figure, the reader is referred to the online version of this chapter.)

obtained already at 280 °C (Fig. 3.9). These results suggest that the deoxygenation reactions of triglycerides are, in fact, irreversible at the selected reaction conditions.

Deoxygenation of triglycerides over the investigated catalyst yielded two main classes of organic liquid reaction products and intermediates—hydrocarbons and oxygenates. The ultimate products of deoxygenation were hydrocarbons and oxygenates were the reaction intermediates, in particular, fatty acids (e.g., stearic acid), fatty alcohols (e.g., *n*-octadecanol), and their fatty esters (e.g., stearyl stearate). The shape of the selectivity curves of the two main reaction product classes confirms that hydrocarbons (Fig. 3.10) are the final reaction products and that their formation is preceded by the formation of oxygenated intermediates (Fig. 3.11). The hydrocarbons consisted mainly of *n*-octadecane, *n*-heptadecane, *n*-hexadecane, and *n*-pentadecane. In addition to the organic liquid products, water and gaseous products, such as propane, carbon oxides, and methane, were formed during deoxygenation of triglycerides. The cracking activity of the CoMo catalyst was very low under the experimental conditions since no other gaseous

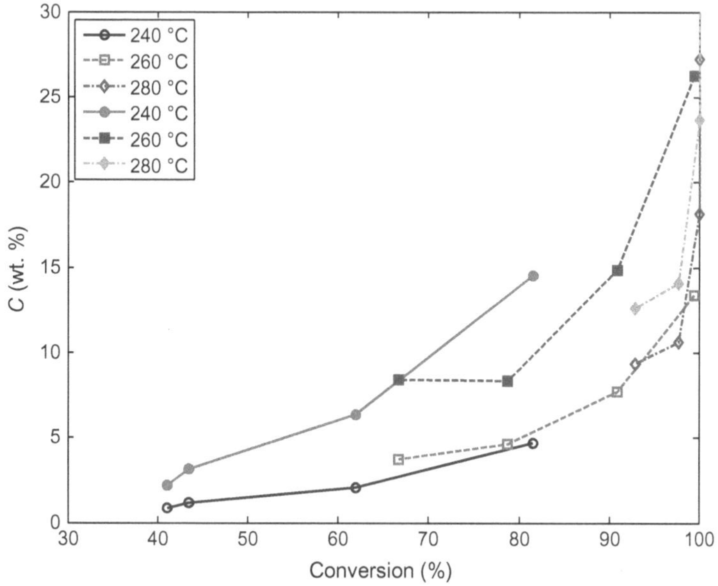

Figure 3.10 Selectivity of alkanes as a function of conversion for the CoMo catalyst at different temperatures. $p=3.5$ MPa, $W/Q_m=0.2$–1 h. (Open points belong to C_{17} and filled to C_{18} hydrocarbons.) (For color version of this figure, the reader is referred to the online version of this chapter.)

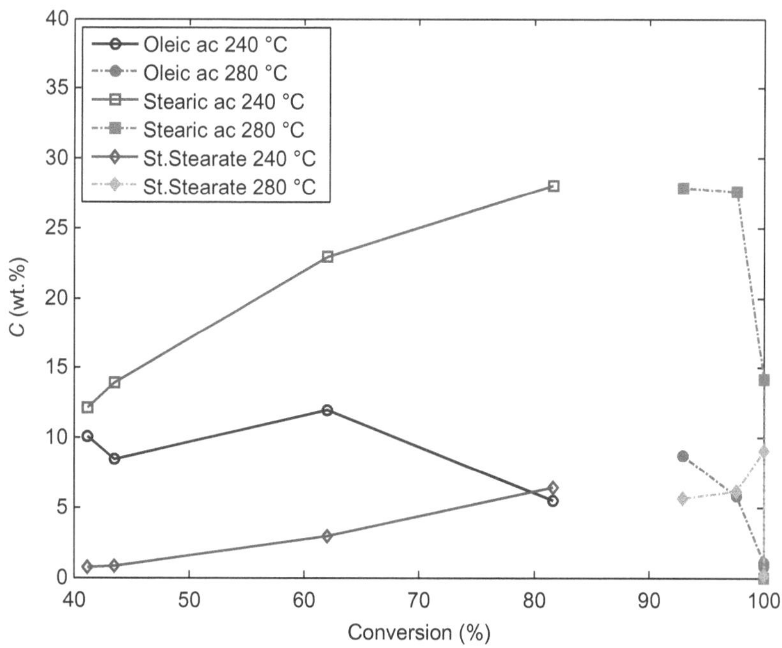

Figure 3.11 Selectivity to oxygenates as a function of conversion for the CoMo catalyst at different temperatures. $p=3.5$ MPa, $W/Q_m=0.2$–1 h. (For color version of this figure, the reader is referred to the online version of this chapter.)

products or short chain liquid hydrocarbons were found in the reaction products. Moreover, the extent of isomerization was also negligible under the investigated conditions since only traces of skeletal isomers were identified in the organic liquid reaction products.

A deeper understanding of the deoxygenation of triglycerides can be obtained from a detailed analysis of both the main product groups, that is, from selectivities to different types of oxygenates and hydrocarbons. The selectivities to *n*-heptadecane and *n*-octadecane provide valuable information about the final reaction step in deoxygenation of triglycerides. Their selectivities are in this case plotted as a function of the total triglyceride conversion (Fig. 3.10). The other hydrocarbons, mainly *n*-pentadecane and *n*-hexadecane, were formed also in the system, but in negligible concentration.

The oxygenated products and intermediates consist of three main types of products—fatty acids (almost exclusively stearic and palmitic acids), fatty alcohols (almost exclusively *n*-octadecanol and *n*-hexadecanol), and their fatty esters, that is, the esters that are formed by the reaction of fatty acids with fatty alcohols that are both reaction intermediates in deoxygenation of triglycerides. The selectivity curves of the three main classes of oxygenated products confirm that fatty acids and fatty alcohols are formed prior to the formation of fatty esters (Fig. 3.11). Fatty alcohols are not depicted here as their measured concentration was very low, presumably due to their quick consumption either by deoxygenation (resulting in formation of hydrocarbons) or by esterification reaction with fatty acids as evidenced by the significant concentration of stearyl stearate (Fig. 3.11).

The overall reaction pathway is schematically depicted in Fig. 3.7. The double bonds in triglycerides, the main components of VOs, are first saturated. Then they are converted into fatty acids that can either (i) undergo further hydrogenation steps to yield fatty alcohols and ultimately saturated *n*-alkanes with an even carbon atoms number (HDO) or (ii) undergo decarboxylation to yield hydrocarbons directly, that is, hydrocarbons with an odd number of carbon atoms (hydrodecarboxylation). HDO leads to elimination of water from triglycerides, fatty acids, fatty alcohols, and fatty esters as a result of reaction with hydrogen. It has been proposed that this reaction occurs via adsorbed enol intermediate (Donnis et al., 2009). HDC consists of decarboxylation (CO_2 elimination) of compounds containing carboxylic functional group and hydrogenation of the unsaturated intermediates of CO_2 elimination. Carbon dioxide is a primary product of this reaction, and it may undergo further reactions, particularly hydrogenation, which is undesired since it causes significant increase in hydrogen consumption. Consequently, there are numerous efforts to achieve selective decarboxylation without CO_2 hydrogenation (Lestari et al., 2008, Simakova et al., 2009, Snåre et al., 2008). The simultaneous presence of both fatty acids and alcohols leads to formation of fatty esters thanks to esterification of fatty acids with fatty alcohols. These esters are, nevertheless, subsequently converted into hydrocarbons. There are two possible initial steps of triglycerides conversion: (i) hydrolysis yielding fatty acids and glycerol and (ii) hydrogenolysis yielding fatty acids and propane. As neither the feed contains water nor the rate of triglycerides conversion increases due to deoxygenation with increasing concentration of water in the product mixture, it is plausible to assume that the reaction is initiated by hydrogenolysis of the C—O bond between the fatty acid chain and the propane (glycerol) backbone of triglycerides.

To evaluate the extent of reactions presented in Fig. 3.7, a set of separate experiments starting with the identified important intermediates was carried out. These experiments included deoxygenation of oleyl alcohol, oleic acid, and oleyl oleate that was synthesized for this purpose.

5.2. Deoxygenation of oleyl alcohol

The simplest reaction pathway of oleyl alcohol hydrogenation implies consecutive hydrogenation to stearyl alcohol followed by its deoxygenation to yield the final product, *n*-octadecane. Alternatively, oleyl alcohol can be deoxygenated first followed by hydrogenation of the unsaturated reaction intermediate to *n*-octadecane. In an attempt to understand the issue of olefin formation, an alternative scheme was tested as well. It involves consecutive hydrogenation of oleyl alcohol to stearyl alcohol, followed by its dehydration to *n*-octadecene and final hydrogenation of *n*-octadecane to *n*-octadecane. As a parallel reaction of oleyl alcohol, direct hydrogenation to *n*-octadecane was used. The second scheme describes the dependence of *n*-octadecane–*n*-octadecene concentration as a function of contact time better (see Fig. 3.12). Arrhenius plot of the corresponding rate constants of the pseudo-first-order kinetics is given in Fig. 3.13. Due to the high reactivity of oleyl alcohol, the experiments were conducted in the temperature range of 220–240 °C in order to distinguish the individual reaction steps involved in transformation of oleyl alcohol into the ultimate product, *n*-octadecane. At the higher temperatures used for rapeseed oil hydrogenation, the scheme presented in Fig. 3.7 is sufficient and, therefore, it was used for the general kinetic model.

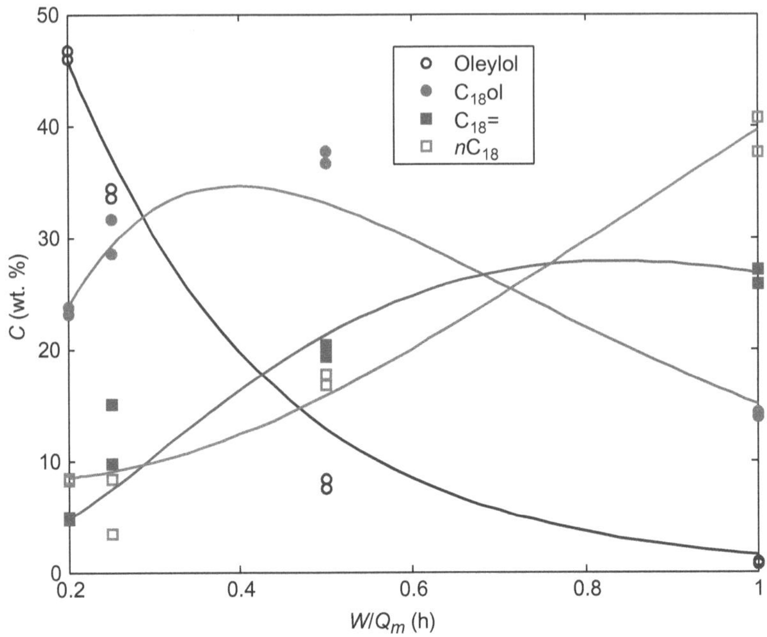

Figure 3.12 Concentration of compounds for oleyl alcohol hydrodeoxygenation at 230 °C and 3.5 MPa (*n*-octadecene formation by dehydration of stearyl alcohol assumed). (For color version of this figure, the reader is referred to the online version of this chapter.)

5.3. Oleic acid deoxygenation

The reaction scheme of oleic acid deoxygenation includes consecutive hydrogenation of oleic acid to stearic acid followed by several parallel reactions: hydrogenation to stearyl alcohol and its subsequent hydrogenation affording *n*-octadecane. Stearic acid produces together with stearyl alcohol an ester (stearyl stearate), which also reacts to hydrocarbons. Moreover, direct decarboxylation of stearic acid is also possible.

From the point of view of selectivity to hydrocarbons and oxygenates like in rapeseed oil deoxygenation, the formation of *n*-octadecane is preferred over HDC to *n*-heptadecane. Temperature enhances very much the formation of the fully hydrogenated product, *n*-octadecane.

5.4. Oleyl oleate HDO

Deoxygenation of oleyl oleate is preceded similar to triglycerides by total saturation of double bonds, that is, by formation of stearyl stearate via either stearyl oleate or oleyl stearate. This fully saturated ester is then split by

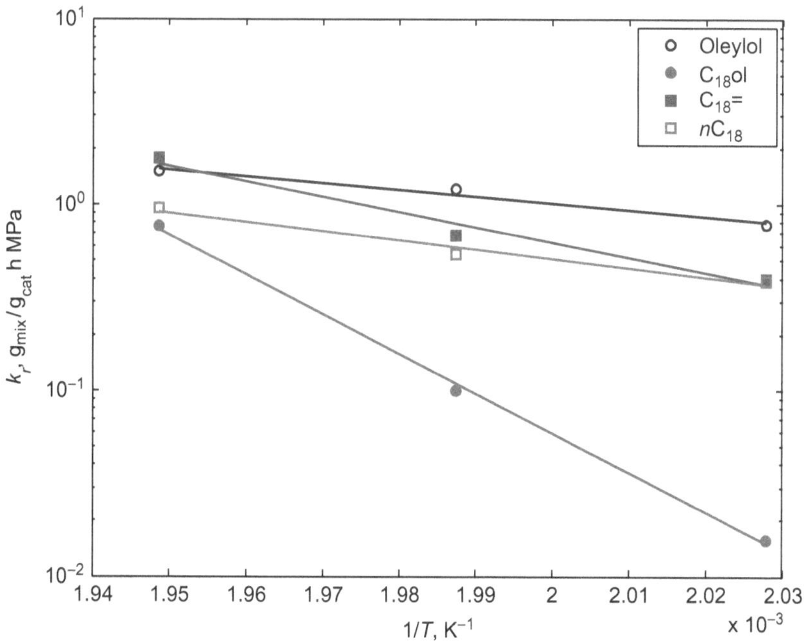

Figure 3.13 Arrhenius plot of oleyl alcohol deoxygenation in the temperature range of 220–240 °C and 3.5 MPa. (For color version of this figure, the reader is referred to the online version of this chapter.)

hydrogenolysis to afford *n*-octadecane and stearic acid that might undergo either further HDO to *n*-octadecane or HDC to *n*-heptadecane. Both results of experiments and kinetic modeling do not support the concept of the possible reversible reaction (hydrolysis) of stearyl stearate to stearic acid and alcohol.

5.5. HDO of rapeseed oil

The complete reaction scheme of deoxygenation of rapeseed oil is presented in Fig. 3.7. As it is not possible to distinguish unsaturated and saturated triglycerides and as it is widely accepted that hydrogenation of olefinic bonds of triglycerides is very fast and thus preceding their deoxygenation, kinetic evaluation starts with the sum of triglycerides. The estimated first-order kinetic parameters are presented in Table 3.8.

5.6. Evaluation of kinetic parameters

The experimental data measured by the method described above in a laboratory trickle-bed reactor using crushed catalyst particles diluted by fine inert SiC particles were processed by a program formulated in Matlab. Simplex

Table 3.8 Pseudo-first-order kinetic parameters for rapeseed oil HDO (Tukač et al., 2012)

Reaction No.	Reaction (initial compound → product)	*k* (280 °C) (g_L/g_{cat} h MPa)	*E* (kJ/mol)	*A* (g_L/g_{cat} h MPa)
1	$TGC_{18} \rightarrow (C_{17}H_{35}COO)_2C_3H_6$	0.987	89.4	2.55×10^8
2	$C_{17}H_{35}COOH \rightarrow C_{17}H_{35}CH_2OH$	1.52	95.7	2.04×10^9
3	$C_{17}H_{35}COOH \rightarrow C_{17}H_{36}$	0.433	47.4	13,000
4	$C_{17}H_{35}COOH \rightarrow C_{18}H_{38}$	2.70	219.2	1.28×10^{21}
5	$C_{17}H_{35}COOH + C_{17}H_{35}CH_2OH \rightarrow C_{17}H_{35}COOC_{18}H_{37}$	3.33	323.1	2.00×10^{31}
8	$C_{17}H_{35}COOC_{18}H_{37} \rightarrow C_{18}H_{38}$	1.36	72.7	1.06×10^7
9	$C_{17}H_{35}COOC_{18}H_{37} \rightarrow C_{17}H_{36}$	0.0574	268.2	5.12×10^{24}
10	$TGC_{18} \rightarrow (C_{17}H_{35}COO)_2C_3H_5OH$	1.43	154.3	7.81×10^{14}
11	$(C_{17}H_{35}COO)_2C_3H_6 \rightarrow C_{17}H_{35}COOH$	2.23	36.8	6695
12	$(C_{17}H_{35}COO)_2C_3H_5OH \rightarrow C_{17}H_{35}COOH$	3.32	41.9	29,100

Nelder–Mead optimization based on minimization of sum of squared deviation of experimental and computed concentration was applied. The kinetic model of reactions supposed the first-order power law rate equations and the reactor differential balance presumed plug flow and pseudo-homogeneous mass and heat transfer conditions without external and internal mass transfer resistance. The differential balance for each compound was solved by the Runge–Kutta method. The parity plot demonstrating the rather good agreement between the experimental and calculated concentrations (Tukač et al., 2012) is shown in Fig. 3.14.

The differential balance of each substrate is introduced in Eq. (3.8), and rate equation depending on concentration and pressure is presented in Eq. (3.9).

$$\frac{dC_i}{d(W/Q_m)} = -r_i \qquad [3.8]$$

$$r_i = k_i C_i P \qquad [3.9]$$

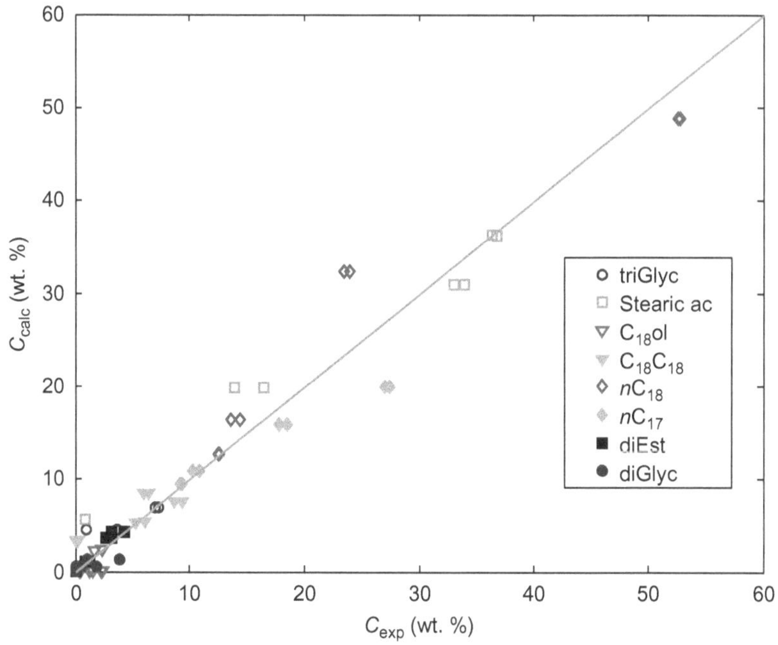

Figure 3.14 Parity plot of experimental and computed concentration of rapeseed oil hydrodeoxygenation at 280 °C, 3.5 MPa and contact time 0.2–1 h (Tukač et al., 2012). (For color version of this figure, the reader is referred to the online version of this chapter.)

The results of kinetic data evaluation are presented in Table 3.8. Chemical reactions are briefly identified by the principal reactant and product. The other columns provide rate constant at 280 °C, apparent energy of activation, and frequency factor of Arrhenius equation for each of the reactions in the model.

Reactions with apparent activation energy lower than 30 kJ/mol are probably influenced by external mass transfer resistance. Also internal mass transfer limitation is possibly of importance if the value of apparent activation energy is below 70 kJ/mol.

The chemical reactions involved in triglycerides deoxygenation can be classified into several groups according to their reaction rate constants. The fastest reactions include the formation of C_{18} alkanes from saturated and unsaturated alcohols, followed by saturation (hydrogenation) of double bonds in unsaturated acids, alcohols, and esters. These are followed by reactions involved esterification and in di- and triglyceride hydrogenolysis. Decarboxylation of acids seems to be the slowest reaction in the system.

6. DEACTIVATION ASPECTS

Long-term stability of catalytic activity is essential for any commercial catalytic process that does not involve continuous catalyst regeneration. From a practical point of view, the cycle length of HDO catalyst should be similar to that of conventional hydrotreating, that is, in the case of stand-alone technology, it should be in the range of 12–24 months that is the typical cycle length of middle distillate hydrotreating units. More importantly, in the case of VO (or generally triglycerides) coprocessing with petroleum-derived feedstocks, typically AGO, the impact on AGO hydrotreating cycle length should be minimized as to not affect the schedule of a given refinery shut-down/maintenance plan. The lifecycle length of a catalyst depends primarily on three factors, namely, feedstock properties, operating conditions, and catalyst properties including its composition. As the focus here is on refinery processing of triglycerides, either stand-alone or coprocessing the obvious choice for coprocessing units is conventional hydrotreating catalyst, while stand-alone units could utilize either conventional hydrotreating catalysts or other types of catalysts, for example, supported noble metals such as Pd on carbon that has been demonstrated to be an efficient deoxygenation catalyst (Kubičková et al., 2005; Murzin et al., 2006; Snåre et al., 2006).

As the triglyceride feedstock is concerned, the catalyst activity regardless of the catalyst type discussed above is affected primarily by the feedstock composition. The main issue to be addressed is the concentration of contaminants originating from VO production. The naturally occurring components having potentially a negative effect on catalyst activity include generally phosphorus, sodium, potassium, calcium, and magnesium. Some other impurities can also be found depending on the oil origin; for example, algae oils often viewed as the bio-feedstock of future are rich in nitrogen compounds. The concentration of the impurities depends not only on the VO origin but also on its processing prior to its hydrotreating. Typically, the highest concentration of impurities is in raw oils, that is, after their pressing and/or extraction from oil seeds. The presence of these impurities is connected to phospholipids, that is, triglycerides in which one of the fatty acid moieties is replaced by phosphoric acid. The additional charge of phosphoric acid moiety in phospholipids is compensated by the alkali metal cations and/or cations of the alkaline earths metals. For example, in neat rapeseed oil, the concentrations were in the range from tens to hundreds of ppm, which are significant amounts for any catalytic process.

The effect of these impurities on catalytic activity of the CoMo catalyst used in hydrotreating of rapeseed oil (310 °C, 3.5 MPa, 2 h^{-1}) has been investigated recently (Kubička and Horáček, 2011). Different rapeseed oil-derived feedstocks available from a rapeseed oil producer were processed. The concentration of impurities ranged from unit ppm or sub-ppm levels found in refined food grade rapeseed oil up to tens to hundreds ppm in neat rapeseed oil or the so-called trap grease. Thanks to using triglyceride streams in different stages of processing, the proportions of phosphorus, on the one hand, and alkali and/or alkaline earths metals, on the other hand, varied making it possible to elucidate the effect of phosphorus and alkali and/or alkaline earths metals, respectively.

The tested feedstocks can be classified into three groups: (i) feedstock with cations–anions charge balance, (ii) feedstocks with excess of phosphoric acid moiety in respect to alkali and/or alkaline earths metals cations, and (iii) feedstock with excess of alkali and/or alkaline earths metals cations in respect to phosphoric acid moiety (Kubička and Horáček, 2011). The slowest deactivation was observed for the first type of feedstock and feedstocks with very low concentration of impurities, that is, units of ppm or even sub-ppm level, despite being of type (iii). On the other hand, the fastest and most severe deactivation was observed when the feedstock with large excess of phosphorus with respect to the available cations was used (type (ii)). In this case, the catalyst was completely deactivated within the first 48 h on stream. The origin of this behavior lied in the large concentration of phosphoric acid moiety which was not compensated by alkali and/or alkaline earths metals cations and was hence acidic. The present phospholipids decomposed under the reaction conditions and phosphoric acid were created and acted as an oligomerization/polymerization catalyst of the unsaturated fatty acid moieties in triglycerides (Kubička and Horáček, 2011). Consequently, very fast coking of the catalyst occurred leading to virtually complete loss of catalyst activity.

On the other hand, the large concentration of phosphoric acid moiety, the charge of which was compensated by alkali and/or alkaline earths metals cations, did not lead to fast deactivation. In fact, the catalyst performance with this feedstock was similar to the performance of the same catalyst using food grade rapeseed oil over the test period that lasted nearly 150 h on stream. Nonetheless, the analyses of the catalyst bed revealed that phospholipids decomposed as well and phosphates of the corresponding alkali and/or alkaline earths metals were deposited just upstream the catalytic bed. As these were not acidic, they could not catalyze the oligomerization reaction.

Nevertheless, it could be anticipated that the reactor would get plugged due to a gradual buildup of phosphates originating from alkali and/or alkaline earths metals containing phospholipids (Kubička and Horáček, 2011). It can be, hence, concluded that the presence of phosphorus is detrimental for the deoxygenation technology regardless of the type of phospholipids and phosphorus has to be eliminated from the feedstock. This process is indeed the first step in upgrading of VOs obtained by pressing and/or extraction to produce food grade product and is called degumming.

Alkali and/or alkaline earths metals present in the feedstock in excess were also shown to deteriorate the catalyst performance in comparison with high-quality food grade VO. A plausible explanation provided suggested that deposition of these metals on catalyst surface induced electronic effects limiting hydrogenation (Kubička and Horáček, 2011). Moreover, deactivation of hydrotreating and hydrocracking catalysts has also been reported previously (Cable et al., 1985). In addition, alkali and/or alkaline earths metals molybdates, tungstanates, and aluminates were indicated by thermodynamic calculations as compounds that could be formed under hydrotreating conditions. Unfortunately, neither their stabilities nor their activities under hydrotreating conditions are known (Furimsky and Massoth, 1999). It can be speculated that if these compounds are indeed formed, their activity will be lower than that of the active components of hydrotreating catalysts, that is, metal sulfides.

The importance of the impurities can be further inferred from the description of the NexBTL technology developed and commercialized by the Finish Company Neste Oil. Any feedstock that is processed in a NeBTL unit undergoes pretreatment in order to ensure suitable quality feedstock. The pretreatment consists of degumming using phosphoric acid to eliminate phospholipids and alkali metals and/or alkaline earths metals cations. Furthermore, sodium hydroxide is used in the pretreatment, plausibly to neutralize free acids that could cause corrosion problems upstream in the technology. The soaps formed as a consequence of neutralization are removed by washing using demineralized water (Hodge, 2006).

In addition, the degree of unsaturation of the triglycerides is of importance not only because it is directly connected to the hydrogen consumption but also because the high degree of unsaturation, that is, two or more double bonds per fatty acid chain, makes the fatty acid chains prone to cyclization reactions. This has been demonstrated by da Rocha Filho et al. (1993), who have used a series of variously unsaturated VOs in hydrocracking experiments (NiMo/Al_2O_3 catalyst, 360 °C, 14 MPa, hydrogen, 2 h). The oils containing linoleic and linolenic fatty acid moieties in large concentration (it totals above 60 wt.%) yielded significant amounts of cycloalkanes (~15 wt.%) and also some alkylaromatics (about 4 wt.%) while oils consisting of saturated fatty acid moieties afforded almost exclusively alkanes. The concentration of cyclic compounds (both cycloalkanes and aromatics) was less than 3 wt.%. When monounsaturated fatty acid moieties (typically oleic acid) prevailed in the feedstock, a moderate yield of cycloalkanes was observed (ca. 8 wt.%; da Rocha Filho et al., 1993).

These conditions might seem too extreme and hence irrelevant for hydrotreating applications. However, the authors (da Rocha Filho et al., 1993) have investigated also the effects of reaction temperature and pressure on formation of cyclic products in hydrocracking of buriti oil (consisting of 80 wt.% oleic acid and 17 wt.% palmitic acid moieties). When pressure was decreased from 14 to 7 MPa at a reaction temperature of 360 °C, which is a common hydrotreating temperature in hydrotreating of middle distillates, the formation of cycloalkanes increased by nearly 40%. The formation of alkylaromatics remained low (it increased from 0.2 to 0.3 wt.%). The increase in reaction temperature had a more profound effect. At 14 MPa and 1-h reaction time, the concentration of cycloalkanes increased from 5 wt.% to more than 20 wt.% upon increase in temperature from 360 to 395 °C, which can be in some hydrotreating units the maximum operating temperature. At the same time, the concentration of aromatics increased from 0.1 to 0.8 wt.% (da Rocha Filho et al., 1993).

To provide more fundamental information reflecting the formation of cycloalkanes and aromatics as a function of the degree of unsaturation, a series of C_{18} fatty acids was also used in hydrocracking ($NiMo/Al_2O_3$ catalyst, 360 °C, 14 MPa, H_2, 2 h; da Rocha Filho et al., 1993). The concentration of alkylcycloalkanes afforded by hydrocracking of stearic, oleic, linoleic, and linolenic acid, that is, of fatty acid with 0, 1, 2, and 3 double bonds in their molecule, increased from <1 wt.% found for stearic acid to ca. 12, 19, and 30 wt.% for oleic, linoleic, and linolenic acid, respectively. For the same sequence of fatty acids, the concentration of alkyl aromatics was increased in the following order: zero, traces, 0.8 and 6.8 wt.%.

Thus, it can be inferred from these observations that with the increasing degree of unsaturation of the triglycerides in an HDO feedstock, the propensity of the feedstock to formation of carbonaceous deposits will increase depending on the operating conditions. The significant formation of cycloalkanes and alkylaromatics from linolenic acid (ca. 30 and 6.8 wt.%, respectively) indicates that the concentration of fatty acid moieties with more than two double bonds needs to be monitored closely in any incoming feedstock as a measure to avoid increased carbonaceous deposits formation. Moreover, the high degree of unsaturation of triglycerides might plausibly cause problems in the feedstock preheating section due to their lower thermal stability as compared with saturated triglycerides.

Apart from the possible adverse effect of significant degree of unsaturation, catalyst deactivation may originate also from deoxygenation products, particularly water and carbon oxides. In this context, a specific problem relates to the use of sulfided catalysts. It is well known that these catalysts are used under reductive conditions. Consequently, the catalysts are susceptible to reduction and loss of their catalytic activity. When treating sulfur-containing feedstocks, this is generally not an issue as the concentration of hydrogen sulfide, the product of hydrodesulfurization, in gas phase is sufficient to keep the catalyst in sulfided form, that is, prevent its desulfidation/reduction. However, the sulfur content of majority of triglyceride feedstocks is very low. For example, rapeseed oil contains approximately 2–3 ppm of sulfur, which is not sufficient to keep a hydrotreating catalyst in its active form. As a result, gradual catalyst reduction and loss of hydrogenation activity occur. This has been demonstrated in long-term experiments focused on deoxygenation of rapeseed oil (Kubička and Horáček, 2011).

The deoxygenation activity of the CoMo catalyst, expressed as the extent of oxygen atom elimination from liquid phase products, dropped from 100% (complete deoxygenation) to ca. 75–80% after the first 144 h on stream when refined sulfur-free rapeseed oil was processed. However, when sulfur was introduced in the feedstock (in the form of dimethyl disulfide) to ensure suitable concentration of hydrogen sulfide in the gas phase, the deoxygenation activity of the CoMo catalyst was stable over the whole test length (ca. 260 h). At the end of the test, the extent of oxygenation was still above 96% (Kubička and Horáček, 2011). Moreover, there were periodic changes in the delivery of hydrogen sulfide source, that is, between short periods (2 h) when rapeseed oil containing dimethyl disulfide was fed in the reactor, and there were longer periods (ca. 50–100 h) when sulfur-free rapeseed oil feedstock was used. These experiments indicated that the loss of activity when sulfur-free feedstock was used was partially reversible, that is, that the deoxygenation activity decrease due to processing sulfur-free rapeseed oil can be stopped by adding dimethyl disulfide in the feedstock and the activity can be even temporarily restored (Kubička and Horáček, 2011).

Obviously, this is an issue only in stand-alone HDO units as during coprocessing of triglycerides with petroleum feedstocks, such as AGO, there is a sufficient amount of sulfur presents to ensure stable catalyst performance. The technology scheme of the NexBTL process (Hodge, 2006) indicates that hydrogen sulfide exits the hydrotreater where only pretreated feedstock consisting of VOs is fed. The origin of hydrogen sulfide is not given. Owing to the readily available hydrogen streams containing hydrogen sulfide (any gas outlet of a conventional hydrotreater), it can be speculated that for hydrotreating/deoxygenation of triglycerides, such a stream is used before it is scrubbed to remove hydrogen sulfide. It is a much simpler alternative than addition of dimethyl disulfide. Alternatively, one could imagine that even in a stand-alone unit, VO is hydrotreated together with a petroleum fraction containing sulfur that is easily hydrotreated. As the NexBTL scheme suggests that the liquid product is directly fed into an isomerization unit, it seems more plausible that the option with hydrogen sulfide containing hydrogen is used. It is also clear that if the stand-alone unit was to use a noble metal type of catalyst, the question of sulfur addition would be redundant.

On the other hand, water and carbon oxides as the ultimate deoxygenation products apart from the desired hydrocarbons can affect negatively the catalytic activity of both noble metal and metal sulfide catalysts. For example, deactivation of hydrogenation catalysts by carbon monoxide is well known. However, its effect is greatly dependent on the type of catalyst and in particular on reaction conditions. From the mechanistic point of view, carbon monoxide is a reaction intermediate that is either formed by hydrogenation of carbon dioxide, that is, the primary product of the decarboxylation reaction pathway, or as a product of decarbonylation that has also been proposed as a viable reaction route in deoxygenation of triglycerides and fatty acids (Donnis et al., 2009; Kubičková and Kubička, 2010). The options that are applied in the commercialized technologies to deal with the inhibition effect of carbon monoxide are discussed in detail in Section 7 that covers the commercial technologies. Apart from purging, pressure swing adsorption (PSA) or methanation can be applied to eliminate carbon monoxide from the recycle hydrogen stream.

Carbon dioxide does not constitute a major problem as it is scrubbed together with hydrogen sulfide from the hydrogen stream. It is also hydrogenated in the deoxygenation reactor which leads to increased hydrogen consumption and formation of undesired carbon monoxide. Several studies have focused on the effect of water formed in course of the deoxygenation reaction on the deoxygenation activity of sulfided catalysts (Laurent and Delmon, 1994b,c). Laurent and Delmon (1994c) reported that water had only a very weak inhibiting effect on majority of various reactions taking place during deoxygenation of phenols and ketones. An exception was hydrolysis of carboxylic esters that was favored by water addition (Laurent and Delmon, 1994c). In another study, the same authors have found that water caused a two-thirds of decrease (in less than 60 h on stream) of the catalytic activity during deoxygenation, but the hydrogenation–hydrogenolysis selectivity remained unaltered (Laurent and Delmon, 1994b). Water caused a small loss of catalyst ($NiMo/\gamma Al_2O_3$)-specific surface area which was connected with a recrystallization of γ-alumina supported into hydrated boehmite phase. More importantly, water caused partial oxidation of nickel sulfide species which was suggested to be the reason behind the observed loss of catalytic activity (Laurent and Delmon, 1994b). Two plausible interpretations were put forward by the authors: (i) nickel sulfide is oxidized to nickel sulfate species that form an inactive layer on the nickel sulfide particles and (ii) water favors migration of nickel in the alumina lattice where the inactive nickel aluminate species are formed (Laurent and Delmon, 1994b). Nonetheless, the effect was not observed when water in the feed was replaced by oxygenated organic compounds. It should be noted here that the treatment was rather severe (360 °C, 7 MPa, dodecane/water = 170/10 ml/ml, 1–5 days) which might be the reason why there was no effect of organic oxygenates that were hydrotreated only for 150 min (13.6 wt.% of dibenzofuran or 4-methylphenol in dodecane). In contrast, the molybdenum sulfide phase was virtually unaffected (not oxidized) by the water or oxygenates treatment (Laurent and Delmon, 1994b).

7. COMMERCIAL STATUS OF HDO

The intensive interest in renewable fuels that are fully compatible with the current automotive fuels as well as with the modern combustion engine technology resulted in fast development and commercialization of several deoxygenation technologies by several companies. In principle, they can be divided into two major categories, namely, stand-alone units and coprocessing units. Stand-alone units are such technologies that process neat VOs (or, in general, triglyceride feedstocks), and their product is renewable (or green) diesel fuel, which can then be used for blending with diesel fuel fulfilling the legislative requirements. As it will be discussed later on, due to isomerization, the renewable diesel outperforms conventional diesel as it is sulfur- and aromatics-free. On the other hand, coprocessing technology refers to hydrotreating of mixtures containing AGO and VOs. The product is then a mixture of conventional diesel and renewable diesel fuel. From the economic point of view, the stand-alone units are definitely more investment intensive, while the coprocessing units require significantly less investments to be made to adapt the existing units to operate efficiently with the new feedstock. The trade-off is connected with the process flexibility. While the stand-alone units are very flexible in terms of feedstock type and quality, and the product properties can be rather easily adjusted, the coprocessing units lack this flexibility as they need to take into account also the HDS process (with the aim to achieve diesel fuel with less than 10 ppm S) and it might be difficult to obtain suitable low-temperature properties of the final product as will be discussed also in this section.

7.1. Stand-alone deoxygenation technologies

The most advanced stand-alone deoxygenation technologies include processes NexBTL be Neste Oil and Ecofining by UOP and ENI. The main features of these will be discussed in the following paragraphs followed by the assessment of the main challenges in coprocessing.

The NexBTL process consists of three essential stages: feedstock pretreatment, hydrotreating, and isomerization (Fig. 3.15). The pretreatment stage is an indispensable part of the process as it ensures high quality of feedstock that can be fed into the catalytic units, particularly the hydrotreating unit which precedes isomerization. As a result, any contaminants of renewable triglyceride-based or -related feedstocks that could affect catalyst performance are removed. Typical contaminants that are being removed in the NexBTL process include phosphorus and metals (Ca, Mg). The processes involved in feedstock pretreatment consist of degumming by H_3PO_4, neutralization by NaOH, and washing by demineralized water. The pretreated (refined) oil is fed in the hydrotreating unit where the

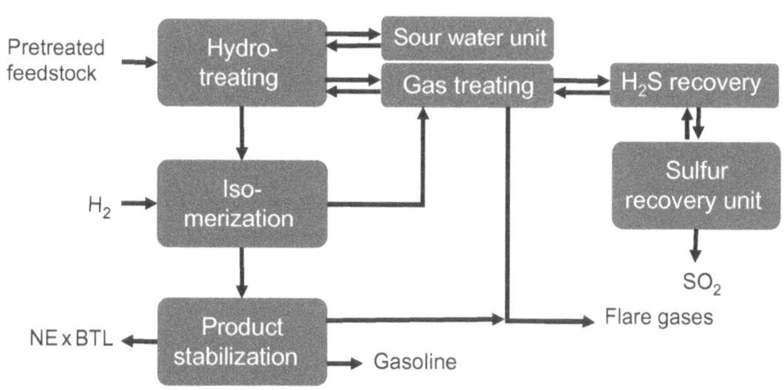

Figure 3.15 Schematic flow scheme of the NexBTL process (Hodge, 2006).

deoxygenation takes place. The catalyst is supplied by Albemarle (Green Car Congress, 2012a), and it is the most probably a sulfided NiMo or CoMo catalyst.

The deoxygenation step of the NexBTL process is followed by isomerization, which is an essential part of the technology as it modifies the chemical structure of the deoxygenation product. The ultimate deoxygenation products, *n*-alkanes, are mildly isomerized to afford a mixture of iso-alkanes that have far better cold flow properties than the original *n*-alkanes. Without this modification, the renewable diesel fuel would not be suitable for use in diesel engines in mild or cold climates as it will be demonstrated later in this chapter when the fuel properties of renewable diesel will be discussed.

Neste Oil has actually pioneered the renewable diesel technology, and currently, four units with the total combined nominal capacity of nearly 2 million tons per year are operated worldwide.

The feedstock portfolio of NexBTL process consists of crude palm oil (<50%); waste and side streams, such as waste animal fat, palm oil fatty acid distillate, and stearin (ca. 40%); and other VOs, such as rapeseed, soybean, and camelina oils (ca. 10%). The given shares of the individual feedstocks were the targets for Neste Oil's procurement of renewable raw materials in 2011. Owing to the sustainability issues that have been raised in connection with production of fuels from renewable raw materials, Neste Oil prioritizes suppliers that follow sustainable cultivation and production practices and have a good greenhouse gas (GHG) balance (Neste Oil, 2012b). Apart from the currently used renewable raw materials, Neste Oil also pays attention to alternative raw materials for production of renewable diesel fuel. These include forest industry by-products, waste fat from fish-processing industry, microalgae oil, and microbial oil (Neste Oil, 2012c).

UOP has developed together with Eni a process called Ecofining that they have been successfully demonstrating since 2008 in 12-kt/yr-demonstration unit in the USA. Commercial units are either completed and awaiting for approval (640 kt/yr in total) or under construction (810 kt/yr), see Table 3.9 for details. The feedstock requirements have evolved similar to Neste Oil—prior to the sustainability question being raised, palm, soybean, and canola oils were considered. Since 2007, however, the focus was shifted toward nonfood and more sustainable sources, such as camelina, jatropha, and algal oils, and cheaper oils and fats, including tallow, yellow grease, and tall oil. The cheaper triglyceride alternatives have the advantage of the lower feedstock price and sustainability since they are typically derived from waste streams in food and wood and forest industries.

Table 3.9 Commercial deoxygenation units and their current status (Frey, 2011; Neste Oil, 2012a)
Technology

Name	Technology developer	Location	Company	Capacity (kt/a)	Status
NexBTL	NesteOil	Finland	NesteOil	190,000	2007
		Finland	NesteOil	190,000	2009
		Singapore		800,000	2010
		Netherlands		800,000	2011
Ecofining	UOP/Eni	UOP	USA	12,000	2008
		Italy	ENI	320,000	Awaiting approval
		Portugal	GALP	320,000	Awaiting approval
		North America	Diamond Green Diesel	490,000	2012—expected
		North America	Undisclosed	320,000	2013—expected
		North America	AltAir Fuels	320,000	Pending

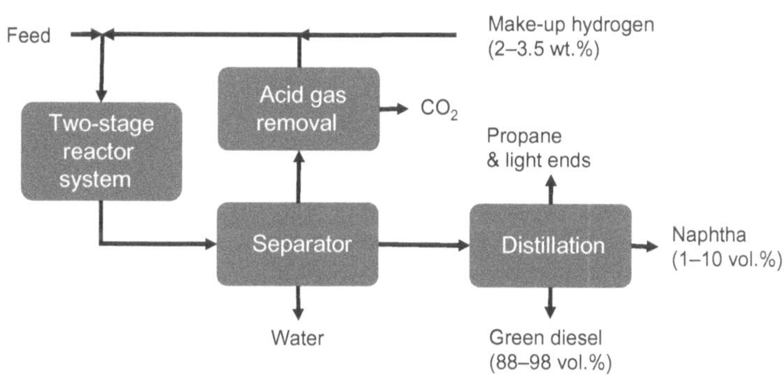

Figure 3.16 Schematic depiction of the Ecofining process (Frey, 2011; UOP, 2012). (For color version of this figure, the reader is referred to the online version of this chapter.)

Their inherent disadvantage is the higher concentration of undesired contaminants and with it connected more demanding preprocessing.

The Ecofining process relies on a two-stage fixed-bed reactor system (Fig. 3.16). Due to the schematic depiction in the available materials, it is not clear whether the reactor system consists of two reactors (a HDO reactor and a hydroisomerization reactor) or a single reactor with a HDO and hydroisomerization stage. The placement of a separator after the reactor system to remove water, an inevitable reaction by-product, indicates that the latter option is most probably used. On the other hand, information provided elsewhere (Kalnes et al., 2008) suggests that the system includes in fact two reactors. In the first one, the feedstock is completely deoxygenated and the light products (water, light hydrocarbons, and carbon dioxide) are immediately separated at reactor pressure. The main product, mixture of *n*-alkanes, is then mixed with additional hydrogen and fed into an integrated catalytic hydroisomerization reactor to afford branched-paraffin-rich diesel (Kalnes et al., 2008). In this way, diesel having suitable cold flow properties is obtained and both reaction steps (deoxygenation and isomerization) are kept at their maximum selectivity. The hydrogen needed can be obtained, for example, by steam reforming of a portion of the light hydrocarbons (Kalnes et al., 2008). Deoxygenation and isomerization catalysts supplied by UOP are employed in the Ecofining process. According to the technical information leaflet, the typical inside battery limits erected costs are between 40 and 60 million USD per a unit with production capacity of ca. 250 kt/yr (UOP, 2012). In order to reduce significantly the capital investment, an existing distillate hydrotreater could be revamped into the Ecofining process.

Depending on the feedstock and the calculation method used, the green diesel offers GHG saving in an excess of 50% relative to petroleum-derived diesel fuel. In addition to GHG savings, the renewable diesel offers further advantages over conventional diesel owing to its superior properties (low aromaticity, low sulfur). As an example, the green diesel from the NexBTL process (100% NexBTL diesel) has been tested extensively. These tests included a 3-year test in buses in the city of Helsinki, Finland; Mercedes Benz trucks tests in Germany; and an arctic trial in Canada. Data on emissions collected in Helsinki showed reduction of particulate matter and nitrogen oxides (NO_x) emissions by 30% and 10%, respectively. The test in Germany confirmed that emissions were reduced significantly in comparison with standard-diesel-driven trucks (Neste Oil, 2012d).

In addition to renewable diesel, renewable jet production is targeted as well. For example, UOP and Eni have announced that they will construct a unit for renewable jet production for AltAir fuels with a capacity of 320,000 t/a (Frey, 2011), and UOP has already performed several successful tests of renewable jet fuel with the major aircraft producers (Boeing, Airbus) and US military demonstrating that replacement of standard jet fuel by renewable jet fuel is feasible. Neste Oil claims that it produces renewable aviation fuel using its NexBTL technology (Neste Oil, 2012e) as a commitment to the European Aviation Biofuels Flightpath that is aimed at increasing the use of aviation biofuels to 2 million tons annually by 2020.

Besides NexBTL and Ecofining, another stand-alone deoxygenation technology, Vegan, was introduced by Axens. The principle of the technology seems to be similar to the discussed technologies. After the hydrotreating step in which VOs are completely deoxygenated follows the second step, which is in this case not called isomerization, but rather mild hydrocracking/dewaxing. Nonetheless, the purpose is the same, that is, to convert *n*-alkanes into iso-alkanes in order to improve the cold flow properties of the final renewable diesel product. However, no information was found whether and where the process is commercially used.

The advantages of renewable (green) diesel in comparison with the traditional biodiesel (FAME), from the application point of view, are rather significant. Apart from the full compatibility with conventional diesel and compliance with the strictest fuel, improved emission characteristics and fossil CO_2 savings have been reported. An overview is given in Table 3.10 (Neste Oil, 2012f).

7.2. Coprocessing deoxygenation technologies

Owing to the apparent simplicity and low investments needed, the coprocessing deoxygenation technology seems to be a popular one. It has been addressed not only by refining companies, such as Petrobras and ConocoPhillips, but also by catalyst manufactures, mainly Haldor Topsoe and Albemarle. As a revamp of existing hydrotreater is sufficient, it is not easy

Table 3.10 Comparison of renewable diesel and biodiesel

NExBTL renewable diesel (HVO)	Traditional biodiesel (FAME)
Can be used in blends in any concentration, that is, 0–100% of the content	Can only be used in blends upto 5–7% of the content[a]
Complies with the strictest quality standards	Biofuel usage requirements cannot be met without compromising fuel quality specifications
Reduces greenhouse gas and tailpipe emissions (PM, NO_x, CO, HC)	Increases NO_x emissions
Offers excellent storability	Must be used before a specific "best before" date
Does not require engine modifications or changes in logistical system	Can cause engine problems

[a]Maximum allowed under the European diesel standard.

to quantify the real extent to which renewable diesel is produced by coprocessing. For example, ConocoPhillips has been produced renewable diesel fuel at its refinery in Whitegate, Cork, Ireland, since 2006, but detailed information about real capacity, content of VO in the hydrotreated feedstock, is not available.

Another example of a coprocessing technology is process H-BIO. It has been developed by the Petrobras research and development center (CENPES). The technology is schematically depicted in Fig. 3.17. The core of the process is catalytic hydrotreatment of diesel fuel blending components, such as straight run diesel, light cycle oil from fluid catalytic cracking (FCC), coker gas oil, and VOs in an existing hydrotreating reactor under controlled reaction temperature and hydrogen pressure (Costa, 2007). Triglycerides in VOs are converted into linear hydrocarbons at very high yields—at least 95% (v/v) to diesel. Owing to the product originating from triglycerides, the quality of refinery diesel gets improved, mainly by increasing the cetane number and decreasing the density and sulfur content (Costa, 2007). The plan of Petrobras was to have the process implemented in at least three of its refineries by the end of 2007. The total VO consumption was about 256,000 m^3/yr, that is, ca. 230,000 t/yr. After implementation of two other refineries, the VO consumption should rise to 425,000 m^3/yr, that is, ca. 390,000 t/yr. The required investment to adjust the production and logistics was estimated to be in the range of USD 23–38 million (Costa, 2007).

However, the implementation was put on hold in 2008 due to high soybean oil prices that should have been the main feedstock. At that time, the refined soy oil price of $180 per barrel by far exceeded a regular diesel price

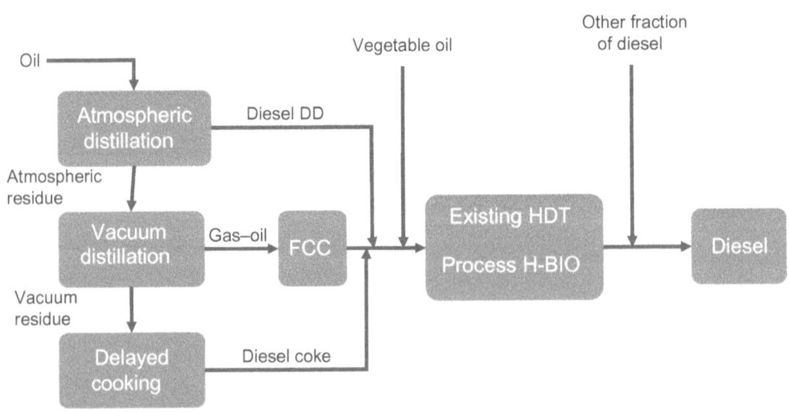

Figure 3.17 A schematic depiction of the H-Bio process (Costa, 2007). (For color version of this figure, the reader is referred to the online version of this chapter.)

of $104 a barrel (Green Car Congress, 2012b). It has been reported that the capacities were planned to be put into operation during late 2009 (World Fuels, 2012). However, no additional information about the current status of the technology and its commercial use was published since the news it was being put on hold due to high soybean oil prices.

Recently, some aspects of coprocessing were discussed in publications by major catalyst manufacturers including general information on the use of their catalysts in refineries for coprocessing. For example, Haldor Topsoe supplies its deoxygenation catalyst to three undisclosed refineries in Australia, Europe, and the USA coprocessing bio-feedstocks in the range of 0–5%, 2–5%, and 10–20%, respectively (Gabrielsen et al., 2011). The feedstocks include typically animal fats, for example, chicken fat in the USA, and other unspecified bio-feedstocks. In Preem, Sweden, Haldor Topsoe has supplied its deoxygenation catalyst for coprocessing of raw tall oil (20–30% of the feed). Apart from coprocessing, Haldor Topsoe catalyst is used by Cetane Energy, USA, to process neat animal fat and algae oil.

The possible pitfalls of revamping industrial hydrotreating or mild hydrocracking units to coprocess renewable feedstocks in significant amounts (up to 30% as in the case of Preem refinery, Sweden) have been discussed by Egeberg et al. (2011). The feed consisted of raw tall oil diesel (RTD) which is produced by esterification of tall oil, a wood processing by-product. There are about 500 kt of tall oil processed every year in Scandinavia (Gabrielsen et al., 2011). The main challenges identified included reaction heat evolution (exotherm, which is a consequence of high oxygen content of bio-feedstocks and also of their often unsaturated nature), corrosion (both up- and downstream of the reactor being consequences of the presence of free fatty acids), hydrogen consumption, and catalyst selection.

The corrosion problem is a consequence of the presence of rosin acids (they are not converted in the tall oil esterification) and unconverted free fatty acids in the RTD. In the process design, the possible problems were minimized by feeding RTD partly into AGO after the heater and partly as a liquid quench after the first catalyst bed. Consequently, the material needed to be changed only in small part of the technology. The liquid quench together with additional hydrogen quenches after second and third catalyst beds helps in controlling the exotherm (see Fig. 3.18). Moreover, due to the liquid quench between the first and second catalyst beds, a higher hydrogen partial pressure can be maintained in the upper part of the reactor (as it is not used for quenching after the first catalyst bed), thus preventing/suppressing condensation and coking reactions leading to catalyst deactivation. Besides, corrosion is suppressed as well (Egeberg et al., 2011).

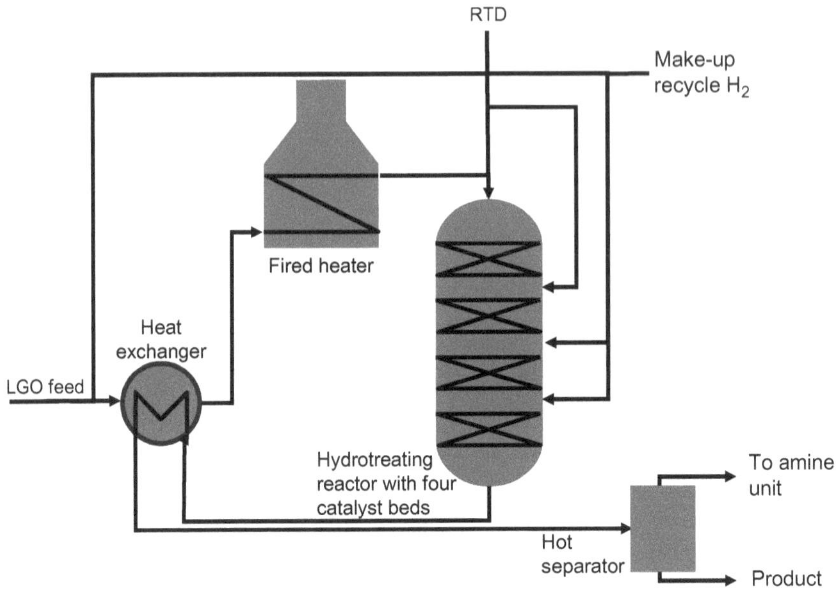

Figure 3.18 Coprocessing of tall oil with LGO in Preem refinery (Egeberg et al., 2011).

Due to the high concentration of oxygen to be removed by hydrotreating, which is significantly higher than typical sulfur concentration in petroleum-derived feeds (typically by an order of magnitude), hydrogen consumption for hydrotreating of bio-feedstocks is significantly higher than in the case of conventional hydrotreating. Obviously, this is also the cause of the increased exotherm when coprocessing bio-feedstocks with conventional petroleum-derived feedstocks. The correct balance between catalyst stability and activity is of the utmost importance (Egeberg et al., 2011). Hence, Haldor Topsoe has designed the catalyst system in the Preem refinery in such a way that RTD is converted in the upper two catalyst beds using catalysts with lower catalyst activity than their latest generation Brim catalysts. Brim NiMo catalyst is then used in the bottom two catalyst beds to ensure final product quality, that is, sulfur content below 10 ppm (Egeberg et al., 2011). The approach is based on the knowledge that deoxygenation reactions are significantly faster than the majority of hydrodesulfurization reactions.

Carbon oxides and methane are typical deoxygenation by-products causing problems in coprocessing units. Although majority of CO_2 from decarboxylation is removed in the amine wash used for H_2S scrubbing.

Downstream CO_2 can form carbonic acid in the presence of liquid water and hence cause corrosion unless the downstream equipment is modified accordingly (Egeberg et al., 2011). CO and methane will build up in the recycle gas, and a substantial purge is necessary to avoid this (Egeberg et al., 2011). Apart from reducing hydrogen partial pressure in the recycle gas loop, CO can act also as catalyst inhibitor. The purge hydrogen-rich gas has to be then treated according to its application. Removal of methane and CO in PSA is one option (Egeberg et al., 2011).

The operating experience in Preem Gothenburg refinery has shown that the reaction exotherm increased, as expected, with the increasing concentration of RTD. While in the absence of RTD the delta T was about 30 °C, it was around 60 °C when the feed contained 30% of RTD. At the same time, the hydrogen consumption increased from ca. 80 to about 140 nm^3/m^3. When the content of RTD was increased further to 50–55%, hydrogen consumption increased to ca. 200–220 nm^3/m^3 (Egeberg et al., 2011). Based on the comparison of different modes of operation (light cut gas oil mode vs. renewable mode), it was concluded that the catalyst was not inhibited either by the renewable feed components or by their products (Egeberg et al., 2011). Unfortunately, no information about the different feedstocks and their products was provided. It has been, however, reported (Gabrielsen et al., 2011) that the renewable diesel product of Preem refinery meets all the requirements of EN 590.

According to Albemarle, coprocessing of VOs in existing hydrotreaters is a straightforward alternative of introducing bio-feedstocks in the diesel fuel range in refineries. Similar issues to those identified in the Preem coprocessing case study have been discussed as well. Despite the relatively simple chemistry and high catalytic activity of the conventional hydrotreating catalyst, coprocessing gives rise to several technological and operational issues including (i) possible formation of CO having an inhibiting effect on the catalyst; (ii) accumulation of CO in recycle gas as it is not removed in scrubbers; (iii) high hydrogen consumption during triglyceride deoxygenation in comparison with HDS of AGO connected with a significant heat release; (iv) formation of *n*-alkanes that affect negatively the cold flow properties of the final diesel fuel; and (v) formation of free fatty acids, reaction intermediates in triglyceride deoxygenation, that can cause local corrosion problems (Melis et al., 2009). Albemarle has proposed a so-called BioSTAX system that is a solution consisting of several different commercial catalysts loaded at a specific place in the reactor and taking into account the constraints of the unit, such as H_2/oil ratio, outlet pressure, capacity of quenching, and intended VO addition (type and amount). The distribution (placement) of catalysts in the reactor reflects the main factors affecting the kinetics of individual feed components conversion. For instance, the requirements of a triglyceride and a dibenzothiophene vary significantly, and their conversion takes place at different positions along the catalyst bed in the reactor, that is, under different reaction conditions (e.g., hydrogen partial pressure). Consequently, different catalysts give optimum performance at a different place in the reactor with respect to the actual feed composition and reaction conditions.

If large guard beds are not available, the inherent impurities of VOs, such as phospholipids and metals, should be removed by refining, bleaching, and deodorization. On the other hand, the absence of sulfur, nitrogen, and aromatics has a positive dilution effect when coprocessed with AGO. An increase in WABT (weighted average bed temperature) of ca. 20 °C has been reported for the CoMo catalyst when coprocessing rapeseed oil (10 wt%) in AGO. When H_2/oil and pressure were increased from 300 Nm^3/m^3 and 40 bar, respectively, to 500 Nm^3/m^3 and 60 bar, the WABT increase dropped to ca. 11 °C in order to reach 10 ppm S at the outlet. The observed phenomenon was attributed to the inhibiting effect of CO, a deoxygenation by-product originating from fatty acids decarbonylation and/or from CO_2 hydrogenation (Melis et al., 2009). The inhibition is particularly severe for the direct desulfurization pathway, that is, it is more pronounced with CoMo than with NiMo catalysts. The lower impact on NiMo catalysts was verified experimentally; at 50 bar and H_2/oil of 200, an WABT increase of less than 5 °C was reported. Moreover, the large WABT increase observed for CoMo catalysts might be detrimental for the stability of catalytic activity due to considerable faster deactivation rates (Melis et al., 2009). The issue of CO-induced inhibition could be augmented with time since it is not removed by scrubbing and can be eliminated in a typical setup only by purging. This may be a limiting factor decisive for the maximum concentration of VO to be coprocessed. For higher VO concentrations, different solutions might be necessary, for example, use of selective membranes or passing of scrubbed, that is, H_2S free, recycle gas through a PSA unit. Due to the significant extent of hydrogenation reactions during deoxygenation of VOs, including double-bond hydrogenation, hydrogenation of the carboxylic group, and CO and CO_2 hydrogenation to methane, the reaction is highly exothermic leading to an increase in temperature along the catalyst bed (increase in Delta *T*). For instance, Albemarle reported an additional hydrogen consumption of 10–20 Nm^3/m^3 and a further Delta *T* increase of 12–21 °C in comparison with typical operations when coprocessing 5% of rapeseed oil. It follows that hydrogen consumption and quenching capacities are of crucial importance and optimum catalyst design is essential for limiting hydrogen consumption and affording acceptable hydrotreater cycle lengths (Melis et al., 2009).

In conclusion, to achieve the optimum performance, the extent of different deoxygenation reactions (HDO, decarboxylation/decarbonylation, methanation) needs to be optimized as well. This can be accomplished by selection of a suitable catalyst combination. The main products, *n*-alkanes, are another limiting factor for the amount of VO to be coprocessed. Their increasing concentration worsens the cold flow properties which might be an issue. For instance, coprocessing of 5% rapeseed oil can result in a decrease in cloud point of 5–6 °C, which is unacceptable typically for the winter diesel fuel grades (Melis et al., 2009).

7.3. Deoxygenation product quality

The final deoxygenation product quality is critical for its end use, that is, either for its blending in the diesel fuel pool in the case of stand-alone deoxygenation units or for the quality of the product of VO and AGO coprocessing. Obviously, the stand-alone option offers a better flexibility as the coprocessing alternative is limited by additional constraints, such as sulfur content of the final product. As discussed in the section describing different technologies, there are ways how to minimize the negative impacts of VOs on AGO hydrotreating during coprocessing. To meet specific fuel targets, the solutions can be either based on technology selection, improvements, and modifications, or on product treatment/modification. A conventional product treatment used for refinery products is the application of various additives enhancing diesel fuel properties. In the case of deoxygenation that affords as primary products *n*-alkanes having poor cold flow properties, the use of depressants might tackle the issue. The application of two industrially, widely used flow improvers (Infineum R-288 and Keroflux® 3566, BASF) for blends consisting of conventional petroleum diesel, having properties in accordance with EN 590, and mixture of hydrocarbons obtained by deoxygenation of rapeseed oil was investigated by Šimáček et al. (2010).

The results clearly demonstrated that neither of the two investigated flow improvers was able to improve the cold flow properties sufficiently. While the sensitivity of conventional diesel was high—the cold flow plugging point (CFPP) dropped from −12 °C to below −25 °C already when 200 mg/kg of flow improver (either Infineum R-288 or Keroflux® 3566) was added, there was hardly any response when the diesel fuel contained 5 wt.% of hydrocarbons from deoxygenation—the CFPP dropped from −12 to ca. −15 °C. The CFPP of diesel fuels containing 10–30 wt.% of hydrocarbons from deoxygenation did not decrease with addition of flow improvers in the range of 100–500 mg/kg more than by 1–2 °C (Šimáček et al., 2010). This clearly demonstrates that a technological solution affecting the composition of deoxygenation products, namely, shifting it from *n*-alkanes to a mixture of *n*- and iso-alkanes, is crucial.

There are several options; the most important being selective hydroisomerization of *n*-alkanes that is used in the NexBTL and Ecofining processes (Frey, 2011; Neste Oil, 2012a) and has also been the topic of recent research activities (Sebos et al., 2009). Another option, mild hydrocracking/dewaxing, was proposed by Axens (process Vegan) and Haldor Topsoe and also investigated experimentally (Šimáček and Kubička, 2010; Šimáček et al., 2011). The beneficial effect of hydrocracking of VOs on the cold flow properties of hydrocarbon mixtures afforded by this process was demonstrated by Šimáček et al. (2011). Hydrocracking of sunflower oil of 18 MPa yielded liquid hydrocarbons with *n*-alkane concentration decreasing from 67 to 20 wt.% as a result of reaction temperature increase from 360 to 420 °C. The concentration of iso-alkanes and cycloalkanes in the products obtained at reaction temperature above 400 °C exceeded 60 wt.%. Consequently, their cold flow properties were considerably improved. For instance, the cloud point of the hydrocarbon product obtained at 360 °C (*n*-alkanes/(iso-alkanes + cycloalkanes)—65/31 wt.%) decreased from 19 to −11 °C when the reaction temperature was increased to 420 °C (*n*-alkanes/(iso-alkanes + cycloalkanes)—20/74 wt.%; Šimáček et al., 2011). At the same time, the distillation curve remained within the EN 590 specification requirements and the cetane index dropped from 91 at 360 °C to 65 at 420 °C which is well above the requirements for diesel fuel (>46 units; Šimáček et al., 2011).

When the products obtained at 420 °C were blended in conventional diesel fuel in concentrations in the range of 10–50 wt.%, the CFPP values were in the range of −15 to −12 °C. As a result of commercial flow improvers (Infineum R-288 and Infineum R-291), the CFPP dropped below −20 °C and there were no differences between neat diesel fuel and diesel fuel blends containing product of sunflower oil hydrocracking. The results indicate that the sensitivity of these diesel fuel blends depends on the total content of long-chain *n*-alkanes (>C_{17}); when it exceeds a certain concentration, the commercial flow improvers become ineffective. On the basis of the results reported by Šimáček et al. (2011), this threshold can be estimated to be around 6–7 wt.% of long-chain *n*-alkanes (>C_{17}) in the final diesel fuel blend.

8. CONCLUSION

While deoxygenation of VOs is a new technology to the refining industry, it can build on the experiences gathered during operation of hydrotreating (HDS) units. Analysis of the thermodynamic, hydrodynamic, and kinetics aspects points out the differences between HDS and deoxygenation technologies that have to be taken into account when developing further deoxygenation technology. The successful examples from both stand-alone as well as coprocessing technologies demonstrate that the technology has a future in the portfolio of refining technologies.

The future development should be focused on two particular directions: further catalyst development and utilization of more sustainable feedstocks. From the point of view of catalyst development, it would be certainly beneficial if more selective catalysts, for example, decarboxylation catalysts, found their application in the stand-alone units limiting, thus, the need for catalyst sulphiding. Moreover, the industry would certainly benefit from catalysts facilitating both deoxygenation and isomerization in one step (reactor). From the point of view of feedstock, the technology has to be focused on nonfood sources of triglycerides that are produced in a sustainable way. Moreover, the need for their preprocessing (purification) should be minimized as this would result in a more cost–effective production, which is a key to avoiding too large dependence on legislation and subsidies.

ACKNOWLEDGMENTS

This publication was created in connection with the project "Unipetrol research and education center" Reg. No. CZ.1.05/2.1.00/03.0071, which is funded through the Operational Program for Research and Innovation Development of the Structural Funds (specifically the European Regional Development Fund) and the state budget of the Czech Republic.

REFERENCES

Andersson K, Hell M, Löwendahl L, Schöön NHJ: Diffusivities of hydrogen and glyceryl trioleate in cottonseed oil at elevated temperature, *J Am Oil Chem Soc* 51:171, 1974.

AspenTech, aspenONE V7.3, Process Engineering Products, 2011.

Bern L, Hell M, Schöön NH: Kinetics of hydrogenation of rapeseed oil: I. Influence of transport steps in kinetic study, *J Am Chem Oil Soc* 52:182, 1975.

Bezergianni S, Dimitriadis A, Kalogianni A, Pilavachi P: Hydrotreating of waste cooking oil for biodiesel production. Part I: Effect of temperature on product yields and heteroatom removal, *Bioresour Technol* 101:6651, 2010a.

Bezergianni S, Dimitriadis A, Sfetsas T, Kalogianni A: Hydrotreating of waste cooking oil for biodiesel production. Part II: effect of temperature on hydrocarbon composition, *Bioresour Technol* 101:7658, 2010b.

Cable TL, Massoth FE, Thomas MG: A basic study of catalyst aging in the H-coal process, *Fuel Process Technol* 10:105, 1985.

Cai HY, Shaw JM, Chung KH: Hydrogen solubility measurements in heavy oil and bitumen cuts, *Fuel* 80:1055, 2001.

Costa PR: Increasing biodiesel production, *PTQ-Biofuels* 2007:32, 2007.

da Rocha Filho GN, Brodzki D, Djega-Mariadassau G: Formation of alkanes, alkylcycloalkanes and alkylbenzenes during the catalytic hydrocracking of vegetable oils, *Fuel* 72:543, 1993.

Donnis B, Egeberg RG, Blom P, Knudsen KG: Hydroprocessing of bio-oils and oxygenates to hydrocarbons. Understanding the reaction routes, *Top Catal* 52:229, 2009.

Egeberg R, Knudsen K, Nystrom S, Grennfelt EL, Efraimsson K: Industrial-scale production of renewable diesel, *Petroleum Technol Q* Q3:59, 2011.

Fillion B, Morsi BI: Gas–liquid mass-transfer and hydrodynamic parameters in a soybean oil hydrogenation process under industrial conditions, *Ind Eng Chem Res* 39:2157, 2000.

Frey, S. 2011. Honeywell development of green jet fuel technology. IAE Bioenergy ExCo67 meeting, paper 08, Helsinki.

Furimsky E, Massoth FE: Deactivation of hydroprocessing catalysts, *Catal Today* 52:381, 1999.

Gabrielsen J, Egeberg RG, Knudsen KG: *Hydrotreating of triglycerides to make aviation fuels, Biomass 2011*, Maryland, 2011, National Harbor.

Green car congress, 2012. http://www.greencarcongress.com/2008/01/petrobras-h-bio.html. Accessed February.

Green car congress, 2012. http://www.greencarcongress.com/2008/08/albemarle-to-pr.html. Accessed February.

Hodge, C. 2006. California Energy Commission, Workshop on bioenergy. http://www.energy.ca.gov/bioenergy_action_plan/documents/2006-03-09_workshop/2006-03-09_NESTE_OIL.PDF. Accessed March 9.

Huber G, O'Connor P, Corma A: Processing biomass in conventional oil refineries: production of high quality diesel by hydrotreating vegetable oils in heavy vacuum oil mixtures, *Appl Catal A Gen* 329:120, 2007.

Immer J, Kelly M, Lamb H: Catalytic reaction pathways in liquid-phase deoxygenation of C18 free fatty acids, *Appl Catal A Gen* 375:134, 2010.

Kalnes TN, Marker T, Shonnard D, Koers KP: Green diesel production by hydrorefining renewable feestocks, *PTQ-Biofuels* 2008:7, 2008.

Kubička D: Future refining catalysis—introduction of biomass feedstocks, *Collect Czech Chem Commun* 73:1015, 2008.

Kubička D, Horáček J: Deactivation of HDS catalysts in deoxygenation of vegetable oils, *Appl Catal A Gen* 394:9, 2011.

Kubička D, Kaluža L: Deoxygenation of vegetable oils over sulfided Ni, Mo and NiMo catalysts, *Appl Catal A Gen* 372:199, 2010.

Kubička D, Šimáček P, Žilková N: Transformation of vegetable oils into hydrocarbons over mesoporous-alumina-supported CoMo catalysts, *Top Catal* 52:161, 2009.

Kubička D, Bejblová M, Vlk J: Conversion of vegetable oils into hydrocarbons over CoMo/MCM-41 catalysts, *Top Catal* 53:168, 2010.

Kubičková I, Kubička D: Utilization of triglycerides and related feedstocks for production of clean hydrocarbon fuels and petrochemicals: a review, *Waste Biomass Valor* 1:293, 2010.

Kubičková I, Kubička D: Heterogeneous catalysts for production of renewable fuels from triglycerides. In Kubičková I, Kubička D, editors: *Heterogeneous catalysis in biomass to chemicals and fuels*, Kerala, India, 2011, Research Signpost, 978-81-308-0462-0.

Kubičková I, Snåre M, Eränen K, Mäki-Arvela P, Murzin D: Hydrocarbons for diesel fuel via decarboxylation of vegetable oils, *Catal Today* 106:197, 2005.

Laurent E, Delmon B: Study of the hydrodeoxygenation of carbonyl, carboxylic and guaiacyl groups over sulfided CoMo/gamma-Al2O3 and NiMo/gamma-Al2O3 catalysts. I. Catalytic reaction schemes, *Appl Catal A Gen* 109:77, 1994a.

Laurent E, Delmon B: Influence of water in the deactivation of a sulfided NiMo gamma-Al2O3 catalyst during hydrodeoxygenation, *J Catal* 146:281, 1994b.

Laurent E, Delmon B: Study of the hydrodeoxygenation of carbonyl, carboxylic and guaiacyl groups over sulfided CoMo/gamma-Al2O3 and NiMo/gamma-Al2O3 catalyst. II. Influence of water, ammonia and hydrogen sulfide, *Appl Catal A Gen* 109:97, 1994c.

Lestari S, Simakova I, Tokarev A, Mäki-Arvela P, Eränen K, Murzin D: Synthesis of biodiesel via deoxygenation of stearic acid over supported Pd/C catalyst, *Catal Lett* 122:247, 2008.

Lestari S, Mäki-Arvela P, Beltramini J, Lu G, Murzin D: Transforming triglycerides and fatty acids into biofuels, *ChemSusChem* 2:1109, 2009.

Levenspiel O: Chemical reaction engineering, *Ind Eng Chem Res* 38:4140, 1999.

Lüft, M., Spicher, U. 2007. *Optimierung von Kraftstoffstrahlausbreitung für Pflanzenöl. insbesondere natürliches Rapsöl. bei der Verwendung moderner Diesel Einspritzsysteme. BWK 25002—Study,* Germany.

Macleod DB: On a relation between surface tension and density, *Trans Faraday Soc* 19:38, 1923.

Maier WF, Roth W, Thies I, Rague Schleyer PV: Hydrogenolysis, IV. Gas phase decarboxylation of carboxylic acids, *Chem Ber* 115:808, 1982.

Mary G, Chaouki J, Luck F: Trickle-bed laboratory reactors for kinetic studies, *Int J Chem Reactor Eng* 7 R2:1, 2009.

Melis S, Mayo S, Leliveld B: Vegetable oil co-processing in diesel hydrotreaters, *PTQ-Biofuels* 2009:43, 2009.

Mikulec J, Cvengroš J, Joríková L, Banič M, Kleinová A: Diesel production technology from renewable sources—second generation biofuels, *Chem Eng Trans* 18:475, 2009.

Murzin, D., Kubičková, I., Snåre, M., Mäki-Arvela, P., Myllyoja, J. 2006. Method for the manufacture of hydrocarbons, PCT International Application 2006/075057 A2, 2006.

Murzin, D.Y., Mäki-Arvela, P., Simakova, I.L. 2012. Triglycerides and oils for biofuels, Kirk-Othmer encyclopedia of chemical technology, Published Online: 13 January, 2012. http://dx.doi.org/10.1002/0471238961.trigmurz.a01.

Neste Oil, 2012. http://www.nesteoil.com/default.asp?path=1,41,11991,12243,13565. Accessed February, 2012.

Neste Oil, 2012. http://www.nesteoil.com/default.asp?path=1,41,11991,12243,15658. Accessed February, 2012.

Neste Oil, 2012. http://www.nesteoil.com/default.asp?path=1,41,11991,12243,15658,12139. Accessed February, 2012.

Neste Oil, 2012. http://www.nesteoil.com/default.asp?path=1,41,11991,12243,12335,14116. Accessed February, 2012.

Neste Oil, 2012. http://www.nesteoil.com/default.asp?path=1,41,11991,12243,17555. Accessed February, 2012.

Neste Oil, 2012. http://www.nesteoil.com/default.asp?path=1,41,11991,12243,12335. Accessed February, 2012.

Piqueras CM, Tonetto G, Bottini S, Damiani DE: Sunflower oil hydrogenation on Pt catalysts: comparison between conventional process and homogeneous phaseoperation using supercritical propane, *Catal Today* 133–135:836, 2008.

Priecel P, Kubička D, Čapek L, Bastl Z, Ryšánek P: The role of Ni species in the deoxygenation of rapeseed oil over NiMo-alumina catalysts, *Appl Catal A Gen* 397:127, 2011.

Reid R, Poling BE, Prausnitz JM: *The properties of gases and liquids,* ed 5, New York, 2001, McGraw-Hill Co., Inc.

Santacesaria E, Parrella P, Di Serio SNM, Borrelli M: Role of mass transfer and kinetics in the hydrogenation of rapeseed oil on a supported palladium catalyst, *Appl Catal A Gen* 116:269, 1994.

Schaub G, Vetter A: Biofuels for automobiles—an overview, *Chem Eng Technol* 31(5):721, 2008.

Schmidt A, Schomacker R: Partial hydrogenation of sunflower oil in a membrane reactor, *J Mol Catal A Chem* 271:192, 2007.

Sebos I, Matsoukas A, Apostolopoulos A, Papayannakos N: Catalytic hydroprocessing of cottonseed oil in petroleum diesel mixtures for production of renewable diesel, *Fuel* 88:145, 2009.

Šimáček P, Kubička D: Hydrocracking of petroleum vacuum distillate containing rapeseed oil: evaluation of diesel fuel, *Fuel* 89:1508, 2010.

Šimáček P, Kubička D, Šebor G, Pospíšil M: Fuel properties of hydroprocessed rapeseed oil, *Fuel* 89:611, 2010.

Šimáček P, Kubička D, Kubičková I, Homola F, Chudoba J, Pospíšil M: Premium quality renewable diesel fuel by hydroprocessing of sunflower oil, *Fuel* 90:2473, 2011.

Simakova I, Simakova O, Mäki-Arvela P, Simakov A, Estrada M, Murzin D: Deoxygenation of palmitic and stearic acid over supported Pd catalysts: Effect of metal dispersion, *Appl Catal A Gen* 355:100, 2009.

Smejkal Q, Smejkalová L, Kubička D: Thermodynamic balance in reaction system of total vegetable oil hydrogenation, *Chem Eng J* 146:155, 2009.

Snåre M, Kubičková I, Mäki-Arvela P, Eränen K, Murzin D: Heterogeneous catalytic deoxygenation of stearic acid for production of biodiesel, *Ind Eng Chem Res* 45:5708, 2006.

Snåre M, Kubičková I, Mäki-Arvela P, Chichova D, Eränen K, Murzin D: Catalytic deoxygenation of unsaturated renewable feedstocks for production of diesel fuel hydrocarbons, *Fuel* 87:933, 2008.

Tukač, V., Kubička, D., Horáček, J. 2012. Manuscript under preparation.

UOP, 2012. www.uop.com, UOP/ENI Ecofining Process.pdf. Accessed February, 2012.

World Fuels, 2012. http://www.worldfuels.com/wfExtract/exports/Content/e0daa202-f81c-4ff2-c5-ca0aba8.html. Accessed February, 2012.

Wu Y, Khadilkar MR, Al-Dahhan MH, Dudukovic´ MP: Comparison of upflow and downflow two-phase flow packed-bed reactors with and without fines: experimental observations, *Ind Eng Chem Res* 35:397, 1996.

CHAPTER FOUR

Chemical Reaction Engineering of Biomass Conversion

Tapio Salmi
Department of Chemical Engineering, Process Chemistry Centre, Laboratory of Industrial Chemistry and Reaction Engineering, Åbo Akademi, Turku, Finland

Contents

Advances in Chemical Engineering, Volume 42
ISSN 0065-2377
http://dx.doi.org/10.1016/B978-0-12-386505-2.00004-3

Abstract

Catalysis and reaction engineering are the key disciplines in a successful conversion of molecules from biomass to end products. The interaction of kinetics, mass transfer, and flow pattern in the processing of biomass is reviewed. A special attention should be paid on the catalytic multiphase transformation of biomass. Chemical analysis and precise kinetic measurements combined to mathematical modeling are the basic elements in the scientific development of the reaction engineering of biomass conversion. Advanced kinetic concepts, such as cluster kinetics, size-dependent kinetics, can be combined to the models of porous catalyst particles and chemical reactors. Challenges and visions for future reaction engineering of biomass conversion are presented.

NOTATION

a catalyst activity factor
a,b empirical parameters
a_p catalyst particle mass transfer area to reactor volume relation
a_{GL} gas–liquid mass transfer area to reactor volume relation
A mass transfer area
c concentration
d diameter
D diffusion coefficient or dispersion coefficient
D_e effective diffusion coefficient
f metal distribution function in eggshell catalysts
k rate constant
k_L, k_G mass transfer coefficients
K equilibrium constant
l length coordinate
L length
n molar amount
n' molar flow
N flux
p parameter in regression analysis
P pressure
Q objective function in regression analysis
r component generation rate
r radius
R reaction rate
R gas constant, 8.3143 J/mol K
R^2 degree of explanation
s catalyst particle shape factor
t time, min
T temperature
V volume
V' volumetric flow rate
w velocity
w weight matrix in regression

x mole fraction
y variable in regression analysis
z dimensionless length coordinate
α Polanyi parameter or maximum coverage
$\Delta\delta'$ parameter in rate equation
ε porosity
ε specific mixing power
η effectiveness factor
θ surface coverage
μ dynamic viscosity
ν kinematic viscosity
ν stoichiometric coefficient
ρ density
τ residence time or space time
φ deactivation function
ω reciprocal value of site concentration

DIMENSIONLESS PARAMETERS

Pe Peclet number
Re Reynolds number
Sc Schmidt number
Sh Sherwood number

SUBSCRIPTS AND SUPERSCRIPTS

b, B bulk property
F fluid
G gas phase
i component index
j reaction index
L liquid phase
p catalyst particle property
* equilibrium conditions

1. INTRODUCTION

Chemical reaction engineering has traditionally focused on the applications related to energy production from fossil fuels and production of bulk chemicals. In these fields, a lot of research and development effort has taken place in both academia and industry. The breakthrough of numerical methods and computing, along with lumping techniques, has enabled the quantitative treatment of very complex chemical systems in various reactors, such as catalytic fixed beds and fluidized beds. Simultaneously, new concepts of catalytic reactors have emerged, such as monoliths, solid foams, and catalyst fibers (Cybulski and Moulijn, 2006).

The mankind is facing a new challenge: the era of easily accessible fossil raw materials is approaching the end. The resources of fossil raw materials diminish, particularly the access to crude oil. The most important change is, however, caused by the concern of the state of the atmosphere. Because of the alarming climate change, the continuous increase in the carbon dioxide content in the atmosphere caused by the burning of fossil fuels, it is necessary to shift to a more sustainable technology, the use of biomass as a source of fuels, chemicals, and materials.

Chemical engineering, in general, and chemical reaction engineering, in particular, are in key position to carry out this transformation. One of the most important sources of biomass is forest, not only the rain forest and eucalyptus trees growing in the tropical areas of the earth but also the forest growing in the Northern hemisphere, for instance, in Canada, the USA, Russia, and the Fenno-Scandic area. The forest growth per capita in some European countries is shown in Fig. 4.1. As the figure shows, the forest growth per capita is high in the Fenno-Scandic area. The same is valid for Canada and Russia.

In the very past, biomass, especially wood, was used as a source of many chemicals; typical examples were ethanol, turpentine, regenerated fibers

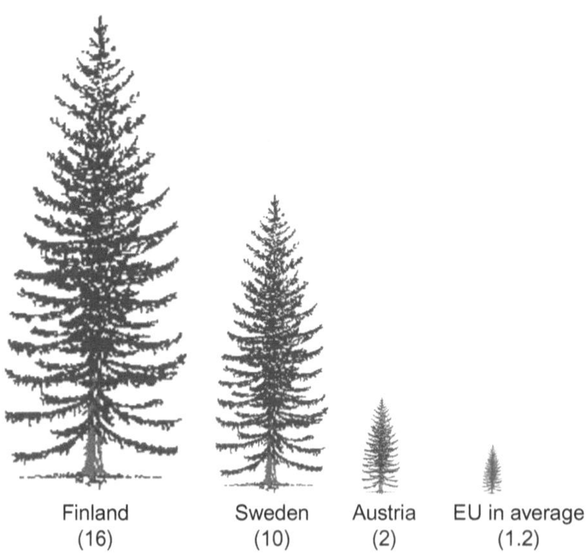

Figure 4.1 Annual forest growth per capita (in m^3) in some European countries. (For color version of this figure, the reader is referred to the online version of this chapter.)

(cellulose xanthate), polymers (e.g., cellulose acetate), and explosives (e.g., cellulose nitrate). However, the access to relatively inexpensive crude oil and natural gas led to a decline of wood-based chemical industry. Few products remained, such as cellulose derivatives.

Now, the concept of biorefinery is making a breakthrough. It implies that biomass will be used as a raw material, instead of natural gas, crude oil, or coal. This development creates big challenges for science and technology, particularly for chemistry and chemical engineering. New catalysts and processes are needed, including new reactors, new separation technologies, and new reaction media. Still, the methodological aspects of chemical engineering, and particularly chemical reaction engineering, remain, and they will be used for the design of biorefineries.

The aim of this overview is to illustrate how the central concepts of chemical reaction engineering can be applied to the transformation of biomass to chemicals and fuel components.

1.1. Chemicals and fuel components from biomass

The two basic routes for the processing of biomass according to a biorefinery concept are shown in Fig. 4.2. This overview chapter focuses on the sugar platform, that is, reactions that preserve the key elements of original structures in biomass, namely, the basic building blocks of cellulose, hemicellulose, and lignin. Conventional chemicals and fuel components can be made in this way, but the sugar platform also opens the perspective for new chemicals as discussed by Gallezot (2011).

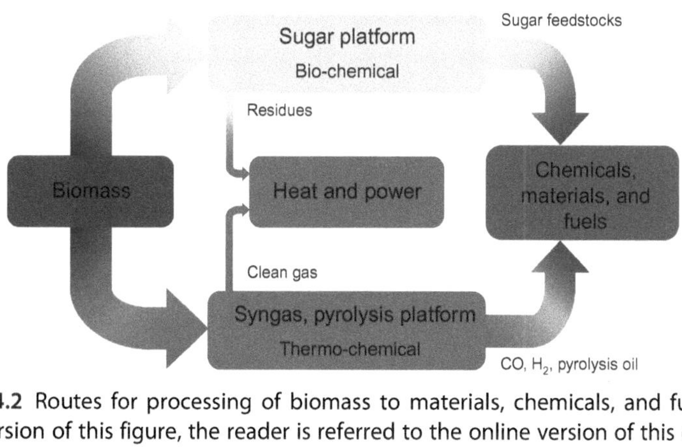

Figure 4.2 Routes for processing of biomass to materials, chemicals, and fuels. (For color version of this figure, the reader is referred to the online version of this chapter.)

In the current industry, wood is processed mechanically to chips, which are processed chemically or mechanically, the ultimate aim being pulp and paper. Roughly speaking, wood consists of cellulose (around 50%), hemicelluloses (10%), lignin (<40%), and extractives (few %). The aim of chemical pulping is to dissolve the lignin material and most part of the hemicelluloses into the cooking liquor. The black liquor originating from the Kraft pulping process contains NaOH and NaHS; lignin is dissolved and cellulose fibers remain for papermaking. It is possible to recover some chemical fractions from chemical pulping, particularly tall oil and turpentine (Sjöström, 1989). Mathematical modeling of the Kraft pulping process has mostly been very empirical, but recently, Grénman et al. (2010) developed a mathematical model which takes into account the chemical and mass transfer processes, as well as the anisotropy and structural changes of the wood material during the progress of chemical pulping.

The turpentine fraction is gasified during the Kraft process, and valuable chemicals, such as α-pinene and carene, can be separated by distillation. They can be further processed by catalytic treatment, such as isomerization and oxidation (Allahverdiev et al., 1998; Corma et al., 2002; Vital et al., 2001).

From the tall oil fraction, valuable chemicals can be separated, such as triterpenoids, fatty acids, and resin acids. Sitosterol, a triterpenoid, is the raw material for a cholesterol-suppressing agent, sitostanol (Ekblom, 2005; Michalson and Devore, 2004). Conjugated fatty acids (conjugated linoleic acid), which are important antioxidants, can be produced by isomerization of fatty acids and fatty acid esters with homogeneous (DeJarlais and Gast, 1971; Frankel, 1970; Mukesh et al., 1988) or heterogeneous catalysts (Bernas et al., 2002a, 2002b). Fatty acids and resin acids can also be saturated through catalytic hydrogenation and used as fuel components after decarboxylation over heterogeneous catalysts, such as supported Pd (Murzin et al., 2006; Snåre et al., 2006).

Some part of the cellulose fraction is taken to make cellulose derivatives, such as methyl and ethyl cellulose, carboxymethyl cellulose, hydroxyethylcellulose, and hydroxypropylcellulose. The preparation of cellulose derivatives takes place as a two-phase reaction: cellulose is pretreated, for example, with alkali, and a reagent is added to get the substitution. Usually, no catalyst is needed (Sjöström, 1989). The recent advances in cellulose derivatives concern the reaction media: by using tailored ionic liquids, it is possible to dissolve cellulose and thus carry out the substitution reaction as a homogeneous liquid-phase process (Heinze et al., 2005), which implies higher reaction rates. The quantitative modeling of cellulose substitution kinetics is described in detail by Salmi et al. (2011).

The use of the lignin fraction is much more cumbersome; the only chemical of a real commercial importance is vanillin, which is obtained by oxidation of the black liquor. The black liquor from the sulfate process contains a lot of hydroxyacids (in pinewood black liquor around 30 wt%), which might be of potential use as a source of chemicals, but their separation from the black liquor mixture is a challenge. The possibilities of catalytic transformation lignin are reviewed by Zakzeski et al. (2010).

Wood chips can be utilized as such to produce bioethanol. The cellulose and hemicellulose material is hydrolyzed in the presence of acids (H_2SO_4, HCl, or HCOOH) or enzymes to get glucose and other monosaccharides (Abatzoglou and Chornet, 1998). Lignin is separated by filtration as a solid residue, and the monosaccharides are fermented to ethanol, which is separated from water and catalyst by distillation. Ethanol can be used not only as an energy source but also as a platform chemical to make ethane and polyethene.

To make chemicals, it is much more straightforward to start from the wood chips, knots, and bark directly rather than from the black liquor produced in the Kraft pulping process. In mechanical pulping, some hemicelluloses, such as glucomannans, can be recovered and processed further by hydrolysis, fermentation, oxidation, and hydrogenation. An even more attractive way is to extract hemicelluloses from wood chips with water—as an example should be mentioned that the larch wood (*Larix sibirica*) contains around 15% and even more arabinogalactan by weight. In an analogous manner, galactoglucomannan and arabinoglucuronexylan can be extracted from the wood material (Willför and Holmbom, 2004; Willför et al., 2002) and hydrolyzed (Kusema et al., 2010; Xu et al., 2008).

The separation of various hemicelluloses and the catalytic treatment of them have a great perspective. Existing and potential products from hemicelluloses are summarized in Fig. 4.3. As the figure reveals, hemicelluloses are rich sources for chemicals and fuel components. Many platform chemicals, such as sugar alcohols, furfurals, and ethanol, can be made from hemicelluloses. The perspective of biopolymers from hemicellulose is broad: they can be used in pharmaceutical, cosmetic, and alimentary industries.

A world-famous product is the sweetener, anticaries, and anti-inflammatory agent, xylitol, which is obtained from glucuronoxylan and arabinoglucuronoxylan (Maloney et al., 1986). The main source of xylanes is birch (*Betula pendula*).The hemicellulose is hydrolyzed to monosaccharides, which are separated, and xylose is hydrogenated over a heterogeneous catalyst

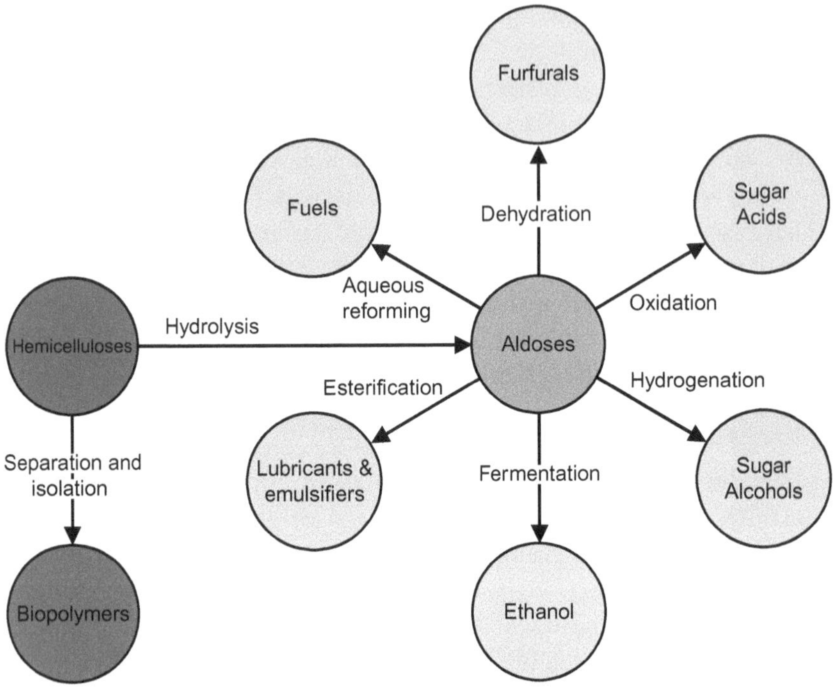

Figure 4.3 Transformation of hemicelluloses to chemicals and fuel components.

to the sugar alcohol, xylitol. The classical catalyst for this purpose has been the Raney nickel (Kuusisto et al., 2006; Mikkola et al., 1999), but supported Ru catalysts have given very promising results (Gallezot et al., 1998; Kuusisto et al., 2007). Sifontes (2012) have recently shown that arabinose, galactose, mannose, rhamnose, and maltose can be successfully hydrogenated to the corresponding sugar alcohols on Ru/active carbon catalysts. In an analogous manner, glucose can be hydrogenated to sorbitol (Gallezot et al., 1998). Sugar alcohols are platform chemicals, which can be used for many purposes, such as the production of fuels via aqueous reforming and preparation of biolubricants through esterification (Alonso et al., 2010).The separation of various hemicelluloses and catalytic treatment are the key issues.

From bark, valuable chemicals, such as betulinol, can be obtained (Sjöström, 1989). Betulinol is used as a health-promoting agent. Holmbom et al. (2002, 2004) discovered that hydroxymatairesinol (HMR) is concentrated in the stems and knots of Scandinavian spruce (*Picea abies*). It can be extracted and transformed catalytically to matairesinol (MAT), which is an antioxidant and anticarcinogenic agent (Eklund et al., 2003; Markus et al., 2006; Sjöholm et al., 2003).

Besides the polysaccharides existing in the biomass originating from wood, two disaccharides appearing in large amounts in the nature should be mentioned: saccharose (mainly in sugar beets and sugar cane) and lactose (in the milk of mammals). Saccharose consists of glucose and fructose units, while lactose consists of glucose and galactose units. They can be transformed to various chemicals, either as such or after hydrolysis (Blecker et al., 2002).

A brief summary of current and potential processes is provided by Table 4.1. As revealed by the table, most of the reactions are hydrolysis,

Table 4.1 Examples of molecules from biomass and their transformation

Reaction	Reactant	Product	Catalyst
Hydrolysis of cellulose (—O—)	Cellulose	Glucose, levulinic acid, formic acid, HMF	Acid, enzyme (HO, HE)
Hydrolysis of hemicelluloses (—O—)	Hemicellulose	Arabinose, galactose, mannose, xylose	Acid, enzyme (HO, HE)
Hydrolytic hydrogenation of cellulose	Cellulose	glucose, sorbitol	Pt (HE)
Hydrolytic hydrogenation of hemicelluloses	Hemicellulose	Sugar monomer, sugar alcohol	Pt (HE)
Hydrogenation of sugars (—CHO)	Arabinose, galactose, glucose, mannose, xylose, lactose +hydrogen	Arabinitol, galactitol, sorbitol, mannitol, xylitol,. . ., lactitol	trad. Raney nickel, Ru/C better (HE)
Oxidation of sugars (—CHO)	Sugars (see above) + oxygen	Sugar acids	Pd, Au, and Au/Pd (HE) catalysts promising
Fermentation of sugars	Mainly glucose	Ethanol	Enzymes (E)
Hydrogenation of phytosterols (C=C)	Sterol (e.g., sitosterol) + hydrogen	Stanol (e.g., sitosterol)	Pd on various supports (C, zeolite, polymer) (HE)
Hydrogenation of fatty acids (C=C)	Unsaturated acid	Saturated acid	Ni, Pt, Pd, Rh, Ru (HE)

Continued

Table 4.1 Examples of molecules from biomass and their transformation—cont'd

Reaction	Reactant	Product	Catalyst
Hydrogenolysis of lignans (—OH)	For example, HMR + hydrogen	For example, MAT	Pd/C, Pd/Cnanofibers, Pd-H-Beta-150 (HE)
Decarboxylation of fatty acids	Carboxylic acid	Alkane, alkene	Pd, Pt (HE)
Ketonization of fatty acids	Carboxylic acid	Ketone	Fe_2O_3,CeO_2 (HE)
Isomerization of fatty acids	For example, linoleic acid	CLA	Rh-phosphine complexes Cr-carbonyls (HO), Ru/C, Ru/Al_2O_3 (HE)
Transesterification	Triglycerides		Fatty acid methyl Esters (FAME) (HO)

HE, heterogeneous catalyst; HO, homogeneous catalyst; E, enzyme; CLA, conjugated linoleic acid; HMF, hydroxymethyl furfural.

hydrogenolysis, hydration, hydrogenation, oxidation, and isomerization reactions, where catalysis plays a key role. Particularly, the role of heterogeneous catalysts has increased in this connection in recent years; therefore, the present chapter concerns mostly the application of heterogeneous solid catalysts in the transformation of biomass. An extensive review of various chemicals originating from nature is provided by Mäki-Arvela et al. (2007).

When looking at the list of catalysts being successful, it is striking to notice that only few catalyst metals seem to be of importance. For the double bond hydrogenation and hydrogenolysis of hydroxyl groups, Pd is both active and selective (Augustine and Reardon, 1969). For the hydrogenation of the carboxyl group (e.g., production of sugar alcohols), sponge nickel (Raney nickel) and Ru on active carbon work well, but Raney nickel is probably going to be replaced by Ru because of environmental and health reasons. Ru is also an active and selective catalyst in the isomerization, in the production of conjugated acids. Oxidation of sugars to sugar acids (Wenkin et al., 1996), for instance, oxidation of lactose to lactobionic acids, is traditionally carried out on Pd catalysts, but the risk is the oxidation of metallic Pd by molecular oxygen, which impairs the activity and selectivity. Recent studies have shown that gold–ceria catalysts could provide a promising alternative for this oxidation process (Tokarev et al., 2006).

Besides the catalyst activity, the selectivity is a central issue in the treatment of the molecules from biomass. A typical example is catalytic hydrogenation of sugar molecules to sugar alcohols. A product selectivity of about 90% is easily achieved on Ni and Ru catalysts, but approaching 100% requires a fine-tuning of the catalyst properties and optimization of the reaction conditions. The selectivity of the catalyst can sometimes be improved with promoters; for instance, Mo is used as a promoter in Raney nickel catalysts. A too high reaction temperature leads to decomposition products, and lack of hydrogen on the catalyst surface promotes the isomerization reaction and enhances catalyst deactivation by fouling. Thus, rather low temperatures (around 100 °C) and moderate hydrogen pressures (40–80 bar) have proven to be the optimal ones for sugar hydrogenations on Ni and Ru catalysts.

The control of pH is an issue for hydrogenation and oxidation processes, particularly for oxygenation of sugars to sugar acids. During the oxidation process, the pH of the solution drops because of the sugar acid formed (Haakana et al., 2004). Increased acidity leads to suppressed catalyst activity. The remedy is a controlled neutralization of the solution during the progress of the reaction (Tokarev et al., 2006).

The role of support is of crucial importance for some of the transformations. A typical example is hydrogenolysis, where an acidity function is needed to accomplish the reaction. Active carbon has proven to be an excellent support material for cases, where Pd and Ru catalysts are used (Table 4.1). The problem with active carbon is that it originates from different sources, having different functional groups attached to the surface. This implies that the reproducibility of the catalyst is not the best possible. Another drawback with active carbon is its rather poor mechanical strength. This has encouraged researchers to investigate alternative support materials, for example, instead of Pd/C catalysts, Pd on zeolite supports can be used in the hydrogenation of the double bond. The recent development of synthetic carbon materials might help solve the problem. Catalyst particles of synthetic carbon, for example, Sibunit, have a good mechanical strength and can be used in fixed beds. Furthermore, it is possible to impregnate the outer layer of the carbon particle only with the active metal and in this way create an eggshell catalyst, which is suitable for the treatment of large organic molecules with a substantial diffusion resistance in the porous catalyst structure. Carbon nanofibers with a reproducible structure can be produced by pyrolysis of a gaseous aromatic component on a metal surface. This kind of nanofibers (Markus et al., 2006) can be impregnated with metals, for example, to prepare a Pd catalyst for hydrogenation of double bonds or hydrogenolysis of hydroxyl groups or a Ru catalyst for the hydrogenation of carbonyl groups.

Catalyst deactivation is a big problem in the treatment of molecules originating from biomass. Leaching, fouling, poisoning, and sintering are, in general, the typical reasons for catalyst deactivation. Catalysts that exhibit leaching of metals are, of course, excluded from the very beginning, as many of the applications concern food, health-promoting chemicals, and pharmaceuticals. Sintering is not a typical phenomenon for catalysts used in biomass transformation, as the reaction temperatures are moderate in the sugar platform (Fig. 4.2). The organic molecules incorporated in the transformations have a tendency to deposit on the surface and initiate coking even at rather low reaction temperatures ($<100^{\circ}C$). The purity of the raw material is a crucial issue. If the reactant molecules originate from the black liqueur, they carry sulfur and sulfurous compounds as impurities, which have a deteriorating effect on the catalyst by irreversible poisoning. Thus, a precleaning of the catalyst is very important. It is applied, for instance, to the hydrogenation of sitosterol to sitostanol: sitosterol is cleaned via adsorption on active carbon before entering the hydrogenation reactor.

An interesting way to retard catalyst deactivation is to expose the reaction mixture to ultrasound. Ultrasound treatment of the mixture creates local hot spots, which lead to the formation of cavitation bubbles, which bombard the solid, dirty surface leading to the removal of carbonaceous deposits (Mikkola and Salmi, 1999). The ultrasound source can be inside the reactor vessel (ultrasound stick), or ultrasound generators can be placed in contact with the wall of the reactor. Both constructions work, and the catalyst lifetime can be essentially prolonged, leading to process intensification. The effects of ultrasound are discussed in detail in the review of Toukoniitty et al. (2008).

1.2. Chemical reaction engineering in the transformation of biomass

Chemical reaction engineering and catalysis are key fields of science in the transformation of biomass components to products. The rational elements of chemical reaction engineering, namely, stoichiometry, thermodynamics, kinetics, heat, and mass transfer, as well as flow modeling, play a decisive role in the design of biorefineries.

A rational scale-up of a production process should be based on correct stoichiometric, thermodynamic, and kinetic model. Chemical kinetics is exciting *per se*, as it yields valuable information about the underlying reaction mechanisms (Murzin and Salmi, 2005). Thus, it is very important that the kinetic information is not obscured by catalyst deactivation and mass transfer effects. To obtain correct stoichiometric information is a challenge itself in case of transformation of organic molecules from biomass, as by-products typically appear in low concentrations in a concentrated environment (the concentrations of reactant and main product being typically 20–60 wt%). Thus, very precise high-pressure liquid chromatography (HPLC) or sophisticated gas chromatography coupled to mass spectrometry (GC–MS) is typically used, and the work is typically completed with NMR analysis. In most demanding cases, coupled GC–NMR and HPLC–NMR techniques are used.

Most of the chemical reactions related to biomass conversion involve a three-phase process: gas, liquid, and solid catalysts are present. Internal and external mass transfer limitations in porous catalyst layers play a central role in three-phase processes. The governing phenomena are well known since the days of Thiele (1939) and Frank-Kamenetskii (1955), but the transport phenomena coupled to chemical reactions are not frequently applied for complex organic systems, but simple—often too simple—tests based on the use of first-order Thiele modulus and Biot number are used. Instead, complete numerical simulations are preferable to reveal the role of mass and heat transfer at the phase boundaries and inside the porous catalyst particles.

It is important to obtain reliable models for catalyst deactivation and to investigate whether it is possible to decouple the deactivation model from the kinetic model or is it necessary to treat the catalyst deactivation as one of the surface reactions on the catalyst (Sandelin et al., 2006).

The reactions of biomass components are often carried out in batch and semibatch reactors, as many of these processes are limited by slow kinetics, which exclude the use of continuous reactors (Trambouze and Euzen, 2002).This implies that time-dependent, dynamic models are required to obtain a realistic description of the process. A typical example is the large-scale production of fatty acid methyl esters (FAME), which is carried out in batch reactors, but recently, it has been proposed that the process could be performed in continuous reactors. An advanced modeling work of FAME reactors has been published by Likozar and Levec (2012).

In all of the transformations indicated, catalysts, particularly heterogeneous catalysts, play the key role. It has turned out that few precious metals (Ru, Pd, Pt, Au) can do most of the transformations needed, but the development of catalyst supports and structures is a challenge. In future, the discontinuous slurry reactor technology should be replaced by continuous

Figure 4.4 Structured catalysts: carbon washcoated steel (CSS) support (left) and active carbon cloth (ACC) support.

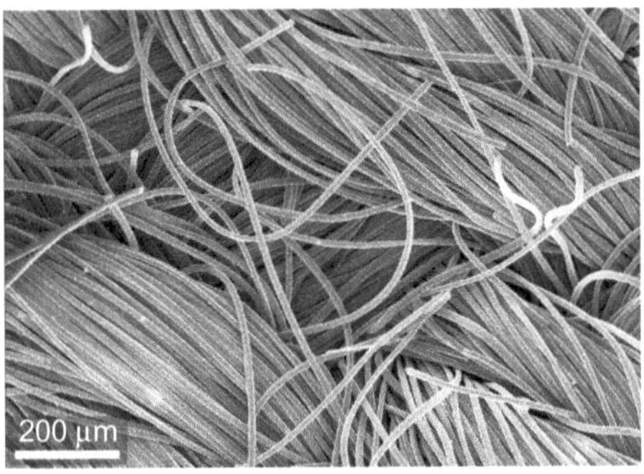

Figure 4.5 SEM image of the active carbon cloth (ACC) support.

reactor technology, in the form of either fixed beds with eggshell catalysts or structured reactors, such as catalyst monoliths, foams, and fibers (Figs. 4.4 and 4.5). These structures enable suppressing the internal diffusion resistance (Cybulski and Moulijn, 2006), which is inevitable in the treatment of large organic molecules originating from biomass. In the design of future biorefineries, it is better to start from the biomass as such—from the macromolecules—and search for technologically and economically feasible solutions instead of trying desperately to recover components from existing chemical pulping processes.

2. REACTION KINETICS

Chemical kinetics, as a discipline, addresses how the reaction rates depend on concentrations, temperature, nature of catalyst, pH, solvent, etc. Chemical kinetics together with other means of studying catalytic reactions, such as spectroscopy of catalysts and catalyst models, molecular modeling, and calculation of the thermodynamics of reactants, intermediates, and products, form the modern basis for understanding chemical processes.

Experimental kinetic investigations are the basis of revealing reaction mechanisms. The following problems can be solved by using a kinetic model:

- choosing the catalyst and comparing the selectivity and activity of various catalysts and their performance under optimum conditions for each catalyst;
- the determination of the main and by-products formed during the process;
- the determination of the optimum sizes and structures of catalyst grains and the necessary amount of the catalyst to achieve the specified values of the selectivity of the product and conversion of the starting material;
- the determination of the short- and long-time stability of the catalyst;
- the determination of the stability of steady states and parametric sensitivity; that is, the influence of deviations of all parameters on the steady-state regime and the behavior of the reactor under unsteady-state conditions;
- the study of the dynamics of the process and deciding if the process should be carried out under unsteady-state conditions;
- the study of the influence of mass and heat transfer processes on the chemical reaction rates and product selectivities as well as the determination of the kinetic region of the process;
- selecting the type of a reactor and structure of the contact unit that provide the best approach to the optimum conditions.

The rates of chemical transformations are very often affected by mass transfer processes. The main mass transfer effects in heterogeneously catalyzed processes are internal diffusion inside the porous catalyst particles/layers, external diffusion in the laminar film surrounding the catalyst particles, and—in multiphase reactors—the gas–liquid or liquid–liquid mass transfer resistance at the phase boundary.

The multiscale approach to kinetic and reactor modeling is visualized in Fig. 4.6. For the understanding of the events occurring at macroscopic (i.e., 1-m) scale, description of not only millimeter (shaped catalyst pellet) and micron scale but microscopic understanding (e.g., nanometer scale—active site) is required.

The method which simplifies the kinetic treatment and reduces the number of parameters in the mathematical model is the so-called steady-state approximation (SSA; also known as quasiSSA or pseudoSSA) originally developed by Bodenstein in 1913.

In general, a reaction is considered to be at steady state, if the concentrations of all species in each element of the reaction space (volume for a

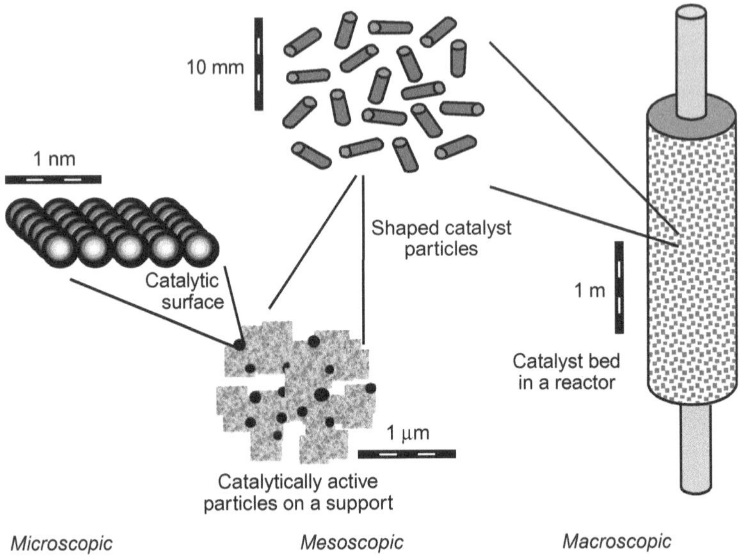

Figure 4.6 Multiscale approach to catalysis—from nanoscale to reactor scale (Chorkendorff and Niemansverdriet, 2003).

homogeneous reaction or surface for a heterogeneous reaction) do not change in time. Such conditions, in general, are fulfilled in open systems, such as continuous tank and tubular reactors and in flow circulation reactors. Grouping the reaction species in two categories: reaction intermediates in one and reactants (substrates) and products in another one, it can be stated, for reaction participants at a steady state, that if a species enters an element of the reaction space, its concentration does not change with time, while for intermediates, the production rate of them is equal to the consumption rate. Bodenstein proposed that there are intermediates in chemical reactions that are present in "inferior" amounts, for example, in much less concentrations than the major species in the mechanism. If this condition is met, the rate of change of the concentration of the intermediate X can be considered negligible. The application of the SSA to heterogeneous catalysis does not require that the surface concentrations are low but implies that they do not essentially change with time.

Another frequently used approximation in catalytic kinetics is the quasiequilibrium approximation. Now imagine that one step in a multistep catalytic process is much slower than the other ones. In such a case, the other steps can be considered as being close to the equilibrium, as their forward and backward rates will be close to each other, for example, r_+/r_- approaches 1. The quasiequilibrium approach, utilized very often, limits the description of catalytic kinetics as it discards transient processes. It is, however, frequently used to obtain simplified and useful expressions for reaction rates. Typically, the surface reaction rate is assumed to be rate limiting, whereas the adsorption and desorption steps are presumed to be rapid. The approach should, however, be supported by experimental evidence obtained by measurements of the reaction and adsorption rates. In the case of complex reactions, several steps are possible and a rigorous treatment is required comprising a multistep rate control, as the quasiequilibrium approach cannot be applied.

2.1. Classical rate equations: Uniform catalyst surfaces

The rate equations depend heavily on the particular chemical case of interest. Some common features in the biomass transformation reactions are of notable interest. Typically, the process involves a reaction between a large organic molecule and a much smaller reagent molecule (e.g., oxygen or hydrogen). In this case, the overall rate is often controlled by the surface reaction step, whereas the adsorption and desorption steps are assumed to be rapid. Further assumptions are needed, concerning the details of the reactant adsorption. Both competitive and noncompetitive adsorption models have been proposed and used to describe the kinetics. A noncompetitive adsorption model can be justified by the size difference of the reacting molecules (e.g., Wärnå et al., 2006). A general overview of kinetic models is provided by Salmi et al. (2004) and Murzin and Salmi (2005), where also a semicompetitive adsorption model is discussed. It has been successfully applied to sugar hydrogenation (Mikkola et al., 1999).

Some generalization of classical kinetic models is possible. For a bimolecular and reversible reaction $A+B=C+D$, the rate takes a form

$$R=\frac{\text{kinetic factor}\times\text{driving force}}{(\text{adsorption term})^{n}} \quad [4.1]$$

where the adsorption term includes $K_A c_A$, etc., for molecular adsorption or $(K_A c_A)^{1/2}$ and the power in denominator corresponds to the number of species in the rate-determining steps, while the driving force is $(1-c_C c_D/K_{eq}\, c_A c_B)$. The adsorption of molecules can be of competitive nature, but in cases when the molecules represent very different sizes, such as catalytic hydrogenation of organic components, noncompetitive adsorption is assumed. A rather general rate equation comprising the rate limitation by the surface reaction step as well as both competitive and noncompetitive adsorption can be written as

$$R_j = \frac{k_j\left(\prod c_i^{\alpha i} - \prod c_i^{\beta i}/K_j\right)}{\left(\sum K_k c_k^{\gamma k} + 1\right)^m \left(\sum K_l c_l^{\delta ll} c_l + 1\right)^n} \quad [4.2]$$

For competitive adsorption, either $m = 0$ or $n = 0$. Of course, rate equations of this kind should be derived case by case, starting from a plausible mechanism.

Recently, models with semicompetitive adsorption have been applied in the hydrogenation of aromatics and sugar molecules (Salmi et al., 2004). The principle of semicompetitive adsorption is that the large organic molecules always leave some empty space, that is, vacant interstitial sites, where the smaller molecule (typically hydrogen and oxygen) can adsorb. Compared to the classical model, the semicompetitive adsorption model includes one additional parameter, namely, the maximum coverage of the organic molecule (Cabrera and Grau, 2008a, 2008b).

2.2. Nonuniform catalyst surfaces

The most often used approach to treat catalytic kinetics is the Langmuir model of uniform surfaces (Section 2.1). This concept presumes that all the surface sites are identical and the binding energies of the reactants are the same independent of the surface coverage. The interactions between adsorbed particles may be neglected.

Ideal adsorbed layer is then considered to be similar to an ideal solution with rapid surface diffusion, allowing thus an application of mass action law with the introduction of surface concentrations and concentrations of free sites into rate expressions of elementary steps. Kinetic models based on uniform catalyst surfaces are treated in many standard textbooks (e.g., Boudart and Djéga-Mariadassou, 1984; Butt, 1999; Fogler, 1999; Froment and Bischoff, 1990; Murzin and Salmi, 2005; Smith, 1990).

It is, however, now well established that the mean field assumption breaks down and surface reconstruction frequently occurs (Somorjai, 2000), adsorbed molecules change the structure of the surface layer and catalytic properties (Ertl, 2000), and reaction rates are dependent on spatial arrangement (Lombardo and Bell, 1991). Two different assumptions are generally used for the description of real adsorbed layers: either surface sites are different or there is a mutual influence of adsorbed species. The first case is defined as biographical nonuniformity and the second one as induced nonuniformity. On biographical nonuniform surfaces, a certain distribution of properties is considered. Such a nonuniformity can be either chaotic, when adsorption energy on a given site is independent of the neighbor site, or discrete. However, if in an elementary surface reaction only one adsorbed particle is involved, the difference in these distributions cannot be observed (Temkin, 1979). The physical nature of the biographical (intrinsic or *a priori*) nonuniformity can be attributed to the differences in the properties of the different crystal faces, the occurrence of dislocations, defects, and other disturbances.

An explicit expression for reaction kinetics in case of biographic non-uniform surfaces could be derived only in few cases, which, in fact, has limited the possibility of a wider application of the *a priori* nonuniform surface concept. One of such cases is a two-step sequence, which addresses the fact that in heterogeneous catalysis it is one of many reaction intermediates, which is the most abundant, while all others are present at the surface at much inferior concentration levels. The rate equation for an evenly non-uniform surface for the two-step sequence of the overall reaction $A+B=C+D$ is given by

$$R=\frac{k(c_Ac_B-c_Cc_D/K)}{(k_1c_A+k_{-2}c_D)^m(k_{-1}c_C+k_2c_B)^{1-m}} \quad [4.3]$$

where m varies from 0 to 1, being very often equal to 0.5. In the case of an irreversible reaction, Eq. (4.3) is simplified to a power-law model. The concept of biographically nonuniform surfaces does not allow one to obtain closed analytic solutions for the reaction rate in case of complex kinetics.

2.3. Size-dependent kinetics: Cluster size

The progress in the physical methods of nanomaterials characterization, computational methods, as well as the discovery of the remarkable reactivity of specifically designed nanoparticles (e.g., gold) resulted in the revival of the interest to the problem of structure sensitivity, that is, dependence of the rate on the particle size (Bell, 2003; Henry, 2000; Narayanan and El-Sayed, 2008; Schlögl and Abd Hamid, 2004). Such variations are observed typically in the domain of 2–20 nm. The dependence of the turnover frequency (TOF) per exposed active site as a function of the nanoparticle dimension is given in Fig. 4.7. Different explanations, for example, quantum size effects, metastability of nanostructures under reactions conditions, defect sites, metal–support interface chemistry to name a few, have been advanced in the literature for the explanation of particle size effects. Size-dependent kinetics on a quantitative level is very rarely described in the literature. It is

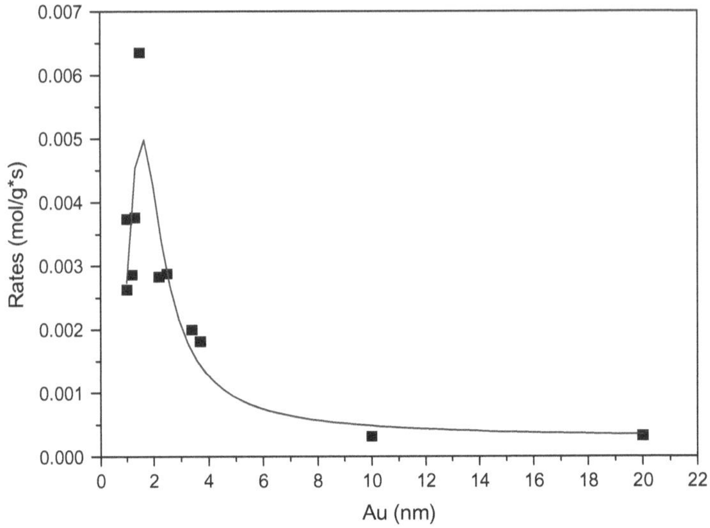

Figure 4.7 Cluster size effect in biomass conversion: oxidation of arabinose on gold (Simakova et al., 2011). (For color version of this figure, the reader is referred to the online version of this chapter.)

interesting to note that recently size-dependent adsorption of nanocrystal surfaces has been discussed (Jiang and Lu, 2008; Zhang et al., 1999).

The explanation for size-dependent adsorption utilized by Jiang and Lu (2008) is based on the application of Laplace–Young equation, which takes into account changes in the interface free energy, namely, the increase in the chemical potential of the active phase due to excessive surface energy and an increase in the internal Laplacian pressure with the particle size decrease. According to the model of Jiang and Lu (2008), the equilibrium adsorption constant increases as the material size decreases, which is in agreement with the experimental data (Zhang et al., 1999) of adsorption of some organic acids on anatase.

Thermodynamic analysis of the effect of the nanoparticle size on the adsorption equilibrium and the reaction rates was performed by Parmon (2007). This kind of approach to the catalytic kinetics, when the rate is governed by adsorption, can explain only a decrease in the TOF as the particle size increases. In order to explain another type of behavior, that is, the TOF increase for large particle sizes, which corresponds to the occurrence of a rate (per mass) maximum, a concept of the increase in the chemical potential upon adsorption with the crystal size increase was utilized by Murzin (2009). Two-step and Christiansen sequences as well as Langmuir–Hinshelwood mechanisms were considered (Murzin, 2009), showing that for catalytic reactions over nanoparticles not only the rates but also the reaction orders can be different from those obtained for large nanoclusters. Comparison with experimental data for arabinose oxidation to arabinonic acid on supported gold catalysts was utilized to explain the nanoparticle size effect on the kinetics (Fig. 4.7).

According to the cluster size kinetics approach, the following rate expression (reaction A+B=C+D, per catalyst weight) can be obtained for the two-step sequence (Parmon, 2007)

$$R(r)=\frac{k(c_A c_B - c_C c_D/K)e^{(2\alpha-1)\Delta\delta'/r}}{(k_1 c_A + k_{-2}c_D)e^{\alpha\Delta\delta'/r} + (k_2 c_B + k_{-1}c_C)e^{(\alpha-1)\Delta\delta'/r}}\frac{3}{r\rho} \qquad [4.4]$$

where ρ is the density of catalytic phase, α is the Polanyi parameter, r is the metal cluster radius, and $\Delta\delta'$ reflects both intrinsic (i.e., excess of surface energy with particle size increase) and induced (i.e., excess of surface energy due to external stress exerted by adsorbed molecules) changes in the chemical potential. For an irreversible two-step sequence, in case of $\Delta\delta' > 0$, for large values of the nanoparticle size, a first-order behavior in reactant A follows. If, in addition, $k_2 c_B \gg k_{-1} c_C$, a zero order with respect to B is observed. On the contrary, a small cluster size leads to zero order in A and first order in reactant B, showing the dependence of kinetic regularities with the metal cluster size. It is thus important during scale-up from the laboratory to pilot and production scale that the materials remain essentially the same of the same or similar size of active clusters; otherwise, not only TOF but also the kinetic regularities might differ substantially.

2.4. Size-dependent kinetics: Reactant size

Historically, the theory of catalytic kinetics was developed based on large-scale processes, such as ammonia and methanol synthesis, or production of sulfuric acid. Later on, heterogeneous catalysis started to be widely used in the field of classical organic chemistry, for example, for the production of fine chemicals. In general, the kinetic analysis is limited to the cases when it is assumed that a given surface intermediate needs only one site for adsorption, neglecting totally the possible multisite adsorption (Frennet and Hubert, 2000). Moreover, in the catalysis involving complex organic molecules with different functional groups, not only the number of sites but also the mode of adsorption is an important issue. The treatment of kinetics of hydrogenolysis (Bernas et al., 2009) and hydrogenation (Cabrera and Grau,

2008a, 2008b; Salmi et al., 2004) on uniform catalyst surfaces involves the concepts of multicentered species.

The concept of semicompetitive adsorption of a large molecule and a smaller one implies that the maximum coverage of the large molecule is less than one, and thus some interstitial sites always remain accessible for the smaller molecule (such as hydrogen). For example, for irreversible catalytic hydrogenation of a large organic molecule (A) with molecular hydrogen (H), the following rate equation can be derived for the rate control of the surface reaction:

$$R = \frac{kK_AK_Hc_Ac_H\alpha((1-\alpha)K_Ac_A+1)}{((1-\alpha)K_AK_Hc_Ac_H+K_Ac_A+K_Hc_H+1)^2} \qquad [4.5]$$

where α is the maximum coverage of A (<1). For the limit case $\alpha=1$ and the standard Langmuir–Hinshelwood model is obtained from Eq. (4.5). Generalizations of this concept are presented in Salmi et al. (2004) and Cabrera and Grau (2008a, 2008b).

2.5. Modeling of catalyst deactivation

The majority of catalysts used in biomass conversion are subject to deactivation, for example, deterioration of the activity with an increasing operation time. The time scale of deactivation depends on the type of process, and it can vary from a few seconds to months and years. A typical example of catalyst deactivation in a continuous fixed bed reactor is displayed in Fig. 4.8, which shows hydrogenation of arabinose to arabitol on supported ruthenium particles. The reactant conversion decreases quite rapidly in time-on-stream but approaches a limit activity.

The main causes of deactivation in heterogeneous catalysis can be poisoning, fouling, thermal degradation (sintering, evaporation) initiated by the often high temperature, mechanical damage, and corrosion/leaching by the reaction mixture. Catalyst poisoning is defined as deactivation by strong adsorption of, usually, impurities in the feed. Poisoning depends upon adsorption strength of such poisons relative to the other species competing for catalytic sites. Adsorbed poisons may not only block active sites but also change the electronic or geometric structure of the surface as well. Surface restructuring by poisons can also take place. The poison may affect the nearest neighbor metal atoms and possibly its next nearest neighbor atoms and modify their abilities to adsorb and/or dissociate reactant molecules. Fouling is associated with surface covering by a certain deposit, which is

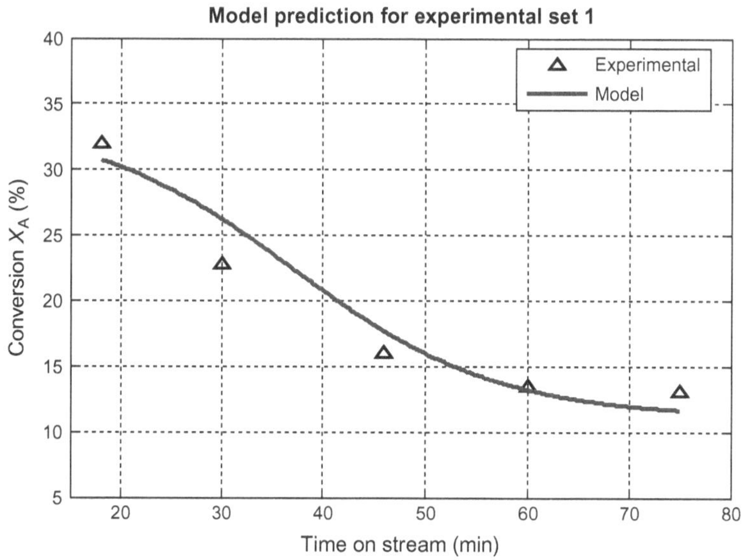

Figure 4.8 Catalyst deactivation in arabinose hydrogenation over supported ruthenium in a laboratory-scale trickle bed reactor: axial dispersion model with catalyst deactivation (Durante, 2011). (For color version of this figure, the reader is referred to the online version of this chapter.)

quite often hydrogen-deficient carbonaceous material (i.e., coke). Coking makes the active sites inaccessible.

Catalyst leaching is extremely important, particularly, for liquid-phase reactions. Often the metals leached are responsible for the high catalytic activity in some reactions, which are considered to be heterogeneously catalyzed, but are in fact homogeneous. Leaching is particularly a problem in oxidation catalysis owing to the strong complexation and solvolytic properties of the oxidants (H_2O_2, RO_2H) and/or the products (H_2O, ROH, RCO_2H, etc.).

Mechanical deactivation is due to strong stresses of packed catalysts beds during start-ups, shut-downs, and catalyst regeneration. Thermal degradation is a physical process leading to catalyst deactivation because of sintering, chemical transformations, evaporation, etc. Sintering is the loss of catalyst active surface due to crystallite growth of either the support material or the active phase.

Due to the industrial importance of deactivation, various kinetic models that account for the deactivation have been advanced. Probably, the most frequently used approach is based on different empirical and semiempirical equations. However, increasingly, the nature of deactivation is considered as a constituent part of the reaction scheme. This approach provides more possibilities for elucidating deactivation mechanisms, which should be an essential part of any new catalyst development. In the simplest case, the catalytic activity is proportional to the number of active sites N_T, intrinsic rate constant, and the effectiveness factor. In the general principles of kinetic modeling, the reaction rate is defined as a function of the intensive variables in the system, typically the concentrations of the components involved and the temperature. In a deactivating system, the initial intrinsic rate on the fresh catalyst is $r_0 = f_1(c_i, T)$. The rate over a deactivated catalyst is $r_t = f_2(c_i, T)$. The activity is defined as the current reaction rate divided by the initial intrinsic rate $a = r_t/r_0$. By carrying out experiments in a fixed bed reactor, the activity changes are observed with time-on-stream. The initial rate r_0 remains unknown, and one has to extrapolate to zero time to obtain this value. This is not always an easy task, particularly for rapid deactivation, because of the influence of the starting up procedure of the experiment.

The catalyst activity is often described by an empirical equation, such as

$$\frac{da}{dt} = -f(c_i, T)\phi(a) \qquad [4.6]$$

where $\phi(a)$ is a deactivation function. Different equations have been proposed in the literature for the deactivation function, for instance, a power-law one $\varphi(a) = a^m$ as well as reversible expressions. Often, the catalyst deactivation is expressed in terms of time-in-stream, but, more correctly, the deactivation function has to be expressed in terms of the deactivating agent itself: the coke precursor or the poison, which means that the amount of coke (or poison) on the catalyst site should be known. The determination of a rate equation for the formation of the coke precursor is thus an integral part of the kinetic study of the process.

Knowledge of the deactivation mechanism deactivation can be explicitly taken into account (Sandelin et al., 2006). For a reaction mechanism,

$$\begin{array}{c} \quad 1 \quad\ \ 2 \quad\ \ 3 \\ A+^* \leftrightarrow A^* \leftrightarrow B^* \leftrightarrow B+^* \\ \downarrow 4 \\ C^* \end{array}$$

where A and B are the reactants, C is the coke, step 2 is reversible, step 4 is irreversible and steps 1 and 3 are in equilibria, the coverage of vacant sites and the rates are

$$\theta_V = \frac{1-\theta_{C*}}{1+K_Ac_A+K_Bc_B}$$

$$r_2 = \frac{(k_{+2}K_Ac - k_{-2}K_Bc_B)(1-\theta_{C*})}{1+K_Ac_A+K_Bc_B}$$

$$r_4 = k_{+4}K_Ac_A\frac{1-\theta_{C*}}{1+K_Ac_A+K_Bc_B}$$

Deactivation can be treated using the general framework of the theory of complex reactions, simply considering deactivation as an independent route leading to coke on the catalyst surface.

It should be mentioned that the analysis of deactivation kinetics is almost exclusively based on the concept of ideal adsorbed layers and lateral interactions or intrinsic inhomogeneity are very seldom taken into account. An attempt to model deactivation on *a priori* nonuniform surfaces for a catalytic reaction proceeding *via* a two-step mechanism with two additional steps, accounting for reversible and irreversible deactivation, was presented by Pogorelov et al. (1983). It required a rather tedious derivation of an equation for the catalyst activity in the region of medium coverage. Intuitively, it is clear that within the framework of any approach based on intrinsic nonuniform surfaces, more complicated reaction mechanism would impose severe restrictions on the derivation of kinetic equations and activity functions. More recently, several aspects of deactivation kinetics over induced nonuniform surfaces were considered by Murzin and Avetisov (1998). When deactivation is not very strong, the activity may exhibit a maximum as a function of time when interactions are of attractive nature. When the deactivation is relatively strong, the nonuniform nature of the surface does not manifest itself, resulting in activity profiles almost independent on the values of interaction parameter.

3. POROUS CATALYST STRUCTURES—REACTION, DIFFUSION, AND CATALYST DEACTIVATION

In any catalytic system, not only chemical reactions *per se* but also mass and heat transfer effects should be considered. For example, mass and heat transfer effects are present inside the porous catalyst particles as well as at the surrounding fluid films. In addition, heat transfer from and to the catalytic reactor gives an essential contribution to the energy balance. The core of modeling of catalytic reactors is the catalyst particle, and the description of simultaneous reaction and diffusion in the pores of the particle should

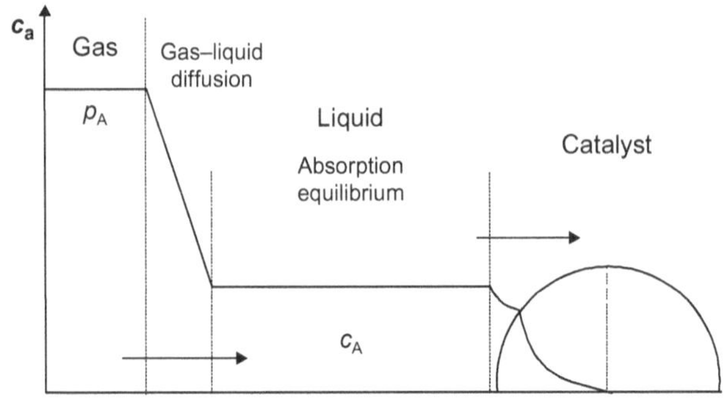

Figure 4.9 Mass transfer processes in three-phase systems.

be accounted for. These effects are completely analogous to reaction–diffusion effects in liquid films appearing in gas–liquid systems.

Gas–liquid diffusion can be essential in homogeneous and enzymatic reactions if the catalyst is dissolved in the fluid phase, while one of the reactants has to be first dissolved. Such a type of diffusion is also present in heterogeneous catalysis for three-phase systems (three-phase catalytic hydrogenations or oxidations) as illustrated in Fig. 4.9.

Physical transport processes can play an especially important role in heterogeneous catalysis. Besides film diffusion on the gas/liquid boundary, there can also be diffusion of the reactants (products) through a boundary layer to (from) the external surface of the solid material and additionally diffusion of them through the porous interior to (from) the active catalyst sites. Heat and mass transfer processes influence the observed reaction rates. The intrinsic rates of catalytic processes follow the Arrhenius law, while the mass transfer effects mask such a pronounced dependence, decreasing the apparent activation energy. The diffusion and mass transfer coefficients display a less pronounced temperature dependence.

3.1. Internal mass transfer

Reaction, diffusion, and catalyst deactivation in a porous catalyst layer are considered. In a porous catalyst particle, the reacting molecules must first diffuse through the fluid film surrounding the particle surface. They then diffuse into the pores of the catalyst, where the chemical reactions take place on active sites. The formed product molecules, of course, need to follow the opposite diffusion path. The diffusion resistances in the fluid film around the particle, as well as inside the particle, are the reason for the fact that the concentrations of the reactant molecules inside the particle are lower than those in the main bulk of the fluid. The result, in the case of most common reaction kinetics, is that the reaction rates inside the pores assume lower values than what would be expected with the concentration levels of the main bulk.

Let us study a component i that diffuses into or out from a catalyst particle. The species i has the flux N_i (mol m^{-2} s^{-1}) in an arbitrary position in the particle. Often the flux N_i of species diffuses into or out from a catalyst particle can be described with sufficient accuracy with the aid of the concentration gradient of the component (dc_i/dr) and its effective diffusion coefficient (D_{ei}): $N_i = -D_e(dc/dx)$. In the calculation of the diffusion in a porous media, the effective diffusion coefficient $D_e = D(\varepsilon/\tau)$ is applied because the cross-section available for diffusion is smaller than the geometric cross section (porosity ε) and the catalyst has irregular pore structure (expressed via tortuosity τ) as illustrated in Fig. 4.10. The approach is called random pore model. More advanced pore models are discussed by Koci et al. (2010).

Reaction, diffusion, and catalyst deactivation in a porous catalyst layer can be considered in detail as follows. A general model for mass transfer and reaction in a porous particle with an arbitrary particle geometry can be written as follows:

$$\frac{dc_i}{dt} = \varepsilon_p^{-1}\left(r_i\rho_p f(r) - r^S\frac{d(N_i r^S)}{dr}\right) \quad [4.7]$$

where the component generation rate (r_i) is calculated from the reaction stoichiometry,

$$r_i = \sum_j v_{ij} R_j a_j \quad [4.8]$$

where R_j is the initial rate of reaction j and a_j is the corresponding activity factor; $f(r)$ is the radial distribution function of the active sites: $f(r) = 1.0$ for conventional catalyst pellets but an increasing function for eggshell catalysts. The form of the function can be determined experimentally with inductive current plasma analysis, and various mathematical functions can be fitted to the data, for instance, $f = f_0(r/R)^\alpha$.

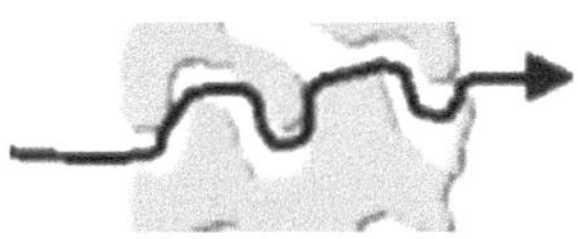

Figure 4.10 Illustration of porosity and tortuosity.

The catalyst activity factor (a_j) is time dependent. Several models have been proposed, depending on the origin of catalyst deactivation, that is, sintering, fouling, or poisoning. The following differential equation can represent semiempirically different kinds of separable deactivation functions,

$$\frac{da_j}{dt} = -k'_j\left(a_j - a_j^*\right)^n \quad [4.9]$$

where k'_j is the deactivation parameter and a'_j is the asymptotic value of the activity factor—for irreversible deactivation, $a_j^* = 0$. Depending on the value of the exponent (n), the solution of Eq. (4.3) becomes

$$a_j = a_j^* + \left(a_{0j} - a_j^*\right)e^{-k_j t} \quad [4.10]$$

and

$$a_j = a_j^* + \left((a_0 - a_*)^{1-n} + k'_j(n-1)t\right)^{1/(1-n)}, \quad n \neq 1 \quad [4.11]$$

Some special cases of Eq. (4.4) are of interest: for irreversible first ($n=1$)- and second ($n=2$)-order deactivation kinetics, we get $a_j = a_{0j}e^{-k_j t}$ and $a_j = a_{0j}/\left(1 + a_{0j}k'_j t\right)$, respectively.

In case that the effective diffusion coefficient approach is used for the molar flux, it is given by $N_i = -D_{ei}(dc_i/dr)$, where $D_{ei} = (\varepsilon_p/\tau_p)D_{mi}$ according to the random pore model. A further development of Eq. (4.1) yields

$$\frac{dc_i}{dt} = \varepsilon_p^{-1}\left(\rho_p f(r)\sum \nu_{ij}R_j a_j + D_{ei}\left(\frac{d^2 c_i}{dr^2} + \frac{s}{r}\frac{dc_i}{dr}\right)\right) \quad [4.12]$$

The boundary conditions of Eq. (4.5) are

$$\frac{dc_i}{dr} = 0 \quad \text{at}\, r = 0 \quad [4.13a]$$

and

$$D_{ei}\left(\frac{dc_i}{dt}\right)_{r=R} = k_{Li}(c_i - c_i(R)) \quad \text{at}\, r = R \quad [4.13b]$$

where c_i is the bulk-phase concentration of the component. The initial condition at $t=0$ is $c_i(r) = c_i$, that is, equal to the bulk-phase concentration. The model presented here is valid for a monodisperse distribution of catalyst particle sizes; it is, however, a straightforward task to expand the model to comprise a distribution of catalyst particle sizes, as shown by Leveneur et al. (2009).

The molecular diffusion coefficients in liquid phase can be estimated from the correlations provided by Wilke and Chang (1955) for organic solutions and by Hayduk and Minhas (1982) for aqueous solutions, respectively. An extensive comparison of the estimation methods for diffusion coefficients is provided by Wild and Charpentier (1987). The solvent viscosities needed in the correlations were obtained from the empirical equation based on the experimental data, $\ln \mu = A + B/T$, where the parameters A and B are experimentally determined constants. The gas-phase component solubility in the organic reaction mixture can be obtained from correlations collected by Fogg and Gerrard (1992), for example, $\ln(x_G^*) = A + B/T$, where x_G^* is the mole fraction of dissolved gas in equilibrium and A and B are experimentally determined constants.

3.2. External mass transfer

External mass transfer effects are expected to play an important role in the conversion of biomass, particularly, in continuous reactors, as the long residence times are needed for many of the reactions listed in Table 4.1. Liquid–solid and gas–liquid mass transfer coefficients are of particular importance.

The liquid–solid mass transfer coefficient can be estimated from correlations based on Sherwood (Sh), Reynolds (Re), and Schmidt (Sc) numbers,

$$Sh_i = 2 + aRe^{\alpha} Sc_i^{\beta} \quad [4.14]$$

where $Sh = k_{Li} d_p / D_i$ and $Sc = \mu/(\rho D_i)$. If the values of Re and Sc numbers are known, then the Sherwood number can be calculated and consequently the mass transfer coefficient can be determined.

Calculation of the Reynolds number depends on the reactor system. For stirred tanks, the Reynolds number is calculated by using the turbulence theory of Kolmogoroff,

$$Re_d = \left(\frac{\varepsilon d}{\nu^3}\right)^{1/3} \quad [4.15]$$

If there is only transfer of mass from the bulk to the external surface of the catalyst, the analysis of the mass transfer coefficient can be done as follows. The Sherwood number is $Sh = k_{Li} d / D_i$ containing, besides the diffusion coefficient, also d (the characteristic length) and the mass transfer coefficient. For fixed bed reactors, a lot of correlations have been proposed for gas–liquid and liquid–solid mass transfer coefficients. They are reviewed in the article of

Al-Dahhan et al. (1997). For fixed bed reactors, the classical experimental approach is to determine the dependence of the reactant conversion on the liquid or gas flow rate.

For vigorously stirred slurry reactors, several correlations have been proposed, relating the Sherwood number to the Reynolds and Schmidt numbers. Data collected by Temkin (1977) for various slurry reactors, where high agitation velocities have been used, indicate that the Sherwood number and mass transfer coefficient can be described by the following simplified equation:

$$Sh = 1.0 Re^{1/2} Sc^{1/3} \quad [4.16]$$

$$k_{LS} = \left(\frac{\varepsilon D^4 \rho}{\mu d_{b/p}{}^2} \right)^{1/6} \quad [4.17]$$

where ε denotes the specific mixing power, D is the diffusion coefficient, ρ is the solvent density, μ is the solvent viscosity, and $d_{b/p}$ is the bubble or particle diameter. The crucial issue is the evaluation of the dissipated energy (ε).

A common test to verify if mass transport controls a catalytic reaction in three-phase systems is to vary the rate of agitation. Agitation of a reaction mixture is usually done by stirring, although reactors in the form of shakers are also used for laboratory purposes. The mass transfer coefficient depends on the energy dissipation, the latter is a function of mixing power W, often related to the stirring speed according to $W = K\rho_{liquid} n^3 d^5{}_{(stirrer)}$, where n is the stirring speed, K is a constant, and $d_{stirrer}$ is the stirrer diameter. The dependence of the mass transfer coefficient on stirring speed can be established as k_{LS} is proportional to $n^{1/2}$. If the rate is independent on the agitation efficiency at sufficiently high stirring speed, then it is conventionally assumed that the mass transport effects are minimized.

However, the energy dissipation has a complex behavior, depending on the selection of solvent, stirring rate, liquid volume, and design of the reactor internals. Experiments with a laboratory-scale slurry reactor (Hájek and Murzin, 2004) demonstrated that the dissipation power at high stirring rates in fact does not follow the square root dependence of mass transfer coefficient on the stirring speed. The energy of dissipation can also depend on the liquid load of an autoclave or a shaking reactor. This shows the apparent danger of establishing the kinetic region based exclusively on the results of experiments at various stirring speeds, as the independence of the catalytic activity of the agitation could be explained simply by the fact that a higher stirring speed or shaking frequency does not necessarily influence the specific mixing power. Therefore, experimental verification of the mass transfer effects should be combined with mathematical modeling.

4. REACTOR MODELING AND SIMULATION ASPECTS

Reactor models for three-phase fixed beds and batch reactors are considered here, as these reactors have the highest potential in the moderate-temperature treatment of biomass. As the aim is to describe transient phenomena, the reactor model is dynamic, including the accumulation of substances in different phases. The mass balances included in the reactor model comprise the gas phase, the liquid phase, and the solid catalyst phase. Mass transfer between the gas and liquid phases as well as between the liquid bulk and the catalyst surface is included in the model. The chemical reactions were presumed to take place exclusively on the solid catalyst surface.

An important factor in the description of tubular reactors is the flow pattern. General reasoning and numerical simulations show that the best flow pattern to reveal transient phenomena is plug flow (see Section 1), but in practice, this is not always reached. Therefore, it is important to incorporate deviations from the plug flow in the fixed bed model. The concept of axial dispersion was selected here to describe the nonideal flow pattern, mainly because of the simplicity: only one parameter, the Peclet number is needed to describe deviations from plug flow.

4.1. Continuous fixed beds

4.1.1 Gas- and liquid-phase components

The flows, dispersion effects, and interfacial fluxes are illustrated in Fig. 4.11. The balance equation for a component in the gas-phase volume element can be written as follows:

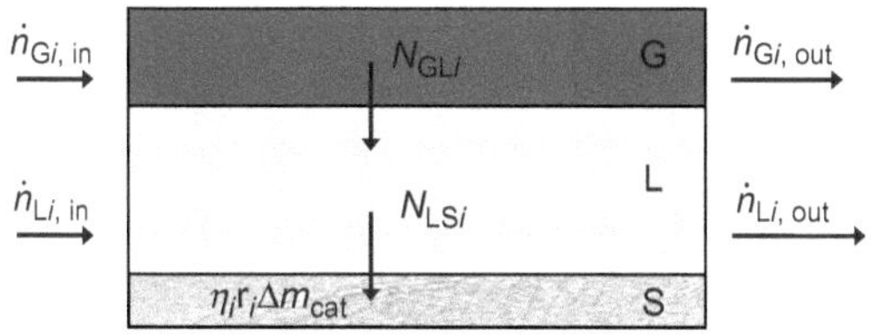

Figure 4.11 The flows and fluxes in a dynamic three-phase fixed bed with axial dispersion. (For color version of this figure, the reader is referred to the online version of this chapter.)

$$\dot{n}_{Gi,\text{in}} + \left(-D_{G,\text{in}}\frac{dc_{Gi}}{dl}A\right)_{\text{in}} = \dot{n}_{Gi,\text{out}} + \left(-D_{G,\text{out}}\frac{dc_{Gi}}{dl}A\right)_{\text{out}} + N_{GLi}A_{GL} + \frac{dn_{Gi}}{dt} \quad [4.18]$$

The mass balance equation for a component in the bulk phase of the liquid is described in an analogous way:

$$\dot{n}_{Li,\text{in}} + \left(-D_{L,\text{in}}\frac{dc_{Li}}{dl}A\right)_{\text{in}} + N_{GLi}A_{GL} = \dot{n}_{Li,\text{out}} + \left(-D_{L,\text{out}}\frac{dc_{Li}}{dl}A\right)_{\text{out}} + N_{LSi}A_{LS} + \frac{dn_{Li}}{dt} \quad [4.19]$$

The symbols are explained in the Notation. By denoting the differences between the inlet (in) and outlet (out) terms by $\Delta(\,)$ and assuming constant values for the axial dispersion coefficients (D_G and D_L), the following general forms of the mass balances are obtained after letting the volume element to shrink and recalling that $dV = Adl$ for a reactor with a constant diameter:

$$\varepsilon_G\frac{dc_{Gi}}{dt} = -\frac{1}{A}\frac{d\dot{n}_{Gi}}{dl} + D_G\frac{d^2c_{Gi}}{dl^2} - N_{GLi}a_{GL} \quad [4.20]$$

$$\varepsilon_L\frac{dc_{Li}}{dt} = -\frac{1}{A}\frac{d\dot{n}_{Li}}{dl} + D_L\frac{d^2c_{Li}}{dl^2} + N_{GLi}a_{GL} + N_{LSi}a_{LS} \quad [4.21]$$

The further treatment of the mass balances depends on the models used for the interfacial fluxes (N_{GL} and N_{LS}) as well as the assumption concerning the flow velocities. The molar flow terms in the above equations can be described with the concentrations and the fluid velocity (w_F),

$$\frac{1}{A}\frac{d\dot{n}_{Fi}}{dl} = \frac{d(w_F c_{Fi})}{dl} \quad [4.22]$$

where F = G or F = L. For a constant fluid velocity, an explicit expression with respect to the concentration is obtained.

If Fick's law and the two-film theory are used in the description of the gas–liquid flux (N_{GL}), the following expression is obtained:

$$N_{GLi} = \frac{c_{Gi} - K_i c_{Li}}{K_i/k_{Li} + 1/k_{Gi}} \quad [4.23]$$

The liquid–solid flux is coupled to the chemical reactions on the catalyst surface. The steady-state assumption for the outer surface of the catalyst particle gives

$$N_{LSi}\Delta A_{LS} + \eta_i r_i \Delta m_{cat} = 0 \qquad [4.24]$$

where the effectiveness (η_i) factor includes the effect of internal diffusion resistance in the pores of the catalyst particle. The details of the modeling of the internal diffusion and the effectiveness factor for porous particles are treated in many references and are thus discarded here. Transient studies for revealing mechanistic details should preferably be carried out in the absence of the internal diffusion limitation, which implies that in such case $\eta_i = 1$ for all components in the above equation and the concentrations of the liquid-phase components are the same throughout the porous particle.

The balance equation for the catalyst particle is divided by the volume element (ΔV) and it becomes

$$N_{LSi}a_{LS} + \eta_i r_i \rho_B = 0 \qquad [4.25]$$

If Fick's law is applied to the liquid–solid (LS) flux for each component,

$$N_{LSi}a_{LS} = k_{LSi}(c_{Li} - c'_{Li})a_{LS} \qquad [4.26]$$

where c'_{Li} denotes the concentration at catalyst surface. The surface concentrations are used in the calculation of the rate (r_i). Thus, the concentrations (c'_{Li}) have in the general case to be calculated iteratively from the balance equation

$$k_{LSi}(c_{Li} - c'_{Li})a_{LS} + \eta_i r_i (c'_{L})\rho_B = 0 \qquad [4.27]$$

where c'_L denotes the vector of the unknown concentrations in the vicinity of the catalyst surface.

An important special case is the operation under kinetic control. In this case, k_{LSi} gets large values, which implies that $c_{Li} \rightarrow c'_{Li}$ and c'_{Li} can be replaced by c_{Li} in the rate expression. Furthermore, $\eta_i = 1$ for rapid internal diffusion. The liquid-phase balance equation becomes now

$$\varepsilon_L \frac{dc_{Li}}{dt} = -\frac{d(c_{Li}w_L)}{dl} + D_L \frac{d^2 c_{Li}}{dl^2} + N_{GLi}a_{GL} + r_i \rho_B \qquad [4.28]$$

An ultimate simplification is obtained for the case that the organic liquid-phase component is nonvolatile ($N_{GLi} = 0$) and the liquid flow rate (w_L) is constant. The dimensionless length coordinate is introduced: $z = l/L$, where L is the total reactor length. Furthermore, the space time is defined as $\tau_L = L/w_L$. The balance equation becomes finally

$$\frac{dc_{Li}}{dt} = -(\varepsilon_L \tau_L)^{-1} \frac{dc_{Li}}{dz} + (Pe_L \varepsilon_L \tau_L)^{-1} \frac{d^2 c_{Li}}{dz^2} + r_i \rho'_B \quad [4.29]$$

where $\rho'_B = \rho_B/\varepsilon_L$ and Pe_L is the liquid-phase Peclet number defined by $Pe_L = w_L L/D_L$.

As revealed by the above equation, the balance contains just two adjustable hydrodynamic parameters, $\tau_L \varepsilon_L$ and Pe_L. For an inert tracer used for the determination of the RTD, the rate term (r_i) is zero in the above equation. Various kinds of boundary conditions have been suggested for the axial dispersion model, the most common ones being Danckwerts' closed boundary conditions, which were used here too.

4.1.2 Adsorbed surface species

For the adsorbed compounds on an ideal, homogeneous surface, the mass balance equations are simply given by the expression

$$\frac{dn_j^*}{dt} = m_{cat} r_j^* \quad [4.30]$$

where n_j^* is the amount of the surface species in the mass element of the catalyst. By introducing the fractional coverages ($\theta_j = c_j^*/c_o^*$), the balance becomes

$$\frac{d\theta_j}{dt} = \omega r_j, \quad \omega = \left(c_0^*\right)^{-1} \quad [4.31]$$

4.1.3 Boundary and initial conditions

The classical approach of the boundary conditions of the axial dispersion model presumes that the axial dispersion is initiated at the reactor inlet and the concentration gradients disappear at the reactor outlet.

$$c_{0i} = c_i - \frac{D_a}{wL} \frac{dc_i}{dz} \quad \text{at} \, z = 0 \quad [4.32]$$

$$\frac{dc_i}{dz} = 0 \quad \text{at} \, z = 1 \quad [4.33]$$

where $D_a/(wL)$ is the reciprocal Peclet number, $Pe = wL/D_a$.

In many practical cases, a departure from the classical treatment is, however, preferred. The outlet boundary condition always suppresses the concentration gradient to zero, which leads to an artificial bending of the concentration profile at $z = 1$ for cases, where the degree of axial dispersion is small and the conversion of the component is far from complete. For such a case, it is better to introduce a finite concentration gradient at the outlet. This can be done either with special functions, as discussed by Salmi and Romanainen (1995), or with an open boundary condition or by introducing a new boundary point beyond the reactor outlet; plug flow is usually assumed to prevail in this imaginary point.

The initial conditions arise from the actual experimental reality. Typically, the concentration profile in the reactor is known in the beginning of the experiment, $c_i = c_i(z)$ at $t = 0$. The changes in the inlet concentrations are also known during the experiment, $c_{0i} = c_{0i}(t)$ at $t \geq 0$.

4.1.4 Residence time distributions

The residence time distribution function $E(t)$ is obtained from pulse experiments—the transient response is obtained from Eq. (4.29) with $r_i = 0$. The residence time distribution function $E(t)$ is calculated from the experimentally recorded responses, after which the $E(t)$ function is obtained by the integration of $E(t)$. The experimental functions are compared to the theoretical ones; the expressions of $E(t)$ and $F(t)$ obtained from the analytical solution of the dynamic, nonreactive axial dispersion model with closed Danckwerts' boundary conditions can be used in the comparison. This approach gave the basic hydrodynamic parameters ($\tau_L \varepsilon_L$ and Pe_L).

4.2. Batch and semibatch operations

The reactors used for transforming biomass are often semibatch and batch reactors equipped with necessary auxiliary equipment for mixing, heating, and cooling as well as for distillation or evaporation of solvents or light reaction products. The reactions typically take place in the liquid phase in the presence of a homogeneous or heterogeneous catalyst. In this section, we summarize the reactor models considering only mass balances.

4.2.1 Liquid phase

The following fundamental assumptions were made in deriving the balance equations for the liquid phase:

- Plug flow and axial dispersion or complete backmixing prevail in gas and liquid phases.
- No reactions take place in the gas phase.
- The reactions proceed in the liquid phase and/or solid phase.

$$\frac{dc_{Li}}{dt} = -\frac{d(c_{Li}w_L)}{dl} + \frac{N_{SLi}a_{SL}}{\varepsilon_L} + \frac{N_{GLi}a_{GL}}{\varepsilon_L} + r_i \quad [4.34]$$

Typically, w_L is approximately constant, that is,

$$\frac{d(c_{Li}w_L)}{dl} = w_L\frac{dc_{Li}}{dl} \quad [4.35]$$

Some special cases of interest are listed below:

a. Nonvolatile component: $N_{GLi}=0$.
b. No homogeneous reactions in liquid phase: $r_i=0$.
c. No heterogeneous reaction: $N_{SLi}=0$
d. High degree of dispersion; complete backmixing:
$d^2c_{Li}/dl^2=0$, $d(c_{Li}w_L)/dl$ is replaced by $(c_{0Li}w_{0L}-c_{Li}w_L)/L$, which is equal to $\left(c_{0Li}\dot{V}_{0L}-c_{Li}\dot{V}_L\right)/V$.
e. No flow, infinite dispersion (or perfect mixing): $d^2c_{Li}/dl^2=0$.

4.2.2 Gas phase

Analogously, the balance equation for the gas phase is of the form

$$\frac{dc_{Gi}}{dt} = \mp\frac{d(c_{Gi}w_G)}{dl} + D_G\frac{d^2c_{Gi}}{dl^2} + \frac{N_{GLi}a_{GL}}{\varepsilon_G}, \quad [4.36]$$

where (−) denotes cocurrent and (+) is for countercurrent flow.

Special cases of interest are:

a. Nonvolatile component, the balance is omitted $(c_{Gi}=0)$, $(N_{GLi}=0)$.
b. High degree of dispersion; complete backmixing.
$d^2c_{Gi}/dl^2=0$, $d(c_{Gi}w_G)/dl$ is replaced by $(c_{Gi}w_G-c_{0Gi}w_{0G})/L$, which is equal to $\left(c_{0Gi}\dot{V}_{0G}-c_{Gi}\dot{V}_G\right)/V$.

4.3. Numerical strategies

The axial dispersion model presented earlier is a dynamic one-dimensional model with respect to the spatial coordinate. For nonisothermal cases with strong heat effects, a two-dimensional model is needed, while in isothermal cases, a one-dimensional model is sufficient. This model consists of a set of coupled second-order partial differential equations (PDEs). The strategy for solving the problem is to discretize the reactor coordinate (z) and integrate the ordinary differential equations (ODEs) in time (t). The number of ODEs is the number of components multiplied by the number of discretization points. Mainly, two kinds of methods are applied on the discretization, the method of lines or orthogonal collocation on finite elements. In using finite differences, expressions are needed for first and second derivatives.

The derivatives which originate from the plug flow term should be described with backward differences, whereas central differences should be used for first and second derivatives originated from diffusion and dispersion. For the first derivative (the plug flow term), for instance, the five-point backward difference formula can be used (Schiesser, 1991). The set of coupled ODEs arise an initial value problem which can be solved numerically. There are several methods for solving ODEs numerically. The ODEs are very often stiff because of the large time-scale difference in reaction rates and deactivation rate. Additionally, the discretization increases the stiffness of the system. For stiff systems, the backward difference method (Gear, 1971) is very suitable and often used. The methods for stiff cases can be compared with nonstiff methods; an often used method for this is the Adams–Moulton method. The application of the model solution is illustrated in connection of the case studies.

5. MODEL SIMULATION AND PARAMETER ESTIMATION METHODOLOGY

In most cases, the kinetic parameters for the reactions involved in the transformation of biomass are unknown and they cannot be predicted from a theory. Therefore, the estimation of kinetic parameters from experimental data is a key issue in the reaction engineering of biomass components.

5.1. Experimental design and parameter estimation

The definition of the objectives is extremely important in kinetic studies, especially how comprehensive the experimental program will be. A heuristic approach includes first collecting all the possible data during the experiments as a function of the parameters which are deemed to be important, that is, concentrations, temperature, pressures, pH, catalyst concentration, volume, etc. Then the rate constants are estimated by regression analysis, and the adequacy of the model is judged based on some criteria (such as, residual sums and parameter significance, which will be discussed further). If the researcher is not satisfied, then additional experiments are performed, followed by parameter estimation and preferably simulations outside the studied parameter domain. The latter procedure provides the possibility to investigate the predictive power of the model. The kinetic model is then gradually improved and the experimental plan is modified, if needed. This process continues until the researcher is satisfied with the kinetic model.

When designing experiments, one should also take care of possible experimental errors, decide which is the minimum set of experiments, and define the necessity of repeated experiments. The minimum number of experiments that must be performed is related to the number of important independent variables that can affect the experiment and to the experimental precision. The frequently used strategy is to carry out first experiments at the extremes (maximum and minimum setting) of the range of the controlled variables. Traditionally, in designing kinetic experiments, it is believed that kinetic data should be obtained under isothermal conditions. The reactants are preheated separately and brought together for the desired reaction to proceed. However, as solid catalysts are involved, in practice the procedure becomes more complicated. With which component to combine the catalyst during preheating and how to pretreat the catalyst; does the pretreatment procedure contribute to catalyst deactivation; are typical questions raised when handling with solid catalysts together with gases and liquids. In practice, many experiments are carried out in such a way that all of the reactants are simply brought together with the catalyst, the mixture is heated up to the reaction temperature, and while heating, the mixture undergoes, at least, some reaction under the rising temperature gradient to the set temperature. Nowadays, this is not a technical problem, as the temperature profile can be followed online and stored on a PC. The temperature dependence of the kinetic model is incorporated in the temperature dependencies of the rate and equilibrium constants, for example, according to the laws of van't Hoff and Arrhenius. All data sets are merged together and the parameters are determined by nonlinear regression, which will be addressed in the following sections. The mutual correlation between the parameters can be suppressed.

A further challenge appears if the reactor equipment itself is nonisothermal. A typical example of this kind is a catalytic fixed bed. If the reaction has considerable heat effects, for instance, highly exothermal reactions, it is difficult to maintain completely isothermal conditions in a tubular fixed bed. Instead, a hot spot appears. Later on, the temperature profiles can be incorporated in the reactor model. Principally, two possibilities exist: to describe the reactor with a complete nonisothermal model including the energy balance or to utilize the experimental temperature profiles more or less directly. The first approach is theoretically correct, but it has the drawback that the heat transfer parameters of the reactor tube need to be determined.

This introduces an additional uncertainty factor to the model, and bottom line, it is not relevant as the kinetic and thermodynamic parameters are of primary interest. Thus, it is better to interpolate in the experimental temperature field, for example, by fitting an empirical model for each temperature profile, $T=f$(reactor length), using, for instance, a polynomial model. The laws of van't Hoff and Arrhenius are incorporated in the thermodynamic and kinetic parameters; all data sets are merged together with the empirical temperature profiles and nonlinear regression is used to obtain kinetic parameters. Nonisothermal data, although rather seldom used in kinetic experiments, indeed, can be used for kinetic analysis. The crucially important issue is to obtain primary data with a large enough concentration and temperature domain, and to record the temperature profiles frequently and precisely. If these requirements are fulfilled, the parameter estimation exercise can be safely carried out under nonisothermal conditions.

The models in chemical kinetics usually contain a number of unknown parameters, whose values should be determined from experimental data. Regression analysis is a powerful and objective tool in the estimation of parameter values. The task in regression analysis can be stated as follows: the value of the dependent variable (y) is predicted by the model; a function (f) contains independent variables (x) and parameters (p). The independent variable is measured experimentally, at different conditions, that is, at different values of the independent variables (x). The goal is to find such numerical values of the parameters (p) that the model gives the best possible agreement with the experimental data. Typical independent variables are reaction times, concentrations, pressures, and temperatures, while molar amounts, concentrations, molar flows, and reaction rates are dependent variables. The parameters to be estimated are usually rate and equilibrium constants, sometimes even mass and heat transfer coefficients. As most models in chemical kinetics and more general in chemical reaction engineering are nonlinear with respect to the parameters, the treatment here is limited to nonlinear regression analysis.

The model equation is written according to $\hat{y}=f(x,p)$; thus, nonlinear regression strictly speaking is valid for algebraic models, but it can be applied to differential models, as well. For differential models, the solution (y) is obtained numerically from the model equation. The objective function to be minimized by regression is defined by:

$$Q(p)=\sum_{i=1}\sum_{s=1}\sum_{t=1}\left[\left(y_{i,s,t}-f\left(x_{i,s,t},p\right)\right)^2\right]w_{i,s,t} \qquad [4.37]$$

where i denotes each component in the reaction mixture, s denotes the different data sets, t refers to the data points, $y_{i,s,t}$ is the experimental value, $f(x_{i,s,t})$ is the corresponding model prediction, and $w_{isj,t}$ is the weight matrix which is used, if experimental scattering varies between different data.

The norm notation in Eq. (4.37) is often used as an abbreviation of the sum of squares:

$$\|y - y_x\|^2 = \sum_{k=1}^{nsets} \sum_{j=1}^{nobs(k)} \sum_{i=1}^{nydata(j,k)} \left(y_{ijk} - y_{x,ijk}\right)^2 \qquad [4.38]$$

where x denotes here the predicted values or in the later case the common mean value. In the sums, *n*set is the number of datasets, *n*obs indicates the number of observations in each set, and *ny*data is the number of observed response components in each dataset at each observation.

The parameter estimation results can be evaluated by a standard statistical analysis, that is, the degree of explanation can be described with the coefficient of determination, the R^2-value.

$$R^2 = 1 - \frac{\|y - \hat{y}\|^2}{\|y - \bar{y}\|^2} \qquad [4.39]$$

In the equation, y is the experimentally observed variable (typically concentration, molar amount, or molar flow), $\hat{y}$ is the prediction of the variable, and $\bar{y}$ is the average of all the data points. This implies that R^2-values approaching 100% are desired.

Various optimization methods exist for the minimization of the objective function (Q). If one has a good initial guess of the parameters, the Levenberg–Marquardt method (Marquardt, 1963) is rapid and efficient, but it typically fails, if the optimization is started far from the minimum. For those cases, the minimization is started with a slower, but a more robust method, such as the simplex algorithm. As the minimum is approached, it is possible to shift to the Levenberg–Marquardt algorithm. Modern software (Haario, 2007) includes this kind of numerical strategy. During parameter estimation, analytical solutions of the reactor models can usually not be obtained; therefore, there is a need to solve algebraic and differential equations by numerical methods.

The numerical solution of the combined reaction–diffusion problem requires a special attention. The PDEs describing the catalyst particle are discretized with central finite difference formulae with respect to the spatial coordinate (Schiesser, 1991). Typically, around 10–20 discretization points are enough for the particle. The ODEs created were solved with respect to time together with the ODEs of the bulk phase. As the system is stiff, the computer code of Hindmarsh (1983) is often used as the ODE solver. In general, the simulations progress without numerical troubles.

5.1.1 Software recommendations

There are several software packages for modeling chemical and biochemical systems. They simulate the kinetics of systems of chemical and biochemical reactions providing tools to fit models to data, estimate parameters, perform simulations and optimizations among other options. In the general purpose software, the task of the user is to develop a mathematical model and rewrite a model in the format expected by the software.

For instance, *Athena Visual Studio* offers an integrated environment for modeling, optimization, estimation, optimal experimental design, and model discrimination of chemically reactive and nonreactive process systems. A user has the possibility to develop models, while the software takes care of the tasks associated with solving the underlying equations and estimating the model adjustable parameters. A convenient graphical user interface is combined with powerful parameter estimation and optimization software, which allows for the analysis of single- and multiresponse experiments, model discrimination, and optimal experimental design. Moreover, several examples are provided illustrating typical cases relevant to chemical reaction engineering. In Athena Visual Studio (2009), the selection of tasks (modeling, estimation, or optimization) is combined with an option to select a variety of mathematical models.

ModEst (from Model Estimation) is a software, which has been designed for parameter estimation of mechanistic mathematical models as well as for experimental design. All the modeling tasks that previously required tailor-made solutions in each project can now be solved in a unified manner in the ModEst environment. ModEst (Haario, 2007) is able to deal with explicit algebraic, implicit algebraic (systems of nonlinear equations), and ODEs. As the standard way to handle PDE systems, the numerical method of lines, which transforms a PDE system to a number of ODEs, is used. In addition, any model with a solver provided may be dealt with as an algebraic system. The basic numerical tools are contained in the well-tested public domain software (Blas, Linpack, Eispack, LSODE). Advanced statistical information including sensitivity data and contour plots is available. Recently, a possibility to evaluate the accuracy of estimated parameters by Markov chain Monte Carlo method is based on the Bayesian approach. The method is incorporated into this software.

6. CASE STUDIES

The methodology which was discussed in previous sections is illustrated for real cases related to components from biomass. Most of the examples concern batch processes, as they are still the dominating ones in biomass conversion.

6.1. Case A: Sitosterol hydrogenation process: From laboratory reactor to plant scale

Cholesterol is found in almost all tissues of animals, particularly in the brain, the spinal column, and gallstones. High level of cholesterol is a leading risk factor for heart diseases. As cardiovascular diseases are one of the leading causes of death, development of functional food to enhance the cholesterol lowering ability of traditional food products is increasing. In Finland, the commercial production of cholesterol-suppressing margarines started in 1995.

Phytosterols have a role in plants similar to that of cholesterol in mammals; for example, they are forming cell membrane structures. More than 40 sterols have been identified, of which β-sitosterol, campesterol, and stigmasterol are the most abundant. These phytosterols and cholesterol share identical ring structures, belonging to the class of 4-desmethylsterols. A less abundant class of related compounds found in plants is the phytostanols. Phytostanols are completely saturated forms of plant sterols and lack the carbon–carbon double bond found in cholesterol and phytosterols. The raw material is derived from vegetable oil processing or wood-pulping processes. In addition to β-sitosterol (ca. 45–47%) (MW = 414.7 g/mol), the feed sterol contains other sterol species, the most frequent of which are campesterol (ca. 25%); stigamsterol (ca. 18%); brassicasterol (ca. 2%); and the target product, sitostanol itself (ca. 2–4%). The sterol feed has a melting point of 135 °C.

Generally speaking, hydrogenation of plant sterols yields the respective stanols by hydrogenating the carbon–carbon double bond and the C=O bond. Some by-products can be formed during the hydrogenation, primarily the ketone and the unsaturated compounds. A proposed route for the hydrogenation of sitosterol and campesterol is sketched in Fig. 4.12 (Wärnå et al., 2006).

It is possible to identify 12 compounds in the reaction scheme; however, campesterol isomer species is not always taken into account if it does not appear in the analysis of the samples. Sitosterol and campesterol were

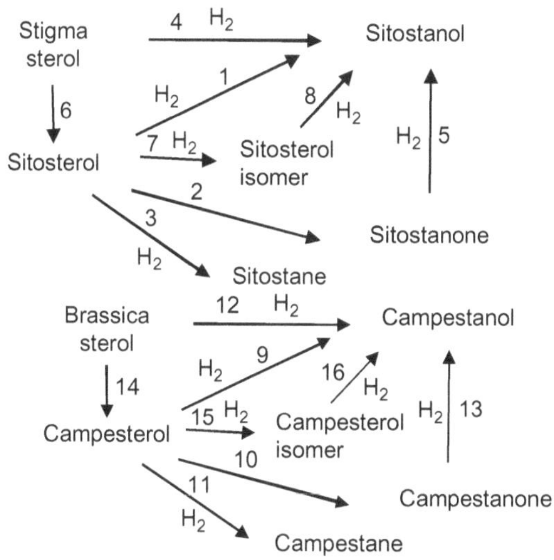

Figure 4.12 Reaction scheme for hydrogenation of sterols.

hydrogenated to sitostanol and campestanol in a batch reactor over 5% Pd/C catalysts (catalyst 1 with the specific surface area = 1010 m^2/g, and catalyst 2 with the surface area = 800 m^2/g Degussa) in an organic solvent (isopropanol). The reactor volume was 100 ml and the liquid volume in the reactor was 500 ml.

A mixture of plant sterols (sitosterol, campesterol, stigmasterol, and brassicasterol) was used as a raw material. Eighteen experiments were performed under different conditions, varying the temperature (70–120 °C) and hydrogen pressure 4–45 bar over a Pd catalyst. The batch reactor was equipped with a heating jacket able to keep the temperature of the reaction fluid constant. The stirring speed was 1000 rpm in all experiments. It was verified that experiments were conducted in kinetic regime, free from the influence of external and internal diffusion. Some initial hydrogenation experiments were performed with stirring speeds from 700 to 1500 rpm. These tests showed that a stirring speed of 1000 rpm is sufficient for the reactor to operate in the kinetic regime. When the system reached the steady state (temperature), reaction was commenced by opening the hydrogen valve. The steady state was set at the time when the temperature inside the reactor was stabilized. Temperature inside the reactor increased just after the reaction was commenced, and then it stabilized again after short time (3 min approximately, depending primarily on the experiment temperature and the mass of catalyst).

Samples were taken along the reaction, and at the same time, temperature and pressure were registered. Hydrogen flow rate was also registered during the experiments. The reaction rate was weakly dependent on the hydrogen pressure. Increased temperature resulted in higher reaction rates. The influence of the vapor pressure of the solvent (1-propanol) had to be taken into account in the calculations, as it lowers the hydrogen partial pressure in the reactor. In the kinetics and reactor modeling, the recorded total pressure in the reactor was corrected by subtracting the vapor pressure of 1-propanol, in order to get the hydrogen pressure. The solubility of hydrogen in sitosterol/1-propanol solution was measured.

The kinetic model was based on the following hypothesis: molecularly adsorbed hydrogen was assumed to react on the catalyst surface with adsorbed organic species. Preliminary analysis of the reaction rate dependence on hydrogen concentration revealed that discrimination between molecular and dissociative hydrogen adsorption cannot be performed in a meaningful way; therefore, molecular hydrogen adsorption was preferred as in principle it gives a possibility to describe first order in hydrogen frequently observed in hydrogenation of carbon–carbon double bonds.

The surface hydrogenation step was assumed to be rate limiting, whereas adsorption and desorption steps were regarded rapid quasiequilibria. Finally, the rate equations were obtained

$$r_j = \frac{k'_j K_A K_{H_2} c_j c_{H_2}}{\left(1+\sum K_i c_i\right)\left(1+K_{H_2} c_{H_2}\right)} \quad [4.40]$$

where index i denotes the components (A-sitosterol, B-sitostanol, C-sitostanone, D-sitostane, E-campesterol, F-campestanol, G-campestanone, H-campestane, I-brassicasterol, J-stigmasterol, K-sitosterol isomer) and index j denotes the reactions (1,...16, Fig. 4.12).

The influence of temperature on the rate and adsorption coefficients can be accounted for with the Arrhenius and van Hoff equations. The rate constants ($k_{0,j}$), adsorption constants ($K_{0,i}$), the activation energies (ΔE_A), and adsorption enthalpies were estimated from the laboratory experiment data by nonlinear regression.

The mass balance for a liquid-phase component (i) in a batch reactor was written as $dc_i/dt = \rho_b r_i$, where r_i is the reaction rate and ρ_B is the catalyst bulk density ($\rho_B = m_{cat}/V_{liq}$). The mass balance is valid in the kinetic regime, that is, in the absence of mass transfer limitations. Hydrogen is present in both the gas and the liquid phases; thus, the mass balance is written as

$dc_H/dt = N_H a_v + r_H \rho_b$, where the flux $N_H a_v$ between the gas and liquid phases is $N_H a_v = k_l a_v (P_G/RT/K - c_H)$. The gas–liquid equilibrium constant K is obtained from gas–liquid solubility data. The proposed kinetic model and the equations of the batch reactor were linked to the MODEST software (Haario, 2007), and the code was compiled and used for the parameter estimation of the data obtained in the experimental work. The rate constants, adsorption coefficients, heats of adsorption, and the activation energies included in the kinetic model were estimated from the laboratory experiments by nonlinear regression with the Levenberg–Marquardt method. The reactor mass balances were solved as a subtask to the parameter estimation with the backward difference method. The parameter estimation routine minimizes the objective function, the sum of square error (Q).

The parameter estimation was performed by applying the estimation procedure directly to the data. The program was executed many times; it has the built-in option to use the last result as a starting condition for the next estimation. Different sets of initial values for the calculation of the model parameters were also tried, finally permitting the convergence of the numerical procedure. The number of components and reactions in the estimation were 11 and 14, respectively, giving 14 rate constants, 14 activation energies, 12 adsorption coefficients, and 12 adsorption energies, altogether 52 adjustable parameters.

Some initial parameter estimations with all 52 parameters revealed that there are too many parameters resulting in large parameter errors. In order to reduce the number of parameters, some assumptions were made, in particular due to the small temperature difference ($\Delta T = 50$ °C) in the experiments, the possible variations for heat of adsorption of hydrogen with temperature were neglected. Furthermore, the adsorption coefficients for sitosterol–campesterol, sitostanol–campestanol, sitonanone–campestanone, sitostane–campestane, and stigmasterol–brassicasterol were set equal to each other which is a reasonable assumption taking into account their structural similarities. With these modifications, the number of parameters was reduced to 36. In order to reduce further the number of parameters, some assumptions were made after initial estimations with all parameters. The reaction rates for reactions 5 and 13 were very low. Therefore, these reactions were considered to be nonsignificant. For reactions 4 and 12 to occur, two double bonds should be hydrogenated simultaneously. In further calculations, they were omitted. The activation energies for the pairs of similar sterols (i.e., sitosterol and campesterol) were set equal to each other ($E_{A9} = E_{A1}$, $E_{A10} = E_{A2}$, $E_{A11} = E_{A3}$, $E_{A12} = E_{A4}$, $E_{A13} = E_{A5}$) also due to structural similarities. The number of kinetic parameters to estimate was now 22. The highest degree of explanation (R^2) after these simplifications was 99.8%.

The kinetic model described very well the concentrations of the reactants and the main products stanols and sterols, although for very minor by-products, stanes and stanones, the fits were not as good as for the main products. The developed kinetic model was used to predict the plant-size reactor behavior. A characteristic feature for a plant-size reactor is the presence of hydrogen mass transfer resistance. In the simulations, only the gas–liquid mass transfer coefficients were fitted.

Experimental data were collected from a plant reactor of 8 m^3 with 25 kg of catalyst. The plant reactor was simulated with the values of kinetic parameters obtained from the laboratory reactor. The simulation of the plant reactor gave the values of the gas–liquid mass transfer coefficient ($k_l a$) equal to 0.34 1/min. For efficiently stirred tank reactors, the typical values are ca. 3–12 1/min[10], demonstrating the low efficiency of gas/liquid mass transfer in the plant reactor. The results displayed in Fig. 4.13 demonstrate the successful application of the adopted approach in process scale-up.

The model can predict the formation of the main products very well (Fig. 4.13). Figure 4.14 shows the calculated concentration of hydrogen in the reaction mixture in the plant conditions. During the first 100 min, the hydrogen concentration is very low because all hydrogen in the liquid phase is immediately consumed by the reaction and thus the gas–liquid mass

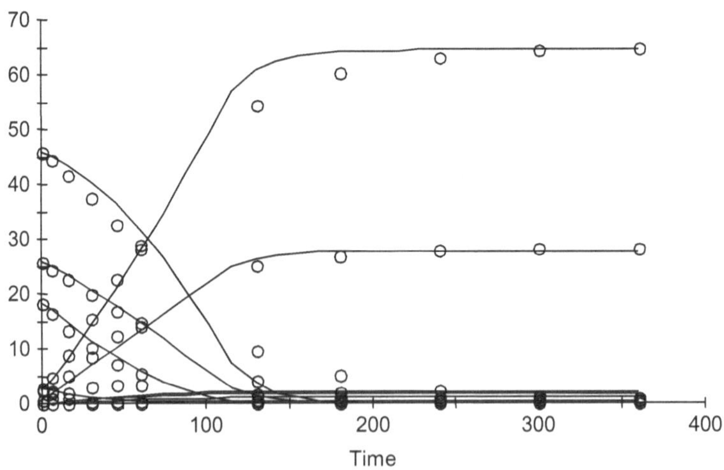

Figure 4.13 Simulation of a plant reactor for sitosterol hydrogenation(4 bar, 70 °C) based on laboratory data. Experimental points (o) from plant reactor and simulated lines.

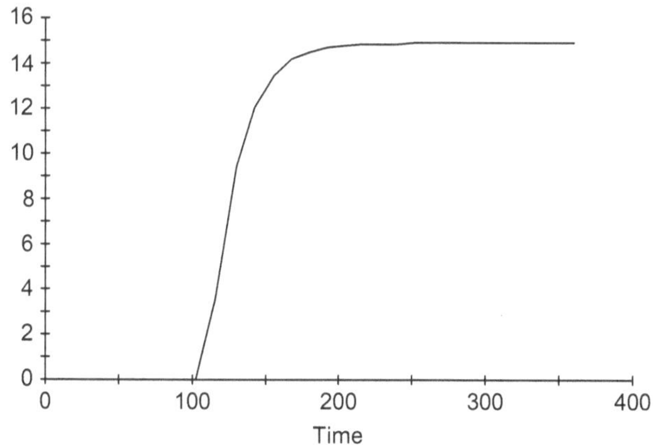

Figure 4.14 Simulation of a plant reactor. Concentration of hydrogen (mol/m^3) in the liquid phase as a function of time (min).

transfer rate controls the overall reaction rate. After 100 min of reaction, less hydrogen is consumed and the hydrogen concentration approaches the saturation level determined by the gas–liquid equilibrium constant.

6.1.1 Catalyst deactivation in sitosterol hydrogenation

Catalytic hydrogenation of β-sitosterol was performed with molecular hydrogen in shaking semibatch reactor vessel, where the hydrogen pressure was kept constant through controlled addition of hydrogen. Besides sitosterol, the raw material contained an analogous phytosterol, campesterol as a minority component as well as a sulfurous poison. The reaction scheme is displayed in Fig. 4.12. In many cases, the superficially very complex reaction scheme displayed in Fig. 4.12 was simplified to two simultaneous reactions, hydrogenation of sitosterol and campesterol to sitostanol and campestanol, respectively.

The liquid-phase components were in batch. 1-Propanol was used as a solvent. The temperature and pressure ranges of the experiments were 60–80 °C and 5.0 bar hydrogen. Thermodynamic calculations revealed that the role of the vapor phase in the system was negligible; thus, the kinetic data could be interpreted as batch reactor data. Samples were taken from the reactor during the experiments by opening the sampling valve. The component concentrations were analyzed by gas chromatography (capillary column, FI detector). Prior to the analysis, the sterol and stanol molecules were derivatized.

Several Pd-based catalysts were screened, such as Pd/C (Sibunit, 4 wt%), Pd/C (Aldrich, 5 wt%), Pd-MCM-41 (5 wt%), Pd-HPS (polystyrene, 5 wt %), Pd-SBA-15NC (5 wt%), and Pd/SiO_2-C. Pd/C (Sibunit, 4 wt%)

turned out to be the best catalyst giving a conversion of sitosterol maximally 95% at a reaction time of 30 min (4 bar and 80 °C). Therefore, this catalyst was selected to systematic kinetic studies which were carried out under diffusion-free conditions, in the kinetic regime.

Extensive deactivation studies were carried out with the Sibunit-based catalyst; for instance, the same catalyst was used in three successive experiments. The results showed that the catalyst activity diminished with about 24% in each experiment. Some typical experimental results are shown in Fig. 4.15.

The mathematical modeling of the hydrogenation process was carried out according to standard principles. The hydrogen concentration was kept constant in each experiment, and the adsorption of hydrogen and the organics were assumed to be of mostly noncompetitive character (Wärnå et al., 2006). Thus, the hydrogen concentration was incorporated in the rate parameters, and equations were directly used in the parameter estimation. A new model was developed for the catalyst deactivation by poisoning. The model equations are briefly summarized below. Reaction step 1 is reactant (A) adsorption, step 2 is surface reaction (hydrogenation of A), and step 3 is poisoning. It can be shown that the following equations are valid for a batch process:

$$\alpha = \frac{c_{0P}}{\rho_B c_0}$$

$$\delta = (k_3/k_1)(1 + k_{-1}/k_2)$$

$$\omega = \frac{\rho_B k_1 c_0}{1 + k_{-1}/k_2}$$

$$d(c_A/c_{0A})/dt = \frac{-\omega(c_A/c_{0A})\left(1 - \alpha\left(1 - (c_A/c_{0A})^{\delta}\right)\right)}{1 + K_A c_A}$$

where A and P refer to the reactant and the poison, respectively. The total concentration of sites is initially c_0. The model parameters were ω, α, and δ as well as K_A, where K_A is the adsorption parameter of sitosterol. The contribution of the adsorption parameter of campesterol was neglected, as its concentration was low compared to that of sitosterol (Fig. 4.15).

The mass balance equations were solved with respect to each liquid-phase concentration during the estimation of the kinetic parameters. The objective function was minimized by a hybrid simplex–Levenberg–Marquardt algorithm implemented in the software Modest (Haario, 2007). The differential equations, that is, the mass balances, were solved numerically during the parameter estimation by using a robust backward difference algorithm for stiff differential equations.

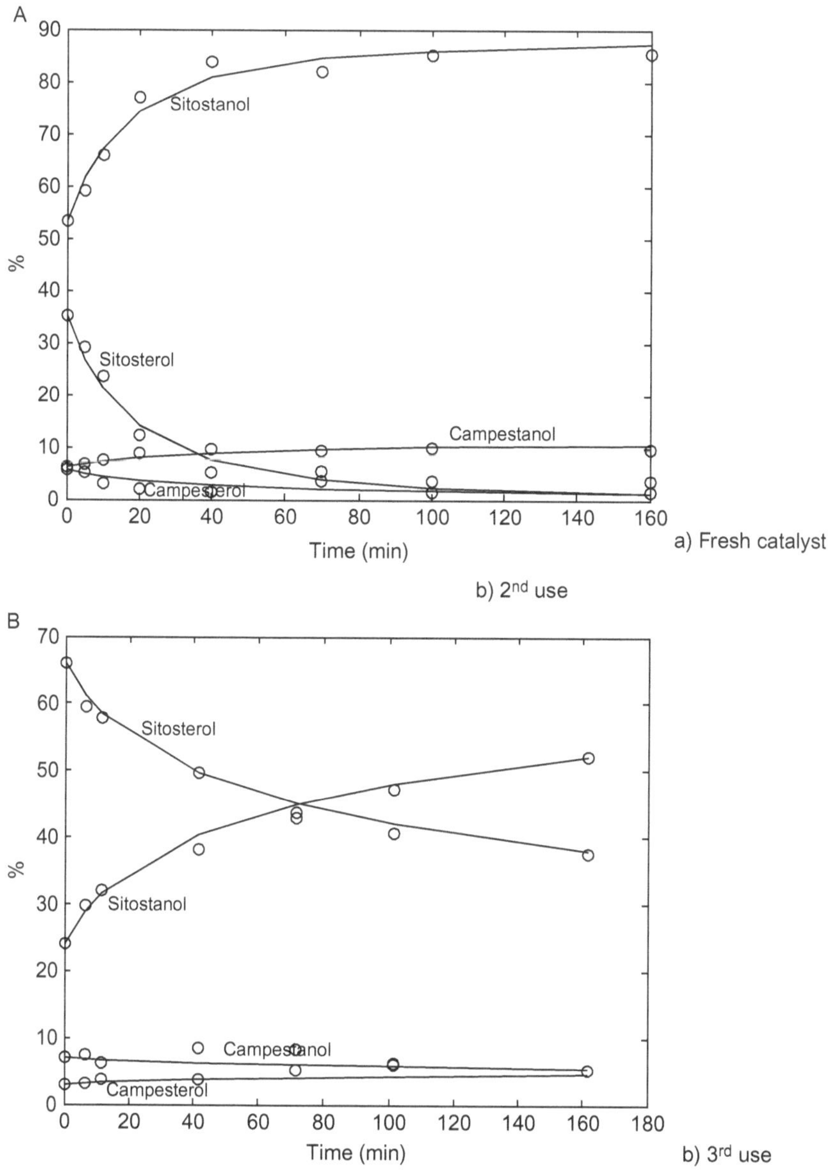

Figure 4.15 Fit of model to experimental data with deactivation model (80°C and 5 bar). Continuous lines represent the model prediction.

The parameters obtained were checked with different methods. The standard statistical analysis gave the standard errors of the parameters as well as their mutual correlations (the correlation matrix). The identification of each parameter was studied by preparing sensitivity plots, in which the objective function was plotted as a function of one of the parameters, while the other parameter values were kept fixed to the value, which gave the objective function minimum.

The fit of the model to the experimental data is demonstrated in Fig. 4.15. As revealed by the figure, the new modeling concept works very well for the hydrogenation of sitosterol and campesterol. The fit of the model to the experimental data is almost perfect, and the parameter estimation statistics is reasonable (Fig. 4.15). Parameter ω contains the initial concentration of sites available (c_0), which decreases in successive runs. For the successive experiments (1, 2, 3, see Fig. 4.15), the ω-parameter for the main reaction, sitosterol hydrogenation, changed with the ratios $\omega_1/\omega_3=4$ and $\omega_2/\omega_3=2$, that is, the catalyst activity declined with 50% after each experiment. The ratio of ω parameters for sitosterol and campesterol hydrogenation ($\omega_{\text{sitosterol}}/\omega_{\text{campesterol}}$) was around 2.2 in all the successive experiments. This result is expected, as the poison occupies sites, which are equally available for both sterols. The parameter δ obtained a quite high value, which reflects the fact that the poison adsorption rate is high, thus giving a high apparent reaction order for the hydrogenation process.

Catalytic reaction kinetics coupled to catalyst poisoning kinetics was considered theoretically starting from elementary steps on the catalyst surface. The kinetic model was applied to batch and plug flow reactors (fixed beds), which were assumed to operate in the kinetic regime. Semianalytical models were derived, which describe the coupled reaction kinetics and poisoning effects. Model simulations revealed the effects of the parameters on the kinetic behavior. For batch reactors, the catalyst poisoning typically increases the experimentally observed reaction order. The simple exponential function $\exp(-k_d t)$ is not correct for the description of poisoning in batch reactors. The model is expected to be very useful in systems originating from renewable sources, as the presence of catalyst poisons can be verified by the modeling approach proposed.

6.2. Case B: Hydrolysis of polysaccharides—From arabinogalactan to monomers

In the acid hydrolysis of arabinogalactane (AG), the two main monosaccharides, arabinose and galactose, (Fig. 4.16) were obtained at temperatures up to 100 °C, pH = 1, 2, and 3, and the initial AG concentration $c_{AG}=0.5$, 2.5,

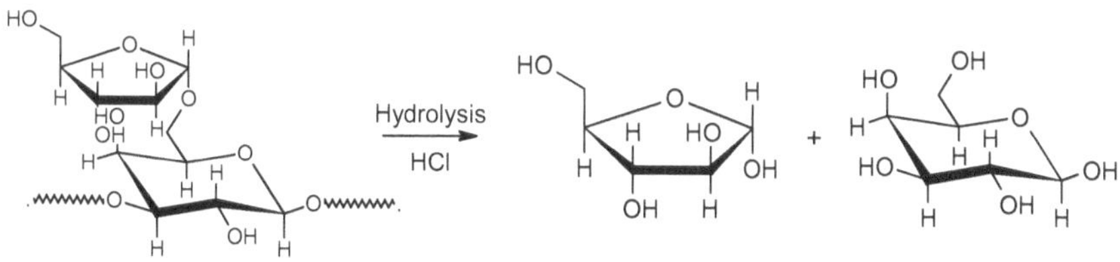

Figure 4.16 Acid-catalyzed hydrolysis of a hemicellulose (arabinogalactane).

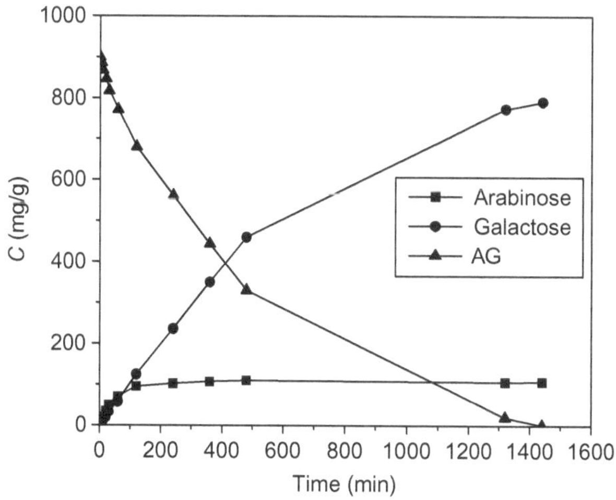

Figure 4.17 Typical kinetic curves of AG hydrolysis to arabinose and galactose at 90 °C and pH 1 (Kusema et al., 2010).

and 5.0 mass%. Typical kinetic curves of the AG hydrolysis and the released monomers are shown in Fig. 4.17. A complete conversion of AG was achieved after 1400 min. It can be noted from the results that arabinose was released totally within 120 min, whereas there was a continuous gradual increase in the galactose concentration during 1400 min. As can be seen, the monomers, arabinose, and galactose were stable at the above-mentioned conditions. There was a sharp decrease in AG molar weight within the first 120 min, which is explained by the presence of the two simultaneous reactions until complete release of arabinose followed by the gradual hydrolysis of galactose residue.

The results of AG hydrolysis at 90 °C and different pH values are shown in Fig. 4.18. The residual AG concentrations were plotted against the reaction times for the different reaction conditions. In each of the experiments, the pH value of the solution was adjusted to 1–3, respectively. The experimental results show a very strong dependence of the hydrolysis rates on the acid concentration (pH). A complete conversion of AG was achieved at pH

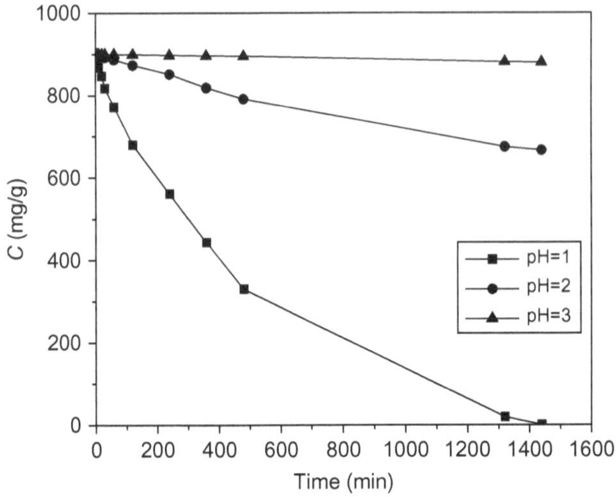

Figure 4.18 Effect of pH on the AG hydrolysis (Kusema et al., 2010).

1 after 1400 min with only 25% and less than 1% conversion at pH 2 and 3, respectively. At pH 1, AG was completely hydrolyzed to monomers achieving the maximum quantities of free sugar units from the polysaccharide. In the case of pH 2, there was almost complete release of arabinose from the polysaccharide with small amounts of galactose monomers leaving a partially hydrolyzed AG. The slowest hydrolysis rates of AG with hardly any galactose sugar units being cleaved from the chain were observed at pH 3. Under these conditions, the obtained results suggest that the lowest acid concentration was not sufficient to hydrolyze the glucosidic bonds of AG. Thus, we conclude here that it is easy to hydrolyze the arabinose unit from the side chains of AG followed by the release of galactose from the main chain.

As expected, the reaction rate increased with temperature. The hydrolysis rate of AG at temperatures lower than 70°C was very slow. At 80°C, only complete release of arabinose was achieved, but partially hydrolyzed galactose residue was left. The conversion of AG was 43%. A complete conversion of AG to monomers was achieved at 90 and 100°C. It can be noted that after the hydrolysis at 100°C, traces of degradation products such as furfural were observed. For this reason, the temperature for AG hydrolysis shall not exceed 100°C.

A kinetic model based on the experimental results was developed. The hemicellulose structure was considered to be composed of two distinct fractions, one that is relatively easy to hydrolyze and the other more difficult. The arabinose residue, because of its easier accessibility might release faster, while the galactose units would be produced more slowly. The kinetic model was therefore set into the reaction scheme described. Applying the first-order kinetics with respect to the functional groups for both reactions, the following rate equations, can be obtained:

$$r_1 = k_1 c_H c_{A,AG}$$
$$r_2 = k_2 c_H c_{G,AG}$$

where r_1 and r_2 are the rates of AG hydrolysis to arabinose and galactose, respectively, k_1 and k_2 are the kinetic rate constants, $c_H = [H^+]$ is the acid concentration, and $c_{A,AG}$ and $c_{G,AG}$ denote the arabinose and galactose fractions in the polysaccharide, that is, easy to hydrolyze (arabinose) and more difficult to hydrolyze (galactose) hemicellulose fractions, respectively. The temperature dependence of the acid hydrolysis was modeled using the Arrhenius equation.

The mass balances of the functional groups (i) in the constant-volume batch reactor are written in the following form:

$$dc_i/dt = r_i$$

The concentrations of arabinose (A) and galactose (G) units liberated in the hydrolysis are thus obtained from the differential equations

$$dc_A/dt = k_1 c_H c_{A,AG}$$
$$dc_G/dt = k_2 c_H c_{G,AG}$$

where $c_{A,AG}$ and $c_{G,AG}$ give the concentrations of the arabinose and galactose units in AG. The following total balance is valid for the monomer units

$$c_A + c_{A,AG} = c_{0A,AG}$$
$$c_G + c_{G,AG} = c_{0G,AG}$$

which implies that the total amounts of the arabinose and galactose units remain constant. This assumption is justified, as the amounts of the degradation products were negligible. After inserting the relations into the differential equations, the expressions

$$dc_A/dt = k_1 c_H \left(c_{0A,AG} - c_A\right)$$
$$dc_G/dt = k_2 c_H \left(c_{0G,AG} - c_G\right)$$

are obtained.

On the other hand, it should be remembered that the initial amounts correspond to the final amounts of arabinose and galactose in the case of complete hydrolysis, that is, $c_{0A,AG} = c_{\infty A}$ and $c_{0G,AG} = c_{\infty G}$. These relations are inserted in the mass balances, which are integrated with the initial condition $t=0$, $c_A=0$, and $c_G=0$. The logarithmic functions are obtained:

$$-\ln\left(1-c_A/c_{\infty A}\right)=k_1 c_H t$$
$$-\ln\left(1-c_G/c_{\infty G}\right)=k_2 c_H t$$

The validity of the proposed model can be checked by plotting the right-hand sides of the equations versus the reaction time. Furthermore, division of the above equations gives

$$\ln(1-c_A/c_{\infty A})=(k_1/k_2)\ln(1-c_G/c_{\infty G})$$

that is, a double logarithmic plot $\ln(1-c_A/c_{\infty A})$ versus $\ln(1-c_G/c_{\infty G})$ should give a straight line, if the model is valid. The slope gives the ratio between the rate constants k_1/k_2 which is independent of the acid catalyst concentration (c_H).The plots obtained with different acid catalyst concentrations coincide as predicted by the model. The test plots gave preliminary values for the rate constants.

The final values of the rate constants, along with their temperature dependencies, were obtained with nonlinear regression analysis, which was applied to the differential equations. The model fits the experimental results well, having an explanation factor of 98%. Examples of the model fit are provided by Figs. 4.17 and 4.18. An analogous treatment can be applied to other hemicelluloses.

6.3. Case C: From sugars to sugar alcohols

Hydrogenation of the carbonyl group of a mono- or disaccharide gives the corresponding sugar alcohol (Table 4.1). Sugar alcohols are important ingredients in pharmaceuticals and alimentary products. They have also a great potential as platform chemicals for biorefineries. On paper, this reaction is simple, and it can schematically be written as

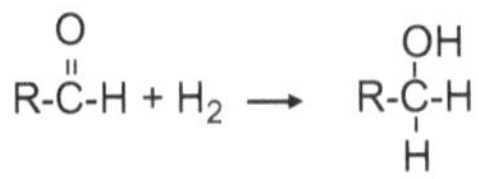

However, the pattern is complicated by several factors. The sugar molecules to be hydrogenated mutarotate in aqueous solutions, thus, coexisting as acyclic aldehydes and ketoses and as cyclic pyranoses and furanoses, reaction kinetics are complicated and involve side reactions, such as isomerization, hydrolysis, and oxidative dehydrogenation reactions. Moreover, catalysts deactivate and external and internal mass transfer limitations interfere with the kinetics, particularly under industrial circumstances.

Catalytic hydrogenation is typically carried out in slurry reactors, where finely dispersed catalyst particles (<100 μm) are immersed in a dispersion of gas and liquid. It has, however, been demonstrated that continuous operation is possible, by using either trickle bed (Gallezot et al., 1998) or monolith technologies. Elevated pressures and temperatures are needed to have a high enough reaction rate. On the other hand, too high a temperature impairs the selectivity of the desired product, as has been demonstrated by Kuusisto et al. (2006). An overview of some feasible processes and catalysts is provided by Table 4.1.

Kinetic experiments were carried out isothermally in autoclave reactors of sizes 300 and 600 ml. The stirring rate was typically 1800 rpm. In most cases, the reactors operated as slurry reactors with small catalyst particles (45–90 μm), but comparative experiments were carried out with a static basket using large catalyst pellets. The HPLC analysis was applied for product analysis. The experimental details are reported in previous papers of our group, for example, by Mikkola et al. (1999) and Kuusisto et al. (2006).

Catalytic hydrogenation of various sugars, such as lactose, maltose, arabinose, galactose, xylose, and mannose, was studied experimentally in a pressurized laboratory-scale batch reactor. Ruthenium on active carbon was used as a catalyst. The main product was always the corresponding sugar alcohol, the selectivity of which typically exceeded 95%. Examples of hydrogenation of arabinose and galactose are shown in Figs. 4.19 and 4.20

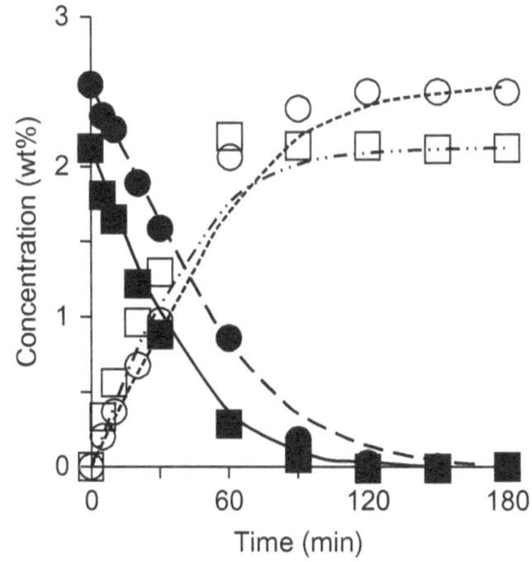

Figure 4.19 Modeling results at 120 °C and 40 bar pressure. Ratio 1:1. ■ L-arabinose, □ L-arabitol, □ D-galactose, ○ D-galactitol. Lines denote model predictions.

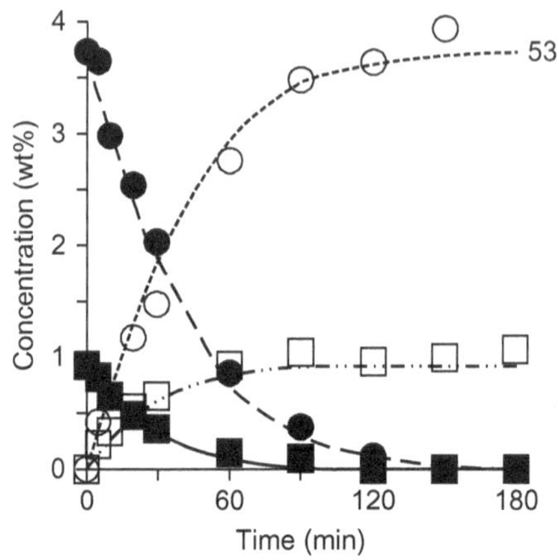

Figure 4.20 Modeling results at 120 °C and 40 bar pressure. Ratio 3.6:1. ■ L-arabinose, □ L-arabitol, □ D-galactose, ○ D-galactitol. Lines denote model predictions.

(Sifontes, 2012). Based on the experimental data, models were developed for simultaneous reaction, diffusion, and catalyst deactivation.

6.3.1 Product distribution analysis—Stoichiometry

Preliminary product distribution plots were prepared based on experimental data representing intrinsic kinetics. The hydrogenation of arabinose and galactose over Ru/C was straightforward: arabinitol and galactitol were the absolutely dominating products with the selectivity close to 100%. The kinetics was described with a Langmuir–Hinshelwood model including rapid competitive reactant adsorption and rate control by the surface reaction between the adsorbed sugar molecule and hydrogen. The kinetic parameters were determined with regression analysis. Examples of the fit of the model are provided by Figs. 4.19 and 4.20 (Sifontes, 2012). The model describes completely the progress of the reaction. The parameters are well defined as shown by the sensitivity plots (Figs. 4.21 and 4.22).

6.3.2 Intrinsic kinetics and diffusion phenomena

Rate expressions based on the principle of an ideal surface, rapid adsorption, and desorption, but rate-limiting hydrogenation steps, were derived. The competitiveness of adsorption and the state of hydrogen on the catalyst surface have been the subject of intensive discussions in the literature (see, e.g., Salmi et al., 2004), but in this case, the simplest possible model was used, noncompetitive adsorption of hydrogen and organics along with molecular adsorption of hydrogen. Previous comparisons (Salmi et al., 2004) have

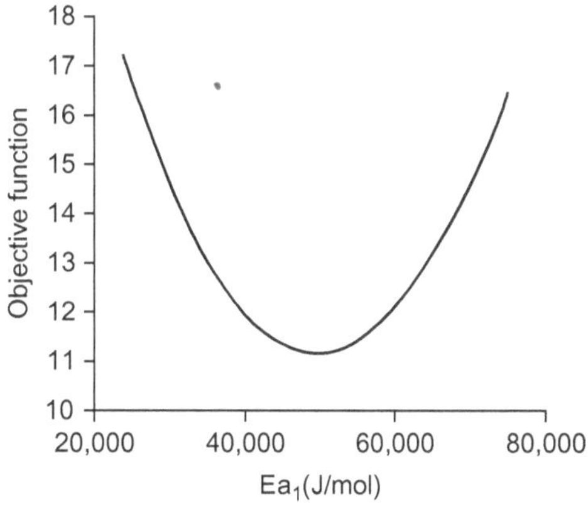

Figure 4.21 Sensitivity analysis for L-arabinose activation energy.

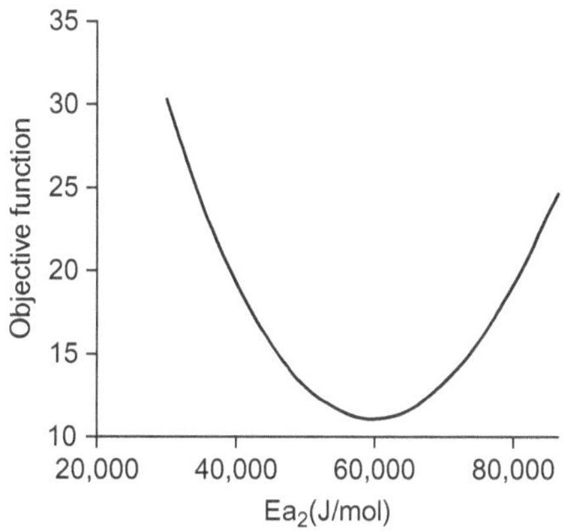

Figure 4.22 Sensitivity analysis for D-galactose activation energy.

shown that this approach gives an adequate description of the sugar hydrogenation kinetics, with a minimum number of adjustable parameters. A thorough comparison of experimental data with modeling results indicated that the model for intrinsic kinetics is sufficient and can be used for simulation of the reaction–diffusion phenomena in larger catalyst particles.

For "small" catalyst particles used in sugar hydrogenation (slurry reactors), one might intuitively assume kinetic control. The simulation results obtained for sugar hydrogenation strongly suggest that this is not the case. The organic molecules are large, having molecular diffusion coefficients of the magnitude 4.0×10^{-9} m^2/s. The porosity-to-tortuosity factor can be rather small ($\ll 0.5$); thus, the effective diffusion coefficient becomes small and internal diffusion resistance becomes an important factor.

The concentration profiles inside the porous catalyst particle were computed by solving numerically the reaction–diffusion model for a porous catalyst layer. Some simulation results are displayed in Fig. 4.23, which clearly reveals the strong diffusion limitation inside the catalyst layer, as the layer thickness (the catalyst particle diameter) exceeds 0.5 mm. On the other hand, if the catalyst layer thickness is 0.2 mm or less, the operation at the kinetic regime, the highest possible hydrogenation rate, is guaranteed. For fixed beds operated in large scale, the catalyst particle diameter should exceed 1 mm, which implies that the process becomes heavily controlled by internal mass transfer as shown by Fig. 4.23. For catalyst particles having suitable diameters for continuous large-scale operation (>2 mm), in order to avoid extensive pressure drop, the internal part of the catalyst is not used because of strong diffusion limitation. Thus, the development of alternative technologies, such as structured reactors, is highly motivated for sugar hydrogenation processes.

In addition, external diffusion resistance is often a crucial factor in sugar hydrogenation: due to inappropriate stirrer design and high viscosity, the risk for lack of hydrogen in the liquid phase is high. A simulations of the hydrogen content in the liquid phase indicated that a complete saturation is achieved at high conversions, while the hydrogen concentration is below saturation in the beginning of the experiment. By increasing the stirring efficiency, the k_L-values can be improved and the role of external mass transfer resistance is suppressed. The hydrogen concentration in the liquid phase plays a crucially important role for the production: in the case of external mass transfer limitation of hydrogen, the isomerization reaction yielding xylulose, which does not need any hydrogen, is favored (Fig. 4.24).

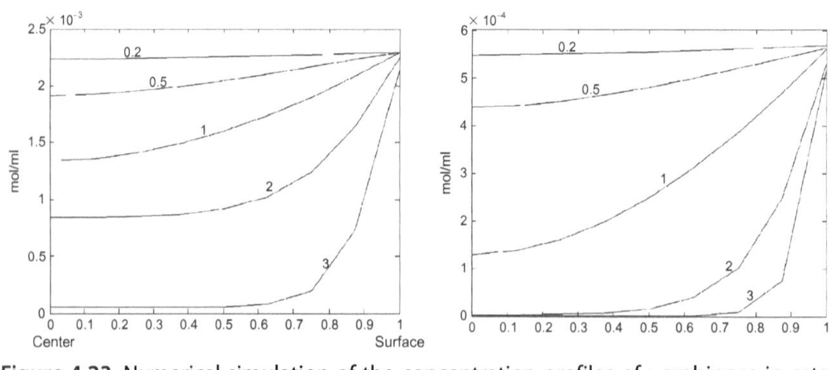

Figure 4.23 Numerical simulation of the concentration profiles of L-arabinose in catalyst layers of different thicknesses (0.2–3 mm). Hydrogenation at 90 °C: thick (50 wt%, left) and dilute (25 wt%, right) solutions of L-arabinose.

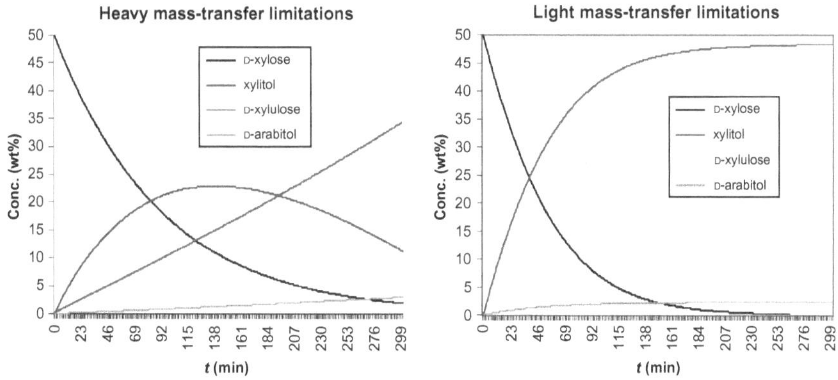

Figure 4.24 External mass transfer resistance—xylose hydrogenation to xylitol and by-products. (For color version of this figure, the reader is referred to the online version of this chapter.)

7. CONCLUSIONS AND PERSPECTIVES

The general principles of chemical reaction engineering, based on the mathematical description of reaction kinetics, mass transfer, and flow patterns in chemical reactors, are applicable on the design of biorefineries. In case that individual molecules originating from biomass undergo chemical transformations, the standard approach of chemical reaction engineering can be used. The recent progress of kinetic modeling has been advantageous for reactions appearing in biomass conversion. Models have been developed, which take into account the cluster size effect and the effect of the size of the adsorbed molecules in catalytic conversions. These kinds of models are already successfully applied to molecules originating from biomass. Molecular modeling can be used to reveal the adsorption stoichiometry, that is, how many active sites on the solid catalyst surface are needed for the adsorption of a particular molecule. This issue is crucially important for the treatment of biomass, as molecules of very different sizes typically coexist and interact on the catalyst surface.

The real challenge is the description of complex mixtures originating from biomass. The reactions of cellulose and hemicellulose have similarities (e.g., hydrolysis of macromolecules as well as transformations of sugar monomers), while the reactions of lignin fractions and various extractives are very different from those of cellulose and hemicelluloses. On the other hand, if a selective fractionation procedure is applied to biomass, the issue becomes easier from a reaction engineering viewpoint: extractives are removed first from woody biomass by leaching with an organic solvent, hemicelluloses can be separated in the next step by hot water extraction, and finally, either cellulose is hydrolyzed or lignin is dissolved.

The dream is that, for instance, the rate constants of the hydrolysis of hemicellulose or hydrogenation of sugar monomers could be related to the structures of the molecules. The future kinetic approach to biomass conversion will be based on the reactivities of functional groups in the modeling of complex mixtures. Advanced lumping techniques will be needed, as they are nowadays applied to hydrocarbon mixtures originating from fossil sources. A lot of physical properties for biomass components should be measured experimentally, such as densities, viscosities, heat conductivities, and polarities of solutions containing biomass. These properties are needed in the modeling of chemical reactors, particularly diffusion phenomena in bulk liquids and porous catalyst layers.

As many of the molecules appearing in biomass are large, the role of internal diffusion is prominent as heterogeneous catalysts are used for chemical transformations. This requires very precise mathematical modeling of simultaneous reaction and diffusion in porous structures. Two issues are of particular importance: the pore structure of the catalyst and the design of the catalyst layer itself. In order to enhance the diffusion of the reactive molecules to the active sites of the catalyst, catalysts with large pores should be preferred. Molecular modeling is of great help to indicate, whether a reactant can fit to the pore structure of a proposed catalyst.

In order to diminish the diffusion resistance, to improve the effectiveness factor, the catalyst layer should be thin. Three principal alternatives exist: the conventional slurry reactor technology with finely dispersed catalyst particles (diameter <0.1 mm), the use of eggshell-type catalyst particles (with the active catalyst layer being in the vicinity of the outer surface of the particle), or to use structured reactors, such as monoliths, fibers, and solid foams. The advantage of structured reactors is that the benefit of high catalyst effectiveness and low pressure drop can be combined, which is not the case with eggshell catalysts. The conventional slurry technology enables a very high catalyst efficiency, but it is very much limited to batch and semibatch reactors. On the other hand, for the manufacture of high-value products, structured catalysts can be implemented in batch reactors, too, as proposed by the group of Moulijn (De Lathouder et al., 2006). The decision on the reactor technology should be based on precise experimental measurements and detailed mathematical modeling, as illustrated below.

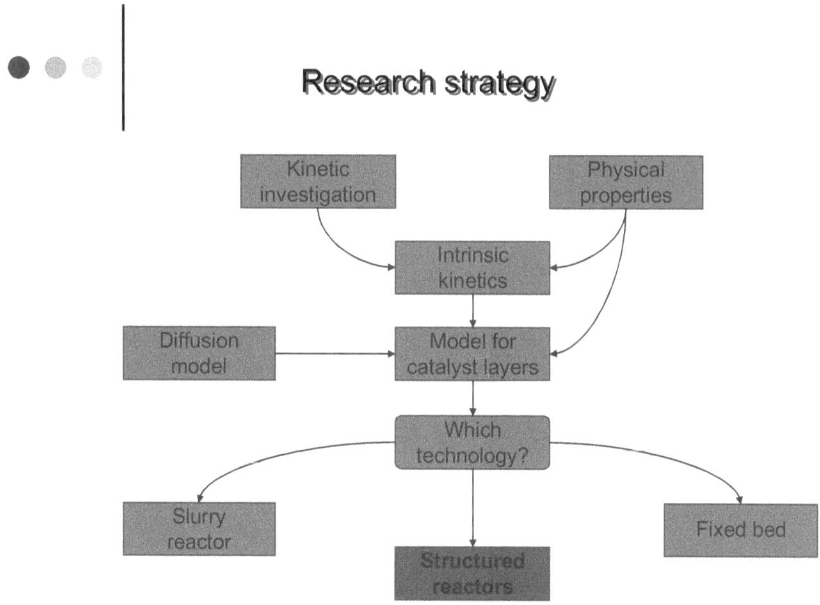

Many tools exist for a precise description of the reaction kinetics mass and heat transfer effects, and they are applicable for biomass conversion, but catalyst deactivation is a dilemma in the treatment of biomass. For instance, the deactivation of decarboxylation catalysts (Table 4.1) is very strong—mainly due to the decarbonylation reaction proceeding simultaneously—and the deactivation hydrogenation catalysts is quite severe, too. Mathematical modeling of deactivation is possible, but still not very predictive. This is, however, not the main point. Catalyst deactivation should be prevented or at least diminished in order to design economically feasible processes based on biomass conversion.

Special techniques, such as ultrasound, can be applied to clean the catalyst surface and refresh the active sites, but they are not a general remedy. For sponge nickel, ultrasound retards the catalyst deactivation in sugar hydrogenation, but for ruthenium catalysts, an opposite effect has been discovered, as ultrasound can contribute to the agglomeration of metal nanoparticles on the catalyst surface (Sifontes, 2012).

It is evident that many of the catalysts developed for the treatment of fossil feedstock do not work in an optimal way when they are exposed to components from biomass. Fundamental research in the reaction mechanisms and catalyst deactivation, including the use of single crystals and other model catalysts, is needed to reveal the chemical and physical reasons for deactivation. In this way, better catalysts can be developed in the future. Active, selective, and durable catalysts can be used in advanced—and sometimes expensive—reactors, such as monoliths, foams, and fiber reactors. The work requires a close collaboration between chemists, physicists, and chemical engineers.

Finally, it has to be stated that there should not be any specific chemical reaction engineering of biomass, but the general principles of chemical engineering should be applied to molecules and mixtures originating from biomass. Without a precise knowledge of chemistry and detailed mathematical modeling of kinetics, thermodynamics, transport phenomena, and flow pattern, a successful biomass conversion, cannot be achieved, but the effort remains Alchemia.

ACKNOWLEDGMENTS

This work is part of the activities at the Åbo Akademi Process Chemistry centre within the Finnish Centre of Excellence Programs (2000–2011) by the Academy of Finland. T. Salmi is grateful for the Academy Professor grant (2009–2013) by the Academy of Finland.

REFERENCES

Abatzoglou N, Chornet E: Acid hydrolysis of hemicelluloses and cellulose: theory and Applications, *Polysaccharides* 1007–1045, 1998.

Al-Dahhan MH, Larachi F, Dudukovic MP, Laurent A: High-pressure trickle-bed reactors: a review, *Ind Eng Chem Res* 36:3292–3314, 1997.

Alonso DM, Bond JQ, Dumesic JA: Catalytic conversion of biomass to biofuels, *Green Chem* 12:1493–1513, 2010.

Allahverdiev AI, Gündüz G, Murzin DY: Kinetics of α-pinene isomerization, *Ind Eng Chem Res* 37:2373–2377, 1998.

Athena Visual Studio: http://www.athenavisual.com/ 2009

Augustine RL, Reardon EJ Jr, : Palladium catalyzed hydrogenation of cholesterol, *Org Prep Proc* 1:107–109, 1969.

Bell AT: The impact of nanoscience on heterogeneous catalysis, *Science* 299:1688, 2003.

Bernas A, Laukkanen P, Kumar N, et al: A new heterogeneously catalytic pathway for isomerization of linoleic acid over Ru/C and Ni/H-MCM-41, *J Catal* 210:354–366, 2002a.

Bernas A, Kumar N, Mäki-Arvela P, et al: Conjugation of linoleic acid over a hydrogen pre-activated heterogeneous catalyst, *Chem Commun* 10:1142–1143, 2002b.

Bernas H, Taskinen A, Wärnå J, Murzin DY: Describing the inverse dependence of hydrogen pressure by multi-site adsorption of the reactant: hydrogenolysis of hydroxymatairesinol on a Pd/C catalyst, *J Mol Catal A Chem* 306:33, 2009.

Blecker C, Fougnies C, van Herck J-C, Chevalier J-P, Paquot M: Kinetic study of the acid hydrolysis of various oligofructose samples, *J Agric Food Chem* 50:1602–1607, 2002.

Boudart M, Djéga-Mariadassou G: *Kinetics of heterogeneous catalytic reactions*, Princeton, New Jersey, 1984, Princeton University.

Butt JB: *Reaction kinetics and reactor design*, ed 2, New York, 1999, Marcel Dekker, Inc.

Cabrera MI, Grau RJ: Methyl oleate isomerization and hydrogenation over Ni/Al_2O_3: a kinetic study recognizing differences in the molecular size of hydrogen and organic species, *J Mol Catal A Chem* 287:24, 2008a.

Cabrera MI, Grau RJ: *Trends Chem Eng* 2008:39–53, 2008b.

Corma CA, Eduardo Domine M, Susarte RM, Rey GF: *MCM-41 type microporous materials containing titanium and their utilization as catalysts in α-pinene oxidation*, WO0054880, 2002.

Chorkendorff I, Niemansverdriet JW: *Concepts of modern catalysis and kinetics*, 2003, Wiley-VCH.

Cybulski A, Moulijn JA: *Structured catalysts and reactors*, Boca Raton, FL, 2006, CRC Press.

DeJarlais W, Gast WL: Conjugation of polyunsaturated fats: methyl linoleate with tris (triphenylphosphine) chlororhodium, *J Am Oil Chem Soc* 48:21–24, 1971.

Durante D: Master's thesis, Åbo Akademi University and University of Padova, Turku/Åbo and Padova, 2011.

Ekblom J: *Process for the preparation of stanol esters*, US 6855837, 2005.

Eklund P, Lindholm A, Mikkola J-P, Smeds A, Lehtilä R, Sjöholm R: Synthesisof (−)-matairesinol, (−)-enterolactone, and (−)-enterodiol from natural lignan hydroxy-matairesinol, *Org Lett* 52:491–493, 2003.

Ertl G: *The Chemical Record 33*, 2000.

Fogg PGT, Gerrard W: *Solubility of gases in liquids*, Chichester, 1992, Wiley.

Fogler HS: *Elements of chemical reaction engineering*, New Jersey, 1999, Prentice Hall International Inc.

Frennet A, Hubert C: Transient kinetics in heterogeneous catalysis by metals, *J Mol Catal A Chem* 163–188, 2000.
Froment GF, Bischoff KB: *Chemical reactor analysis and design,* 2nd ed, New York, 1990, John Wiley & Sons Inc.
Frankel E: Conversion of polyunsaturates in vegetable oils to cis-monounsaturates by homogeneous hydrogenation catalyzed with chromium carbonyls, *J Am Oil Chem Soc* 47:11–14, 1970.
Frank-Kamenetskii DA: *Diffusion and heat-transfer in chemical kinetics,* New Jersey, 1955, Princeton University Press.
Gallezot P, Nicolaus N, Flèche G, Fuertes P, Perrard A: Glucose hydrogenation on ruthenium catalysts in a trickle-bed reactor, *J Catal* 180:51–55, 1998.
Gallezot P: Direct routes from biomass to end-products, *Catal Today* 167:31–36, 2011.
Gear CW: *Numerical initial value problems in ordinary differential equations,* Englewood Cliffs, New Jersey, 1971, Prentice Hall.
Grénman H, Wärnå J, Mikkola J-P, et al: Modelling the influence of wood anisotropy and internal diffusion on delignification kinetics, *Ind Eng Chem Res* 49:9703–9711, 2010.
Hájek J, Murzin DY: Liquid-phase hydrogenation of cinnamaldehyde over a Ru-Sn sol-gel Catalyst, 1.Evaluation of mass transfer via a combined experimental/theoretical approach, *Ind Eng Chem Res* 43:2030–2038, 2004.
Hayduk W, Minhas BS: correlations for prediction of molecular diffusion coefficients in liquids, *Can J Chem Eng* 60:295, 1982.
Henry CR: Catalytic activity of supported nanometer-sized metal clusters, *Appl Surf Sci* 164:252, 2000.
Holmbom B, Eckerman C, Eklund P, et al: Knots in trees—a new rich source of lignans, *Phytochem Rev* 2:331–340, 2004.
Holmbom B, Eckerman C, Hemming J, Reunanen M, Sundberg K, Willför S: *A method for isolating phenolic substances or juvabiones from wood comprising knot-wood,* WO 02/098830, 2002.
Haakana T, Kolehmainen E, Turunen I, Mikkola J-P, Salmi T: The development of monolith reactors: general strategy with a case study, *Chem Eng Sci* 59:5629–5635, 2004.
Haario H: *MODEST User's Guide 6.0,* Profmath Oy. Helsinki, Finland, 2007.
Heinze T, Schwikal K, Barthel S: Ionic liquids as reaction medium in cellulose functionalization, *Macromol Biosci* 5:520–525, 2005.
Hindmarsh AC: ODEPACK—a systematized collection of ODE-solvers. In Stepleman R, et al, editor: *Scientific computing,* Amsterdam, 1983, IMACS/North Holland Publishing Company, pp 55–64.
Jiang Q, Lu HM: Size dependent interface energy and its applications, *Surf Sci Rep* 63:427–464, 2008.
Koci P, Novak V, Stepanek F, Marek M, Kubicek M: Multi-scale modeling of reaction and transport in porous catalysts, *Chem Eng Sci* 65:412–419, 2010.
Kusema BT, Mäki-Arvela P, Xu C, et al: Kinetics of acid hydrolysis of arabinogalactans, *Int J Chem React Eng* 8(44):1–16, 2010.
Kuusisto J, Mikkola JP, Sparv M, et al: Hydrogenation of lactose over sponge nickel catalysts—kinetics and modeling, *Ind Eng Chem Res* 45:5900–5910, 2006.
Kuusisto J, Tokarev AV, Murzina EV, et al: From renewable raw materials to high-value added fine chemicals—catalytic hydrogenation and oxidation of D-lactose, *Catal Today* 121:92–99, 2007.
Leveneur S, Wärnå J, Salmi T, Murzin D, Estel L: Interaction of intrinsic kinetics and internal mass transfer in porous ion-exchange catalysts, *Chem Eng Sci* 64:4101–4114, 2009.
De Lathouder KM, Bakker JJW, Kreutzer MTS, Wallin A, Kapteijn F, Moulijn JA: Structured reactors for enzyme immobilization: amonolithic stirrer reactor for application in organic media, *Chem Eng Res Dev* 84:390, 2006.

Likozar B, Levec J: Effect of temperature, hydrodynamic conditions, phase ratio and catalyst concentration on biodiesel production: kinetics and mass transfer modeling. In *Final program A6.2, CHISA 2012, Prague, August*, 2012.

Lombardo SJ, Bell AT: A review of theoretical models of adsorption, desorption and reaction of gases on metal surfaces, *Surf Sci Rep* 13:3–72, 1991.

Marquardt DW: An algorithm for least squares estimation of nonlinear parameters, *J Soc Ind Appl Maths.* 11:431–441, 1963.

Markus H, Mäki-Arvela P, Kumar N, et al: Reactions of hydroxymatairesinol over supported palladium catalysts, *J Catal* 238:301–308, 2006.

Maloney MT, Chapman TW, Baker AJ: An engineering analysis of the production of xylose by dilute acid hydrolysis of hardwood hemicellulose, *Biotechnol Prog* 2:193, 1986.

Michalson ET, Devore JD: *Process for catalytically hydrogenating phytosterols*, US Pat. 6673951, 2004.

Mikkola J-P, Salmi T, Sjöholm R: Modelling of kinetics and mass transfer in the hydrogenation of xylose over Raney nickel catalyst, *J Chem Technol Biotechnol* 74:655–662, 1999.

Mikkola J-P, Salmi T: In-situ ultrasonic catalyst rejuvenation in three-phase hydrogenation of xylose, *Chem Eng Sci* 54:1583–1588, 1999.

Mukesh D, Narasimham S, Deshpande VM, Ramnarayan K: Isomerization of methyl linoleate on supported ruthenium–nickel catalyst, *Ind Eng Chem Res* 27:409–414, 1988.

Murzin DYu, Avetisov AK: Deactivation kinetics over induced nonuniform surfaces with linear steps of surface reactions, *Chem Eng Sci* 53:2469–2474, 1998.

Murzin D, Salmi T: *Catalytic kinetics,* Amsterdam, 2005, Elsevier, ISBN: 0-444-51605-0.

Murzin DYu: Thermodynamic analysis of nanoparticle size effect on catalytic kinetics, *Chem Eng Sci* 64:1046–1052, 2009.

Murzin DYu, Kubickova I, Snåre M, Mäki-Arvela P, Myllyoja J: *Method for the manufacture of hydrocarbons*, WO 2006075057, 2006.

Mäki-Arvela P, Holmbom B, Salmi T, Murzin D: Recent progress in synthesis of fine and specialty chemicals from wood and other biomass by heterogeneous catalytic processes, *Catal Rev—Sci Eng* 49:197–340, 2007.

Narayanan R, El-Sayed MA: Some aspects of colloidal nanoparticle stability, catalytic activity and recycling potential, *Top Catal* 47:15–21, 2008.

Parmon VN: Thermodynamic analysis of the effect of nanoparticle size of the active component on the adsorption equilibria and the rate of heterogeneous catalytic processes, *Doklady Phys Chem* 413:42, 2007.

Pogorelov VV, Avetisov AK, Vigdorovich FL: *Kinet Katal* 24(472):1983, 1983.

Salmi T, Murzin DYu, Mikkola J-P, et al: Advanced kinetic concepts and experimental methods for catalytic three-phase processes, *Ind Eng Chem Res* 43:4540–4550, 2004.

Salmi T, Romanainen J: A novel boundary condition for the axial dispersion model, *Chem Eng Proc* 34:359–366, 1995.

Salmi T, Damlin P, Mikkola J-P, Kangas M: Modelling and experimental verification of cellulose substitution kinetics, *Chem Eng Sci* 66:171–182, 2011.

Sandelin F, Salmi T, Murzin D: Dynamic modelling of catalyst deactivation in fixed bed reactors: skeletal isomerization of 1-pentene on ferrierite, *Ind Eng Chem Res* 45:558–566, 2006.

Schiesser WE: *The numerical method of lines: integration of partial differential equations,* San Diego, California, 1991, Academic Press.

Schlögl R, Abd Hamid SB: Nanocatalysis: mature science revisited or something new?, *Angew Chem Int Ed* 43:1628–1637, 2004.

Sifontes V: Doctoral thesis, Åbo Akademi University, Turku/Åbo, 2012

Simakova O, Kusema B, Campo B, et al: Structure sensitivity in L-arabinose oxidation over Au/Al_2O_3 catalysts, *J Phys Chem C* 115:1036–1043, 2011.

Sjöholm R, Eklund P, Mikkola J-P, Lehtilä R, Södervall M, Kalapudas A: Preparation of matairesinol from hydroxymatairesinol, WO 03/057209, 2003.
Sjöström E: *Puukemia—teoreettiset perusteet ja sovellutukset,* Espoo, Finland, 1989, Otakustantamo.
Smith JM: *Chemical engineering kinetics,* ed 3, Singapore, 1990, McGraw-Hill.
Snåre M, Kubickova I, Mäki-Arvela P, Eränen K, Murzin DYu: Heterogeneous catalytic deoxygenation of stearic acid for production of biodiesel, *Ind Eng Chem Res* 45:5708–5715, 2006.
Somorjai GA: The development of molecular surface science and surface science of catalysis: the Berkeley contribution, *J Phys Chem B* 104:2969–2979, 2000.
Temkin MI: *Kinet Katal* 28:493, 1977.
Temkin MI: The Kinetics of some industrial heterogenous catalytic reactions, *Adv Catal* 28:173–287, 1979.
Thiele EW: Relation between catalytic activity and size of the particle, *Ind Eng Chem* 31:916–920, 1939.
Tokarev A, Murzina E, Kuusisto J, Mikkola J-P, Eränen K, Murzin D: Kinetic behaviour of electrochemical potential in three-phase heterogeneous catalytic oxidation reactions, *J Mol Catal* 255:199–208, 2006.
Toukoniitty B, Murzin D, Salmi T: Utilization of electromagnetic and acoustic irradiation in enhancing heterogeneous catalytic reactions, *Trends Chem Eng* 11:1–37, 2008.
Trambouze P, Euzen J-P: *Les réacteurs chimiques,* Paris, 2002, Editions Technip.
Wenkin M, Touillaux R, Ruiz P, Delmon B, Devillers M: Influence of metallic precursors on the properties of carbon-supported bismuth-promoted palladium catalysts for the selective oxidation of glucose to gluconic acid, *Appl Catal A* 148:181–199, 1996.
Wild G, Charpentier J-C: *Diffusivité des gaz dans les liquides,* Paris, 1987, Techniques d'ingenieur.
Vital J, Ramos AM, Silva IF, Castanheiro JE: The effect of α-terpienol on the hydration of α-pinene over zeolites dispersed in polymeric membranes, *Catal Today* 67:217–223, 2001.
Wilke CR, Chang P: Correlation of diffusion coefficients in dilute solutions, *AIChE J* 1:264–270, 1955.
Willför S, Sjöholm R, Laine C, Holmbom B: Structural features of water-soluble arabinogalactans from Norway spruce and Scots pine heartwood, *J Wood Sci Technol* 36:101–110, 2002.
Willför S, Holmbom B: Isolation and characterization of water-soluble polysaccharides from Norway spruce and Scots pine, *J Wood Sci Technol* 38:173–179, 2004.
Wärnå J, Flores Geant M, Salmi T, et al: Modelling and scale-up of sitosterol hydrogenation process: from laboratory slurry reactor to plant scale, *Ind Eng Chem Res* 45:7067–7076, 2006.
Xu C, Pranovich A, Vähäsalo L, et al: Kinetics of acid hydrolysis of water-soluble spruce O-acetyl galactoglucomannans, *J Agric Food Chem* 56:2429–2435, 2008.
Zakzeski J, Bruijnincx PCA, Jongerius AL, Weckhuysen BM: The catalytic valorization of lignin for the production of renewable chemicals, *Chem Rev* 110:3552–3599, 2010.
Zhang H, Lee Penn R, Hamers RJ, Banfield JF: Enhanced adsorption of molecules on surfaces of nanocrystalline particles, *J Phys Chem B* 103:4656–4662, 1999.

CHAPTER FIVE

Chromatographic Fractionation of Lignocellulosic Hydrolysates

Jari Heinonen, Tuomo Sainio

Lappeenranta University of Technology, Department of Chemical Technology/Laboratory of Industrial Chemistry, Lappeenranta, Finland

Contents

Abstract

Monosaccharides (glucose, xylose, etc.) are valuable platform chemicals that can be produced from lignocellulosic (polysaccharide containing) biomasses via hydrolysis. Their cost-effective recovery from lignocellulosic biomass hydrolysates and further

Advances in Chemical Engineering, Volume 42
ISSN 0065-2377
http://dx.doi.org/10.1016/B978-0-12-386505-2.00005-5

purification requires sophisticated separation technology. Adsorption and electrolyte exclusion chromatography can be used for the fractionation of hydrolysates containing mineral acids. The recovered hydrolysis acid can be recycled, which saves and reduces chemicals consumption. Chromatographic separation can be done either batchwise, using steady-state recycling chromatography, or using continuous simulated moving bed chromatography. The latter two process options offer significant increase in productivity when compared to the batchwise fractionation process. In this chapter, chromatographic and adsorptive fractionation techniques for the treatment of acidic lignocellulosic hydrolysates are reviewed. The relevant physical phenomena affecting the separation efficiency as well as various industrially applicable process options are discussed.

1. INTRODUCTION

Monosaccharides (e.g., glucose and xylose) are valuable raw materials for biofuel and chemicals manufacturing. Monosaccharides can be obtained from different types of biomasses; some of which are mainly used for human nutrition (corn, sugar beet) (Hamelinck et al., 2005; Takerzadeh and Karimi, 2007; Taherzadeh and Karimi, 2008). One source for monosaccharides, that does not create competition between food manufacturing and chemical production, is lignocellulosic biomass (e.g., wood, straws, forest residues) which is the most abundant biomass type on Earth (Blommel and Cortright, 2008; Heinonen and Sainio, 2010, 2012; Heinonen et al., 2012a; Parikka, 2004; Xie et al., 2005).

Monosaccharide-based biofuels, such as ethanol and butanol, derived from lignocellulosic biomasses can be used to substitute crude oil-derived fuels currently used for transportation and other purposes. However, it is clear that, on a sustainable level, lignocellulosic biomass-derived fuels are not enough to satisfy the need for transportation fuels. Therefore, a better option would be to use the monosaccharides derived from lignocellulosic biomasses as precursors in manufacturing of chemicals that are currently produced by the petrochemical industry (see, e.g., Blommel and Cortright, 2008; Burk, 2010; Colmenares et al., 2011; Ezeji and Blaschek, 2010; Kunkes et al., 2008; van Leeuwen et al., 2012; Nghiem et al., 2010; Pedersen, 2008; Weber et al., 2010). This would lead to lower green house gas emissions, waste amounts (due to higher level of utilization of the raw materials), and economical savings. The monosaccharides are highly important and valuable platform chemicals due to their high reactivity and oxygen content (Blommel and Cortright, 2008; Kunkes et al., 2008; McMurry, 2008).

Recovery of monosaccharides from lignocellulosic biomasses is a multistep process. This is because these raw materials contain, instead of monosaccharides, polysaccharides (cellulose and hemicelluloses) that must be cleaved to monosaccharides in a process known as hydrolysis.

Hydrolysis of lignocellulosic biomasses can be conducted either as mineral acid catalyzed or as enzymatically catalyzed process (Farone and Cuzens, 1993, 1995; Galbe and Zacchi, 2002; Hamelinck et al., 2005; Heinonen and Sainio, 2010; Heinonen et al., 2012a; Iranmahboob et al., 2002; Taherzadeh and Karimi, 2008). Enzymatic hydrolysis is a selective process in which no by-products are formed. However, it is slow method and typically only cellulose is hydrolyzed by the enzymes (Camacho et al., 1996; Duff and Murray, 1996; Galbe and Zacchi, 2002; Gurgel et al., 2012; Kristensen et al., 2009; Sun and Cheng, 2002; Torget et al., 2000). In acid hydrolysis, however, both cellulose and hemicellulose chains of the lignocellulosic biomass are cut into monosaccharides. Acid hydrolysis can be catalyzed with either dilute or concentrated acid, usually with sulfuric acid as catalyst. The negative aspects of acid-catalyzed hydrolysis are high acid consumption and corrosion problems in the concentrated acid hydrolysis, and formation of by-products mainly in the dilute acid hydrolysis (Clausen and Caddy, 1993; Duff and Murray, 1996; Farone and Cuzens, 1993, 1995; Iranmahboob et al., 2002; Hamelinck et al., 2005; Takerzadeh and Karimi, 2007; Taherzadeh and Karimi, 2008).

In order to improve the economics of the acid-catalyzed hydrolysis processes, the hydrolysis acid should be recovered and recycled. In addition, the by-products formed in the hydrolysis, especially in the case of dilute acid hydrolysis, should also be recovered selectively because of high value of these chemicals.

The recovery of the hydrolysis acid and removal of by-products can be accomplished successfully using chromatography and adsorption. The separation is based on a chromatographic technique known as electrolyte (or ion) exclusion. In this technique, strong electrolytes are separated from nonelectrolytes and weak electrolytes using a strong ion-exchange resin as a stationary phase. The strong electrolytes are excluded from the resin due to electrostatic repulsion, while the nonelectrolytes and weak electrolytes are adsorbed to the resin (Dorfner, 1991; Helfferich, 1995; Wheaton and Bauman, 1953).

In the chromatographic fractionation of concentrated acid lignocellulosic hydrolysates, gel-type strong acid PS–DVB cation-exchange resins in acid (H^+) form are used as an adsorbent. The hydrolysates formed in the acid hydrolysis process are fractionated into hydrolysis acid fraction, monosaccharide fraction, and by-products fractions (Heinonen and Sainio, 2010, 2012). The hydrolysis acid fraction is led through a concentration step back to the hydrolysis. This recycling lowers the chemical consumption (costs) of the process considerably because the need for fresh acid is reduced (Heinonen and Sainio, 2010, 2012; Nanguneri and Hester, 1990).

Chromatographic fractionation of lignocellulosic hydrolysates brings about interesting phenomena. This is due a cooperative effect of sulfuric acid on the sorption of the other components present in the system. This cooperation, however, complicates the separation to some extent (Heinonen et al., 2012a, 2012b; Neuman et al., 1987). Chromatographic fractionation of lignocellulosic hydrolysates can be accomplished using batchwise chromatography, steady-state recycling chromatography, or simulated moving bed (SMB) chromatography.

Although the by-products formed in the acid hydrolysis process can be separated selectively using electrolyte exclusion chromatography, it may be more economical to separate them in a separate adsorption unit (Heinonen and Sainio, 2010, 2012; Sainio et al., 2011). Adsorbents such as activated carbon, polymeric adsorbents, zeolites, and ion-exchange resins have been successfully used in the selective recovery of hydrolysis by-products from lignocellulosic hydrolysates (Chandel et al., 2007; Nilvebrant et al., 2001; Ranjan et al., 2009; Sainio et al., 2011; Sun et al., 2011; Weil et al., 2002; Wang and Feng, 2010; Xie et al., 2005).

After the fractionation steps, the monosaccharides are led to biochemical- or thermochemical processing units (see, e.g., Blommel and Cortright, 2008; Burk, 2010; Colmenares et al., 2011; Ezeji and Blaschek, 2010; Heinonen et al., 2012a, 2012b; Kunkes et al., 2008; van Leeuwen et al., 2012; Nghiem et al., 2010; Pedersen, 2008; Weber et al., 2010). However, if there is need for selective separation of the monosaccharides, this can also be accomplished after the hydrolysate fractionation using chromatography. Chromatographic fractionation of sugars is widely used in the sweetener industry, for example, for the enrichment of fructose in fructose–glucose syrups. Gel-type strong acid PS–DVB cation-exchange resins in calcium form are usually used as adsorbents in the sugar separations (Bart et al., 1996; Bubnik et al., 2004; Farone and Fatigati, 2003; Howard et al., 1988; Schoenrock, 1991; Lee, 2003; Saari and Hurme, 2011).

Here, the chromatographic and adsorptive separation techniques that can be used for the fractionation of lignocellulosic hydrolysates and recovery of monosaccharides from the hydrolysates are reviewed. In order to clarify the phenomena occurring during the separation, the main separation principles are presented. Short introductions of different process options available for the chromatographic/adsorptive fractionation of lignocellulosic hydrolysates are given. The effects of different variables on the separation performance are discussed. The advantages of using chromatography for the purification of concentrated acid lignocellulosic hydrolysates for fermentative ethanol production are also discussed. In addition, short overviews of the acid hydrolysis processes and the selective separation of monosaccharides from each other are provided.

2. HYDROLYSIS OF LIGNOCELLULOSIC BIOMASS

Lignocellulosic raw materials include wood, forestry industry and agricultural residues (corn and rice straws), bamboo, energy crops (reed canary grass and witch grass) processing waste materials, construction waste, and beet pulp (Duff and Murray, 1996; Galbe and Zacchi, 2002; Hamelinck et al., 2005; Sun and Cheng, 2002; Sun et al., 2011; Takerzadeh and Karimi, 2007; Taherzadeh and Karimi, 2008). These materials consist mainly of cellulose, hemicellulose, and lignin (Table 5.1). In addition to these three main components, small amounts of ash, acids, and extractives can be found from lignocellulosic raw materials (Duff and Murray, 1996; Hamelinck et al., 2005).

Cellulose is the main component of lignocellulosic materials (Table 5.1). It is a linear homopolymer consisting of 2000–27,000 cellobiose (a dimer of two glucose monomers) units. The glucose units are linked together with β-(1,4)-glycosidic bonds (Arslan et al., 2012; Duff and Murray, 1996; Takerzadeh and Karimi, 2007; Taherzadeh and Karimi, 2008; Xiang et al., 2003a, 2003b). Cellulose chains form numerous intra- and intermolecular hydrogen bonds via the hydroxyl groups in the sugar molecules. This results in a highly ordered crystalline structure interspersed with less-ordered amorphous regions (Duff and Murray, 1996; Farone and Cuzens, 1993, 1995; Taherzadeh and Karimi, 2008).

Hemicelluloses surround the cellulose chains. They are amorphous linear heteropolymers with monomeric side chains. Structural units of hemicellulose chains depend on the lignocellulosic raw material. The chains consist of hexose (e.g., glucose, mannose, galactose, and rhamnose) and pentose

Table 5.1 Fractions of cellulose, hemicellulose, and lignin in lignocellulosic biomasses (Hamelinck et al., 2005; Sun and Cheng, 2002; Takerzadeh and Karimi 2008)

	Cellulose (%)	Hemicelluloses (%)	Lignin (%)
Hardwood	40–75	10–40	15–25
Softwood	30–50	25–40	25–35
Wheat straw	30	50	15
Rice straw	32–47	19–27	5–24
Sugarcane bagasse	40	24	25
Switch grass	32–45	25–32	12–18
Grasses	25–40	25–50	10–30

(e.g., xylose and arabinose) sugars, acetic acid, and sugar acids (Arslan et al., 2012; Duff and Murray, 1996; Hamelinck et al., 2005; Takerzadeh and Karimi, 2007; Taherzadeh and Karimi, 2008). Like glucose units in cellulose chains, also the monosaccharides in hemicelluloses are linked together by β-(1,4)-glycosidic bonds (Duff and Murray, 1996; Gonzáles et al., 1986; Taherzadeh and Karimi, 2008; Xiang et al., 2003a, 2003b).

The third major constituent of lignocellulosic raw materials is lignin. It is an aromatic three-dimensional polymer consisting different kinds of phenylpropane units (Duff and Murray, 1996; Takerzadeh and Karimi, 2007; Taherzadeh and Karimi, 2008). Usually, softwoods contain more lignin than hardwoods (Table 5.1). The role of lignin is to strengthen and shield the polysaccharide network (Duff and Murray, 1996).

The polysaccharides (cellulose and hemicellulose) in lignocellulosic biomass can be utilized in producing monosaccharides. For this purpose, the lignocellulosic biomass must be hydrolyzed, that is, the polysaccharide chains must be cleaved to monosaccharides. Hydrolysis can be done using acids or enzymes as catalysts (Galbe and Zacchi, 2002; Hamelinck et al., 2005; Heinonen and Sainio, 2010; Heinonen et al., 2012a; Iranmahboob et al., 2002; Sun et al., 2011; Takerzadeh and Karimi, 2007; Taherzadeh and Karimi, 2008). In acid-catalyzed hydrolysis, sulfuric acid is usually used as a catalyst although the use of hydrochloric acid, sulphurous acid, acetic acid, and phosphoric acid has also been studied (Baniel and Eyal, 2011; Farone and Cuzens, 1993, 1995; Gonzáles et al., 1986; Hamelinck et al., 2005; Hoshino et al., 2007; Iranmahboob et al., 2002; Lee et al., 1999; Lenihan et al., 2010; Sun et al., 2011; Takerzadeh and Karimi, 2007; Taherzadeh and Karimi, 2008). The acid hydrolysis can be done with dilute or concentrated acids. The lower the acid concentration in the hydrolysis, the higher the reaction temperature must be (Lee et al., 1999). Acid-catalyzed hydrolysis is a complex heterogeneous reaction scheme depending not only on the hydrolysis conditions (temperature, hydrolysis acid concentration) but also on physical factors (crystallinity of cellulose, swelling of the lignocellulosic material, shearing forces in the hydrolysis reactor) (Xiang et al., 2003a, 2003b).

In acid hydrolysis, cellulose chains break down into glucose monomers which, in turn, degrade to hydroxymethyl furfural (HMF) (Lenihan et al., 2010; Palmqvist and Hahn-Hägerdahl, 2000a, 2000b; Takerzadeh and Karimi, 2007; Taherzadeh and Karimi, 2008). Eventually, the HMF break downs into formic and levulinic acids (Mussatto and Roberto, 2004; Palmqvist and Hahn-Hägerdahl, 2000b; Taherzadeh and Karimi, 2008). The amorphous regions of cellulose are more easily hydrolyzed than the crystalline regions (Taherzadeh and Karimi, 2008).

Hemicelluloses are more easily hydrolyzed to monosaccharides than cellulose due to the completely amorphous structure (Galbe and Zacchi, 2002; Hamelinck et al., 2005; Lenihan et al., 2010; Takerzadeh and Karimi, 2007; Taherzadeh and Karimi, 2008; Torget et al., 2000). As a consequence, in acid hydrolysis, monosaccharides derived from hemicelluloses are more easily degraded due to longer availability for degradation reactions. The degradation products of hemicellulosic pentose and hexose sugars are furfural and HMF, respectively (Galbe and Zacchi, 2002; Hamelinck et al., 2005; Lenihan et al., 2010; Palmqvist and Hahn-Hägerdahl, 2000a, 2000b; Rafiqul and Sakinah, 2012; Takerzadeh and Karimi, 2007; Taherzadeh and Karimi, 2008; Torget et al., 2000). These can, in turn, degrade to formic and levulinic acids (Lewkowski, 2001; Taherzadeh and Karimi, 2008).

The degradation reactions of the monosaccharides are considerably slower than the release rate of the monosaccharides in concentrated acid hydrolysis conditions (Gurgel et al., 2012; Heinonen et al., 2012a; Iranmahboob et al., 2002; Lee et al., 1999; Taherzadeh and Karimi, 2008; Torget et al., 2000). In dilute acid hydrolysis conditions, the degradation order of the main hemicellulosic monosaccharides is xylose > arabinose > mannose > galactose > glucose (Takerzadeh and Karimi, 2007).

Lignin is highly resistant for chemical and enzymatic modifications and therefore breaks down only partially during acid hydrolysis yielding phenolic compounds (Clark and Mackie, 1984; Delgenes et al., 1996; Palmqvist and Hahn-Hägerdahl, 2000a, 2000b; Taherzadeh and Karimi, 2008). The concentrations of the phenolic compounds in acidic hydrolysates are usually very low due to the resistant nature of lignin and low solubility in acidic medium (Clark and Mackie, 1984; Lenihan et al., 2010; Takerzadeh and Karimi, 2007; Taherzadeh and Karimi, 2008).

2.1. Mechanism of acid-catalyzed hydrolysis

Acid-catalyzed hydrolysis of hemicelluloses is easier than that of cellulose due to the amorphous structure of the former polymer (Arslan et al., 2012; Rinaldi and Schüth, 2009; Xiang et al., 2003a, 2003b). The crystalline regions of the cellulose chains due to the hydrogen bonding between the OH groups complicate hydrolysis: at low temperature and hydrolysis acid concentrations, the prevailing reaction mode is endwise hydrolysis (Xiang et al., 2003a). When temperature or hydrolysis acid concentration is increased above certain level, the crystallinity of cellulose disappears as protons from the acid interfere with the inter- and intramolecular hydrogen bonding (Xiang et al., 2003a, 2003b). With sulfuric acid, the limiting concentration is 65 wt% at room temperature and decreases with increasing temperature. When the acid is diluted with water, cellulose chains reprecipitate but form an irregular structure (Xiang et al., 2003a). Due to the disappearance of the crystallinity of cellulose, the protons of the hydrolysis acid can penetrate into inner layers of cellulose chains and hydrolysis reactions can occur also in the middle of cellulose chains. This increases the rate of cellulose hydrolysis considerably (Rinaldi and Schüth, 2009; Xiang et al., 2003a, 2003b).

The actual hydrolysis reactions in amorphous cellulose and hemicelluloses and in end of crystalline cellulose chains occur as follows (Fig. 5.1).

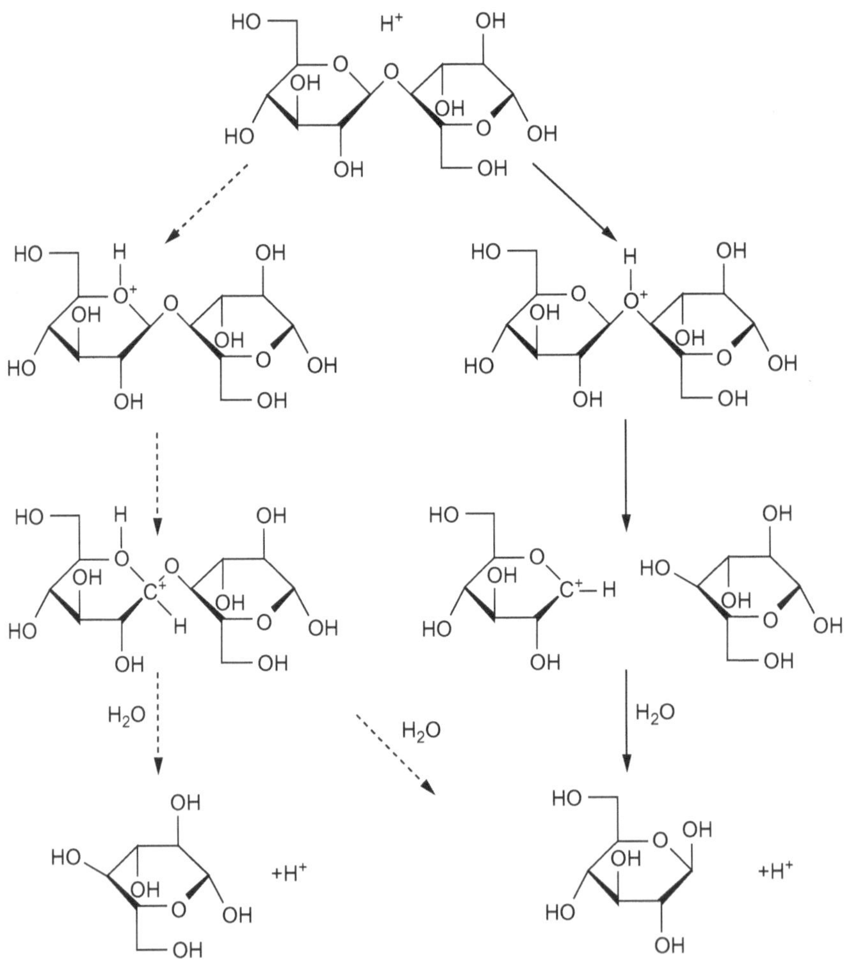

Figure 5.1 Mechanism of acid-catalyzed hydrolysis of β-1,4-glycosidic bonds of polysaccharides. Solid lines mark the dominant reaction mechanism. *Adapted from Xiang et al. (2003b).*

Protons from the hydrolysis acid interact with the glycosidic oxygen atoms that link two sugar units and a conjugate acid is formed. The C—O bond between the sugar molecules in the conjugate acid is then cleaved, and a carbonium ion is formed from the other half of the conjugate acid. The other half of the conjugate acid becomes an end of mono-, or oligosaccharide, or shortened polysaccharide chain, depending on at which part of the chain the cleavage occurred. The carbonium ion reacts rapidly with a water molecule and forms a sugar molecule, and a proton is regenerated (Fig. 5.1) (Rinaldi and Schüth, 2009; Xiang et al., 2003b). The formation of the carbonium ion occurs faster in the end of a polysaccharide chain than in the middle (Xiang et al., 2003b). During acid hydrolysis, monosaccharide yield losses can occur when the carbonium ions undergo condensation reactions with acid-soluble lignin moieties (Xiang et al., 2003a).

The monosaccharides obtained from cellulose and hemicelluloses can undergo further reactions in which furfural and HMF are formed from pentoses and hexoses, respectively (Binder et al., 2010; Lenihan et al., 2010; Lewkowski, 2001; Lima et al., 2010; Taherzadeh and Karimi, 2008). These degradation products are formed when the monosaccharides undergo thermal dehydration reaction in acidic conditions. Protons from the hydrolysis acid act as catalysts also in these reactions (Binder et al., 2010; Lewkowski, 2001). However, temperature, instead of acid concentration, determines the rate of the monosaccharide dehydration reactions: the formation of furfural and HMF is much slower in concentrated acid hydrolysis at low temperature (<100 °C) than in dilute acid hydrolysis at high temperature (>130 °C) (Farone and Cuzens, 1993, 1995; Galbe and Zacchi, 2002; Heinonen et al., 2012a; Lenihan et al., 2010; Lewkowski, 2001; Takerzadeh and Karimi, 2007; Taherzadeh and Karimi, 2008). HMF and furfural can also undergo decomposition reactions in acidic conditions and form formic (from HMF) and levulinic acids (from furfural and HMF) (Lewkowski, 2001; Taherzadeh and Karimi, 2008).

2.2. Concentrated acid hydrolysis

The most common acid used to breakdown polysaccharide chains in lignocellulosic biomasses is sulfuric acid (Clausen and Caddy, 1993; Farone and Cuzens, 1993, 1995; Galbe and Zacchi, 2002; Hamelinck et al., 2005; Heinonen and Sainio, 2010; Heinonen et al., 2012a; Hoshino et al., 2007; Iranmahboob et al., 2002; Janga et al., 2012; Sun et al., 2011; Takerzadeh and Karimi, 2007; Taherzadeh and Karimi, 2008; Torget et al., 2000). Concentrated acid hydrolysis is an old process first reported already in 1819 by Braconnot (Takerzadeh, 2007). The hydrolysis is usually done in two steps at temperatures below 100 °C (Clausen and Caddy, 1993; Farone and Cuzens, 1993, 1995; Heinonen et al., 2012a; Hoshino et al., 2007; Janga et al., 2012; Sun et al., 2011; Taherzadeh and Karimi, 2008). In the first step, the lignocellulosic biomass is mixed with 65–85 wt% sulfuric acid at a temperature of 30–55 °C in order to break down the crystalline structure of cellulose and solubilize the biomass (Clausen and Caddy, 1993; Farone and Cuzens, 1993, 1995; Hoshino et al., 2007; Iranmahboob et al., 2002; Janga et al., 2012; Sun et al., 2011; Ucar and Balaban, 2003; Xiang et al., 2003a, 2003b). Usually, the acid to biomass ratio by weight is 0.3–5. The duration of the decrystallization step is less than 2 h. The actual hydrolysis is commenced after the pretreatment step by diluting the hydrolysis acid to 20–30 wt% and increasing temperature to 80–100 °C. The reaction is continued for 40–480 min (Farone and Cuzens, 1993, 1995; Hoshino et al., 2007; Janga et al., 2012; Sun et al., 2011).

Concentrated acid hydrolysis gives a high total monosaccharide yield: even 80% or higher (Clausen and Caddy, 1993; Duff and Murray, 1996; Farone and Cuzens, 1993, 1995; Hamelinck et al., 2005; Iranmahboob et al., 2002; Sun et al., 2011). For example, Iranmahboob et al. (2002) obtained 78–82 wt% monosaccharide yield from concentrated acid hydrolysis of mixture of softwood and hardwood chips (1:1) with 20–33 wt% H_2SO_4, respectively. Sun et al. (2011) obtained 81.6 % monosaccharide yield in two-step concentrated sulfuric acid hydrolysis of bamboo chips (wood to acid ratio: 0.7).

In concentrated acid hydrolysis, only small amounts of by-products (mainly acetic acid, furfural, and HMF) are formed (Clausen and Caddy, 1993; Farone and Cuzens, 1993, 1995; Hamelinck et al., 2005; Heinonen et al., 2012a; Janga et al., 2012; Sun et al., 2011). Acetic acid is formed already at mild conditions when the acetylate groups in hemicelluloses are cleaved off, and the acetic acid yield does not depend much on the conditions of the hydrolysis. Furfural and HMF are the only furans found in significant quantities in the hydrolysates (Lenihan et al., 2010; Mussatto and Roberto, 2004; Palmqvist and Hahn-Hägerdahl, 2000b; Taherzadeh and Karimi, 2008).

In addition to the three main by-products, also a variety of phenolic compounds originating from lignin and extractives may be formed (Delgenes et al., 1996; Palmqvist and Hahn-Hägerdahl, 2000a, 2003b; Takerzadeh and Karimi, 2007; Taherzadeh and Karimi, 2008). Most of the unreacted lignin precipitates (90% at pH 2.6 (Alén et al., 1985)) during the hydrolysis due to low solubility in acidic conditions (Alén et al., 1985; Fuhrmann et al., 2002; Iranmahboob et al., 2002; Koljonen et al., 2004) and can be separated from the hydrolysates by filtration.

The disadvantages of the concentrated acid hydrolysis process are corrosion problems related to the high acid concentration and high hydrolysis acid consumption. The hydrolysis acid must be removed from the hydrolysate prior to fermentation to ethanol (Galbe and Zacchi, 2002; Heinonen et al., 2011) or butanol (Shapovalov and Ashkinazi, 2008) or manufacturing of other chemicals (Blommel and Cortright, 2012; Burk, 2010; Pedersen, 2012; Rafiqul and Sakinah, 2012). Traditionally, the acid removal has been done by neutralization with $Ca(OH)_2$. This, however, results in high chemical costs of the hydrolysis. In addition, high amounts of $CaSO_4$ are generated (Clausen and Caddy, 1993; Farone and Cuzens, 1993, 1995; Hamelinck et al., 2005), and in the ethanol fermentation, the ethanol yield is relatively low (Palmqvist and Hahn-Hägerdahl, 2000b).

The disadvantages regarding the economy of the concentrated acid hydrolysis have made it unpopular. However, the economics of the process can be increased considerably if the hydrolysis acid would be recycled and reused instead of neutralizing. This can be done by using chromatography as has been shown by Farone and Cuzens (1993, 1995), Heinonen and Sainio (2010, 2012), Laatikainen et al. (2011), Neuman et al. (1987), and Sprinfield and Hester (1999). With chromatography using strong acid cation-exchange resins as adsorbents, the monosaccharides in the hydrolysates can be separated from sulfuric acid and by-products formed during the hydrolysis. This kind of chromatographic separation of electrolytes and nonelectrolytes is known as electrolyte exclusion chromatography (Dorfner, 1991; Helfferich, 1995; Wheaton and Bauman, 1953). Although the fractionation of concentrated acid hydrolysates can be accomplished in one step, it may be more economical to remove the hydrolysis by-products in an additional adsorption step prior to the chromatographic monosaccharide–acid separation (Palmqvist and Hahn-Hägerdahl, 2000b; Sainio et al., 2011).

After the chromatographic purification, the monosaccharide fraction and the by-product fractions are led to downstream processing. The hydrolysis acid fraction is led to a concentration process in which the acid is concentrated to 70 wt% (Farone and Cuzens, 1993, 1995), for example, by using multiple effect evaporator or vapor compression distillation (Gsell, 2004). After the concentration step, the acid is reused in the hydrolysis.

2.3. Dilute acid hydrolysis

Hydrolysis of lignocellulosic biomasses can also be done with dilute mineral acids. Also in this process option, sulfuric acid is usually used as a catalyst (Farone and Cuzens, 1993, 1995; Gurgel et al., 2012; Hamelinck et al., 2005; Iranmahboob et al., 2002; Lee et al., 1999; Takerzadeh and Karimi, 2007;

Taherzadeh and Karimi, 2008; Torget et al., 2000), although the use of, for example, phosphoric acid has also been studied (Lenihan et al., 2010, 2012). Sulfuric acid concentration in the process ranges between 0.05 and 6.0 wt% (Gurgel et al., 2012; Hamelinck et al., 2005; Iranmahboob et al., 2002; Karimi et al., 2006; Lee et al., 1999; Rafiqul and Sakinah, 2012; Takerzadeh and Karimi, 2007; Taherzadeh and Karimi, 2008; Torget et al., 2000; Xiang et al., 2004). Low sulfuric acid concentration means low acid consumption, and therefore, acid recovery is not necessary with this process. The dilute acid hydrolysis is operated at high temperatures (120–260 °C) and at high pressure (>5 bar) in order to compensate the low acid concentration.

Due to the harsh hydrolysis conditions, the hydrolysis reactions occur considerably faster than in the concentrated acid hydrolysis process: reaction times vary between 2 and 10 min (Hamelinck et al., 2005; Karimi et al., 2006; Lee et al., 1999). In traditional one-step dilute acid hydrolysis, the overall monosaccharide yield in the hydrolysis is low. In order to hydrolyze cellulose, harsher conditions are needed than in the hydrolysis of hemicelluloses. The conditions in the one-step dilute acid hydrolysis have to be set so that cellulose chains are hydrolyzed. However, in these conditions, the hemicelluloses are very easily hydrolyzed and the released monosaccharides degrade rapidly to furfural and HMF, lowering the overall sugar yield considerably. In order to minimize the degradation of the monosaccharides originating from the hemicelluloses, the one-step dilute acid hydrolysis is not carried out long enough to completely hydrolyze cellulose; this results in a low glucose yield (~40–60%) (Farone and Cuzens, 1993, 1995; Galbe and Zacchi, 2002; Iranmahboob et al., 2002; Lee et al., 1999; Lenihan et al., 2010; Takerzadeh and Karimi, 2007; Taherzadeh and Karimi, 2008). The monosaccharide yield losses due to degradation reactions can be as high as 30% of the maximum yield (Clark and Mackie, 1984). The traditional one-step dilute acid hydrolysis method has thus become unpopular.

To obtain higher total monosaccharide yield, two-step dilute acid hydrolysis can be used. In the first step, the hemicellulose chains are hydrolyzed with dilute sulfuric acid in "mild" conditions at a temperature between 120 and 190 °C (Hamelinck et al., 2005; Karimi et al., 2006; Takerzadeh and Karimi, 2007; Torget et al., 2000). However, considerable monosaccharide degradation may still occur and the yield of hemicellulosic sugars is typically 70–95% (Galbe and Zacchi, 2002; Karimi et al., 2006; Lee et al., 1999; Takerzadeh and Karimi, 2007). At the end of the first step, the solubilized hemicellulosic monosaccharides are separated from the remaining solids which are taken to the second step and fresh acid is added. The cellulose hydrolysis is done at a temperature between 200 and 260 °C. Although harsher conditions can be used for cellulose hydrolysis in the two-step process, the glucose yield still remains low: only 40–60% (Duff and Murray, 1996; Galbe and Zacchi, 2002; Karimi et al., 2006; Lee et al., 1999; Takerzadeh and Karimi, 2007).

Karimi et al. (2006) studied the effects of temperature (pressure), retention time, and hydrolysis acid concentration on the two-step dilute acid hydrolysis of rice straw. Highest xylose (78.9% of theoretical) and glucose (46.6% of theoretical) yields were obtained when the conditions for the first step were 201 °C (15 bar), 10-min retention time, and 0.5 wt% H_2SO_4 concentration, and for the second step, 234 °C (30 bar), 3-min retention time, and 0.5 wt% H_2SO_4 concentration (Karimi et al., 2006).

Low yield of glucose in dilute acid hydrolysis is mainly due to formation of nonhydrolyzable oligomers from cellulose. Using special reactor designs, the glucose yield can be improved considerably (Xiang et al., 2003b). Xiang et al. (2003b) obtained 90% glucose yield in dilute acid hydrolysis (205–235 °C, 0.07 wt% H_2SO_4) of yellow poplar saw dust in a bed-shrinking flow-through reactor with a quick quenching heat exchanger at the reactor outlet. The high glucose yield was due to quick quenching of the reactor effluent (Xiang et al., 2003b).

Although acid recovery for reuse is not critical in dilute acid hydrolysis, the by-products formed in the hydrolysis must be removed from the hydrolysates. These compounds include furfural, HMF, acetic acid, and possible phenolic compounds (Mussatto and Roberto, 2004; Palmqvist and Hahn-Hägerdahl, 2000a, 2000b; Takerzadeh and Karimi, 2007; Taherzadeh and Karimi, 2008). The by-product removal can be accomplished using chromatography (Heinonen and Sainio, 2010; Laatikainen et al., 2011; Wooley et al., 1998; Xie et al., 2005) or adsorption (Chandel et al., 2007; Nilvebrant et al., 2001; Ranjan et al., 2009; Sainio et al., 2011; Sun et al., 2011; Weil et al., 2002).

2.4. Enzymatic hydrolysis

Cellulose and hemicelluloses can be hydrolyzed highly selectively using enzymes as catalysts. (Duff and Murray, 1996; Galbe and Zacchi, 2002; Gurgel et al., 2012; Kristensen et al., 2009; Peng et al., 2011; Saddler et al., 1983; Sun and Cheng, 2002; Torget et al., 2000). Cellulose hydrolyzing cellulase enzymes can be obtained from many fungi or bacteria including species among *Trichoderma*, *Penicillium*, and *Aspergillus* (Galbe and Zacchi, 2002). The structure of the lignocellulosic biomass hinder the access of the enzymes to the polysaccharides, and therefore, the biomass must be pretreated prior to the enzymatic hydrolysis (Galbe and Zacchi, 2002; Gurgel et al., 2012; Sun and Cheng, 2002). Biomass pretreatment methods include mechanical comminution, pyrolysis, steam explosion, ammonia fiber explosion, CO_2 explosion, ozonolysis, dilute acid hydrolysis, alkaline hydrolysis, oxidative delignification, and biological pretreatment with rot fungi (Duff and Murray, 1996; Galbe and Zacchi, 2002; Sun and Cheng, 2002).

The optimal conditions for enzymatic hydrolysis are temperature of 45–55 °C and pH 4.0–5.0 (Duff and Murray, 1996; Galbe and Zacchi, 2002; Kristensen et al., 2009; Sun and Cheng, 2002). Enzymatic hydrolysis is a slow method when compared to the dilute or concentrated acid-catalyzed hydrolysis: 48–84 h is needed to obtain high glucose yield (Camacho et al., 1996; Duff and Murray, 1996; Kristensen et al., 2009; Sun and Cheng, 2002). The solid content in enzymatic hydrolysis must be low in order to obtain high cellulose conversion: the conversion decreases with increasing solid concentration. For example, Kristensen et al. (2009) obtained 65% conversion of the theoretical value with 5 wt% solid concentration and 43% conversion with 20wt% solid concentration in 84-h hydrolysis. The decrease in the conversion is mainly due to the inhibitory effect of the hydrolysis products (in cellulose hydrolysis cellobiose as an intermediate product and glucose as an end-product) (Galbe and Zacchi, 2002; Kristensen et al., 2009).

3. PROCESS OPTIONS FOR CHROMATOGRAPHIC SEPARATIONS

This chapter describes the main process options used in industrial scale liquid chromatography in fixed beds. The column is filled with adsorbent material such as activated carbon, zeolites, polymeric adsorbents, or ion-exchange resins (Dorfner, 1991; Guiochon et al., 2006; Heinonen and Sainio, 2010; Helfferich, 1995; Wang and Feng, 2010; Wankat, 1990; Wheaton and Bauman, 1953; Eken-Saracoglu and Arslan, 2000). The solid materials used in chromatography and adsorption are usually highly porous and have therefore large surface areas for adsorption (Wankat, 1990). The pore size distribution affects mass transfer resistance (Helfferich, 1995).

3.1. Batch chromatography

The basic implementation mode of chromatography is the batch mode. In this method, an eluent (e.g., water, an organic solvent, or a mixture of these) is pumped through a column continuously and, at certain time intervals, the mixture that is to be fractionated is injected to the column at the column inlet. The components are pushed by the eluent and travel through the packed bed. The components in the injected sample travel with different velocities because they interact differently with the adsorbent material. Components that are strongly adsorbed to the adsorbent propagate slower through the column than the components that are weakly adsorbed. As a consequence, the components are separated from each other and exit the column at different times. The interactions between solute component and the adsorbent material are reviewed in Section 4.1. Distribution of species between liquid and solid phases is described by adsorption isotherms. Components in a mixture can also have cooperative (enhance sorption) or competitive (reduce sorption) effects on the sorption of the other components in the system (Guiochon et al., 2006; Wankat, 1990).

Batch chromatography is commonly used on production scale in biotechnology, sweetener industry, and pharmaceutical industry (Wankat, 1990). On analytical scale, the separation of sample components is complete. This is because of the small column loadings used in analytical chromatography: the concentrations are small, and no interactions between the components occur in the column. On preparative scale, however, the column loadings are large and there may be interactions between the solute components. Due to these reasons, the separation between components on a preparative scale is often not complete: the component profiles overlap each other. For fractionation of binary mixtures, this results in two or three fractions, depending on the purity requirements set for the chromatographic separation. If a mixed fraction (containing both components) is obtained, it can be recycled and fed back to the column (see below).

3.2. Steady-state recycling chromatography

The purity of the product fractions can be increased substantially if the middle (unresolved) part of the chromatogram is recycled back to the column inlet as shown in Fig. 5.2. If a constant amount of fresh feed is introduced to the process on every cycle (e.g., by mixing it with the recycled fraction), the mass that leaves the process eventually becomes equal to the mass of fresh feed. The process thus operates at a periodic steady state.

The benefit of SSR chromatography is increased purity with a given production rate or, alternatively, increased production rate at a given purity level. However, designing such a process is relatively difficult even with the help of a numerical simulation tool. There are five independent operating parameters (the four cut times shown in Fig. 5.2 and the injection

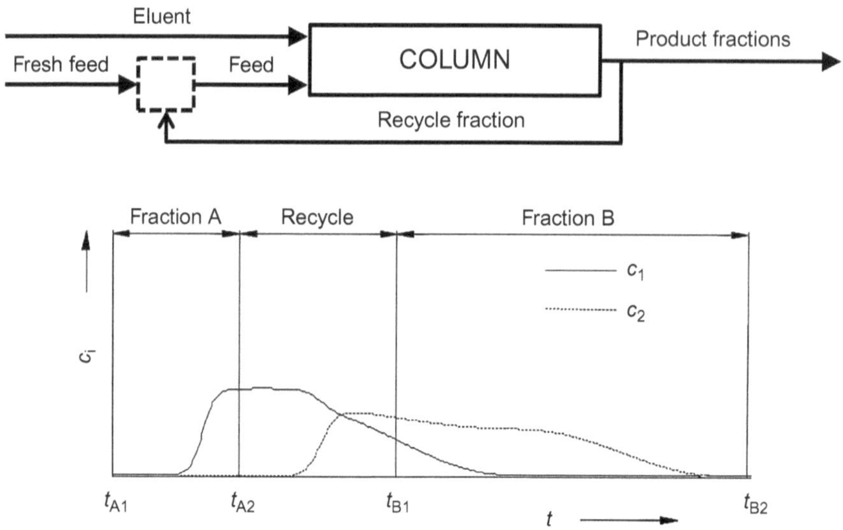

Figure 5.2 Operating principle of steady-state recycling chromatography. (For color version of this figure, the reader is referred to the online version of this chapter.)

volume) that determine the amount of fresh feed processed per cycle at steady state. Nearly optimal values for two of these (t_{A1} and t_{B2}) are easily chosen, whereas choosing properly the size and position of the recycle fraction (determined by t_{A2} and t_{B1}) is not straightforward. This is because the chromatogram at steady state usually differs considerably from that of the first cycle, and cut times that initially fulfill the purity requirements fail to do so at steady state. A design method that guarantees correct purities under ideal conditions (i.e., no axial dispersion or mass transfer resistance) has been derived by Sainio and Kaspereit (2009). In the practically more relevant case of significant dispersion, a short-cut design method by Kaspereit and Sainio (2011) provides excellent first approximations for the cut times.

A further improvement in process performance is possible by shortening the time between two consecutive pulses such that the front of the second pulse overlaps with the tail of the first pulse. At present, there are no design methods, other than fully dynamic numerical simulations, available for such an operating mode.

3.3. Continuous SMB chromatography

There are also continuous countercurrent process schemes for chromatographic separation. When properly optimized, they can yield lower eluent consumption, higher productivity, higher product concentrations, higher product purities, and enhanced utilization of the adsorbent (Guiochon et al., 2006; Kaspereit, 2009). In addition, continuous chromatographic separation process is easy to scale up or down (Guiochon et al., 2006; Katsuo and Mazzotti, 2010).

In practice, a truly continuous chromatographic separation process (true moving bed, TMB) is challenging to implement. This is because continuous countercurrent movement of the solid and liquid phases is difficult to establish at high flow rates. For the solid phase, plug flow conditions are impossible to achieve due to back mixing of the solid phase. In addition, severe particle abrasion and breakage occurs during the movement of the solid phase (Guiochon et al., 2006; Kaspereit, 2009; Wankat, 1990).

Due to the impracticality of the truly continuous chromatographic separation process, a process in which the movement of the solid phase is simulated was developed by Universal Oil Products in 1961 (Broughton and Gerhold, 1961). This process is called simulated moving bed (SMB) chromatographic separation process, and it was initially developed for separations in the petrochemical industry (Broughton and Gerhold, 1961). Today, various SMB processes can be found from the petrochemical industry (Broughton and Gerhold, 1961; Lee and Wankat, 2010; Mata and Rodrigues, 2001), the sweetener (sugar) industry (Airaksinen et al., 2010; Ando et al., 1990; Giacobello et al., 2000; Heikkilä et al., 1989, 1994, 1997, 2000, 2007; Masuda et al., 1993), and the fine chemicals and pharmaceutical industry (Barreto et al., 2008; Goncalves et al., 2008; Juza et al., 2000; Lee et al., 2010; Lehoucq et al., 2000; Pais et al., 1998; Ribeiro et al., 2011; Schulte and Strube, 2001). In this chapter, introductions to the operation of the SMB chromatography and some of its variants are given.

3.3.1 Operation of a four-zone SMB process for binary fractionations

In SMB chromatography, the continuous movement of the solid phase is simulated by periodically shifting the inlet and outlet ports in the direction of the liquid flow. This results in a countercurrent movement of the solid and liquid phases. If the time between the port switching and the column lengths are infinitely short, SMB and TMB processes are equal (Guiochon et al., 2006).

The principle of a conventional SMB system which can be used to separate binary mixtures into two fractions is shown in Fig. 5.3. This system consists of four zones which all have different functions (Guiochon et al., 2006). Each of the zones in SMB system can contain one or more columns. Zone I (Fig. 5.3) is located between the eluent inlet and extract (more retained component) outlet ports. Desorption of the more retained component occurs in

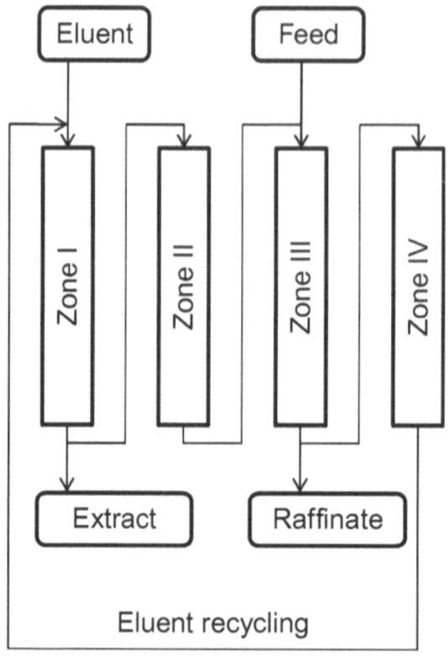

Figure 5.3 Four-zone simulated moving bed chromatographic system for binary separation. Eluent, eluent inlet stream; Feed, feed inlet stream; Extract, outlet stream of the most retained component; Raffinate, outlet stream of the least retained component.

this zone as well as the regeneration of the solid phase. Zone I also prevents the movement of the more retained component from zone I to zone IV (through the recycling stream) and is therefore called buffer zone.

The actual separation occurs in Zones II and III (Fig. 5.3). Zone II is located between the extract outlet and feed inlet ports (Fig. 5.3). Desorption of the less retained component occurs in zone II. The columns between the feed inlet and raffinate (less retained component) outlet ports belong to zone III (Fig. 5.3). Adsorption of the more retained component occurs in this zone. Components of a binary mixture have overlapping profiles in zones II and III.

The last zone in the SMB train, zone IV, is for the adsorption of the less retained component. It contains the columns between the raffinate outlet and eluent inlet ports (Fig. 5.3). It is also a buffer zone: the movement of the less retained component from zone IV to zone I is prevented with zone IV.

The basic four-zone SMB system has two inlet streams (feed and eluent) and two outlet streams (extract and raffinate) (Fig. 5.3). The eluent and the feed solution are continuously fed to the first column of zones I and III, respectively. The less retained component is continuously taken out from the system as raffinate stream at the outlet of the last column of zone III. Similarly, the more retained component is taken out from the system as extract stream at the outlet of the last column of zone I.

When the ports in an SMB system are shifted toward the liquid flow, the solid phase moves to the opposite direction. This means that the first column of zone III becomes the last column of zone II and the first column of zone II becomes the last column of zone I, etc. The separation of components of a binary mixture is achieved by adjusting the flow rates and the port switching time so that the more retained component moves with the solid phase toward the extract outlet and the less retained component with the liquid phase toward the raffinate outlet. In other words, the less retained component should be made to propagate faster than the port switching and the more retained component slower than the port switching.

3.3.2 SMB processes for ternary separations

The standard four-zone SMB can only be used for fractionations of binary mixtures. In many applications, however, the mixtures contain at least three components that need to be separated selectively. For example, in the chromatographic fractionation of lignocellulosic hydrolysates with a strong acid cation-exchange resin (sulfonated PS–DVB resin in H^+ form), three fractions are formed (Heinonen and Sainio, 2012): sulfuric acid, monosaccharide, and by-product (acetic acid) fractions (see Section 5.2). Due to this, one standard four-zone SMB system is not applicable for this fractionation. To overcome the limitations of the conventional SMB, modifications of the SMB have been developed that can be used for complex fractionations (Heikkilä et al., 1989; Jermann et al., 2012; Kaspereit, 2009; Kessler and Seidel-Morgenstern, 2006; Masuda et al., 1993; Paredes et al., 2004). These modified SMB processes operate continuously, that is, all flows are continuous, or semi-continuously, for example, feed is not introduced continuously to the system. The continuous SMBs for ternary fractionations use usually large number of columns, whereas the semi-continuous SMBs can operate with a relatively small number of columns (Heikkilä et al., 1989; Jermann et al., 2012; Kaspereit, 2009; Kessler and Seidel-Morgenstern, 2006; Masuda et al., 1993; Paredes et al., 2004). SMB systems for ternary or multicomponent separations include a cascade of two (or even more) four-zone SMBs (Kaspereit, 2009; Kessler and Seidel-Morgenstern, 2006), Japan Organo process (Agrawal and Kawajiri, 2012; Masuda et al., 1993; Mata and Rodrigues, 2001), multizone SMBs (more than four zones) (Kessler and Seidel-Morgenstern, 2006; Mun, 2011; Paredes et al., 2004), intermittent SMB (Jermann et al., 2012; Katsuo and Mazzotti, 2010), and sequential SMB (see, e.g., Heikkilä et al., 1989).

a. *Four-zone cascade SMB.* The four-zone cascade SMB consists of two SMBs in series: the outlet stream from the first unit containing two components is fed to the second unit. The second unit may be placed directly at the outlet of the first unit, or there may be a collection tank between the units. In the former option, the feed flow rate of the second unit is coupled with the outlet flow rate of the first unit (raffinate or extract) (Guiochon et al., 2006; Kaspereit, 2009). In this option, the feed to the second unit is always more diluted than the feed to the first unit (Kaspereit, 2009). When a collection tank is placed between the units, the feed to the second unit can be concentrated. With two four-zone cascade SMBs, it is more beneficial to do the easier separation in the first unit (Kessler and Seidel-Morgenstern, 2006): the concentrations of feed to the second unit can be altered, and therefore, the separation in the second unit can be made easier.

b. *Multizone SMBs.* In multizone SMBs, the number of zones is higher than four. Five-zone SMB units have been investigated, for example, by Mun (2011) and Paredes et al. (2004). In one option for a five-zone SMB, the unit combines both batch and continuous system (Paredes et al., 2004). One zone of the unit (zone 0) operates as a batch unit and is not connected to the rest of the SMB system. Desorption of the most retained component in the ternary separation system occurs in this zone. The other four zones operate like standard four-zone SMB unit. The feed is introduced continuously to the system (the first column of zone III). Eluent is introduced to the first column of zone I such as in the conventional SMB, and also to zone 0 in order to elute the most retained component (Paredes et al., 2004).

Mun (2011) investigated a five-zone SMB unit for ternary separations, in which all the zones operate continuously in countercurrent manner. This system has three outlet streams: one raffinate (least retained component) and two extract (for the intermediately and most retained components) stream. Instead of continuously feeding the system, partial feeding can be applied to improve the separation of ternary mixtures. The partial feeding can be implemented in numerous ways (Mun, 2011).

Also, SMB systems with more than five zones have been investigated (e.g., Kessler and Seidel-Morgenstern, 2006; Wooley et al., 1998). The cascade of two four-zone SMBs can also be integrated into the same unit creating an eight-zone SMB. In this configuration, part of the recycle stream between the subunits must be taken out of the system in order to accomplish ternary separation. In addition, the solid-phase flow rate is fixed. Eight-zone SMBs have lower productivity and product mass flow than two coupled four-zone SMBs (Kessler and Seidel-Morgenstern, 2006). Wooley et al. (1998)

have presented a nine-zone SMB (four- and five-zone SMBs integrated into the same unit) for the fractionation of dilute acid lignocellulosic hydrolysates into three fractions (see Section 5.4).

c. *Intermittent SMB.* The so-called intermittent (or improved) SMB (I-SMB) was originally developed in the sweetener industry (Katsuo and Mazzotti, 2010) for binary fractionations. It has since been modified for ternary separations (Jermann et al., 2012).

Intermittent SMB usually has one column in each of its four zones and a single switch consists of two steps. Two different configurations exist for separation of ternary mixtures. In the first configuration, the unit is operated like a conventional SMB in the first step with two inlet (feed and eluent) and two outlet (raffinate and extract) streams. However, there is no flow-through zone IV (Jermann et al., 2012; Katsuo and Mazzotti, 2010). The two least adsorbed components are collected in this step from the raffinate and extract streams. In the second step, in the separation of binary mixtures, all the inlets and outlets are closed and the liquid phase is circulated through the columns for adjustment of the relative positions of the concentration profiles (Katsuo and Mazzotti, 2010). In the case of ternary fractionation, the most retained component is taken out from the system in the second step. This component moves with the solid until it is eluted from the column in the first position in the second step, thus allowing the regeneration of the solid phase (Jermann et al., 2012).

In the second process configuration for intermittent SMB for ternary separation, flow passes through all four zones in the first step. However, there is no flow between zones I and II, and zones II and III. The feed is introduced to the first column of zone III, and it is recycled to zone I into which eluent is also fed. The most retained component is eluted in the first step from the last column of zone I, and the intermediately retained component from the last column of zone II. The least retained component is eluted in the second step from the outlet of the last column of zone IV. In this step, zone I is disconnected from the SMB unit and regenerated. The other three zones are connected in series, and fresh eluent is fed to the system (to the first column of zone II) in order to adjust the relative position of the two most retained components and elute the least retained component (Jermann et al., 2012).

d. *Sequential SMB.* Also, the sequential SMB system was developed in the sweetener industry (Heikkilä et al., 1989, 1994). Main applications have been in the recovery of betaine from molasses, that is, separation of betaine from salts, oligosaccharides, and sucrose. Sequential SMB differs considerably from other SMB systems presented above. First of all, the sequential SMB can be operated already with just two columns, while typically the column number is from three to eight (Airaksinen et al., 2010; Heikkilä et al., 1989, 1994, 1997, 2000, 2007).

The most distinct difference between sequential SMB and other SMB systems for ternary fractionations is that the sequential SMB is operated without (simulated) countercurrent movement of the solid and liquid phases. This means that the ports are not switched synchronously in the direction of the fluid flow during a cycle. In principle, sequential SMB is not an MSB system, but a multicolumn batch chromatographic system with recycling.

Another feature of the sequential SMB is that, according to the available patents, the column outlet streams are not split but the whole column outlet stream is taken out of the system or led to the next column (Airaksinen et al., 2010; Heikkilä et al., 1989, 1994, 1997, 2000, 2007). In conventional SMB, on the contrary, part of an outlet stream goes to collection tank and the rest to the next column in the series (Guiochon et al., 2006; Kaspereit, 2009).

One cycle of sequential SMB system consists of three phases: feeding, elution, and recycling. Each of these phases is conducted at least once in a cycle. One cycle of the sequential SMB consists of four to eight stages with one stage being one of the three phases given above (Heikkilä et al., 1989, 1994).

In the feeding phase, the mixture that is to be fractionated is fed to one column of the SMB system. Simultaneously, eluent can be fed to another column(s), and two or three product fractions are taken out of the system. The columns from which the products are withdrawn can be connected to the column (or be the column) into which the feeding is conducted or they may be disconnected from this column. During the elution phase, no feeding of the mixture to be separated is conducted. Eluent is fed to the system, and product fractions are collected from the outlets of various columns in the system (Heikkilä et al., 1989).

In the recycling phase, no eluent or fresh feed is introduced to the system and nothing is taken out of the system. The component profiles are circulated through the system in order to adjust the positions of the profiles relative to the outlet ports. All the columns can be connected in a single loop, or there may be two or more loops in which the liquid phase is circulated. There may even be simultaneously one closed loop in which recycling occurs and an open loop into which eluent is fed and products are taken out: the three phases of the sequential SMB can be done in series or they can overlap (Heikkilä et al., 1989).

The eluent consumption of the sequential SMB system can be reduced up to 70% by introducing parts of the collected fractions (diluted or mixed) back to the unit as an eluent substitute. In addition, product yield and purity may be increased. Proper position and stage of recycling must be determined in a way that the components in the reintroduced fractions meet similar components (Airaksinen et al., 2010).

e. *Japan Organo process.* Another interesting and commercially utilized SMB variation for ternary fractionations (or more complex fractionations) is the Japan Organo process, which is also known as pseudo SMB. The origins of this process are also in the sweetener industry (Masuda et al., 1993). The JO process can easily be built from standard four-zone SMB unit, and it can be run with four columns (one for each zone) (Agrawal and Kawajiri, 2012; Kurup et al., 2006; Lee and Wankat, 2010; Masuda et al., 1993; Mata and Rodrigues, 2001).

The operating principle of the JO process is shown in Fig. 5.4. One cycle consists of two steps. In the first step, the feed stream is introduced to the system and the intermediately retained component is taken out from the system. The SMB column train is disconnected between the zones II and III. The system is operated like a series of batch columns with the exception that eluent is fed to the system from different position than the feed (Fig. 5.4). Feeding of the eluent in the first step is not necessary but can improve the separation (Masuda et al., 1993). The feed mixture is introduced to

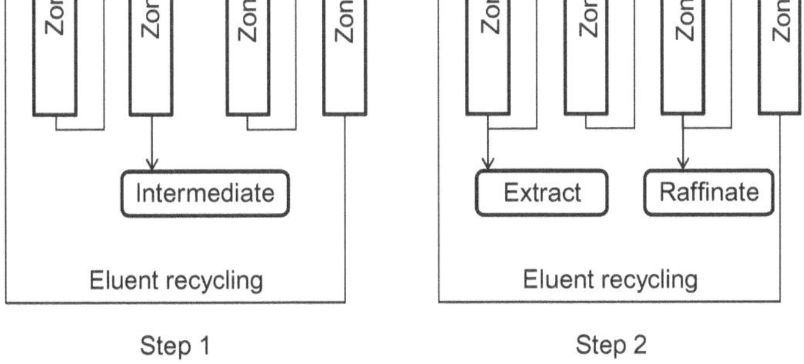

Figure 5.4 Principle of Japan Organo SMB system for ternary fractionations. Eluent, inlet stream of eluent; Feed, inlet stream of feed; Intermediate, outlet stream of the intermediately retained component; Extract, outlet stream of the most retained component; Raffinate, outlet stream of the least retained component.

the system through the inlet of the first column of zone III, while eluent is typically fed through the inlet of the first column of zone I (Fig. 5.4). Flow rates in zones I and II as well as in zones III and IV are equal. The intermediately retained component is taken out from the system at the outlet of the last column of zone II. In the first step, due to the batchwise operation, more than one component can be withdrawn consecutively from the outlet port at the end of zone II. In addition, also the raffinate and extract ports can be open in the first step allowing the withdrawal of the least and most retained components, respectively (Masuda et al., 1993).

In the second step of the JO process, only eluent is introduced to the system (Fig. 5.4). The system is operated like a standard four-zone SMB without the feed stream. Due to this, the flow rates in zones II and III are equal (Agrawal and Kawajiri, 2012; Masuda et al., 1993). The least and most retained components are taken out from the system in raffinate and extract streams, respectively (Agrawal and Kawajiri, 2012; Lee and Wankat, 2010; Masuda et al., 1993; Mata and Rodrigues, 2001). This is continued for several SMB switches before the first step is conducted again. In the end of step two, the intermediately retained component is located upstream from the feed point (in zone II) and can be collected during the next cycle (first step) (Lee and Wankat, 2010; Masuda et al., 1993; Mata and Rodrigues, 2001).

The number of switches in the second step can be equal or different to the number of columns. If they are equal, the feed in the first step is always introduced to the same column (Kurup et al., 2006; Lee and Wankat, 2010). As the number of port switches in the second step increases, the feed amount in the first step can be increased. This is because the effective column length increases with increasing number of port switches, and larger amounts of the least and most retained components can be separated from the intermediate component. However, if the amount of feed introduced to the system is larger than the loading capacity of zone III, the raffinate and extract streams may become contaminated with the intermediate component (Lee and Wankat, 2010).

The minimum duration of the first step is the time needed for the intermediate component to reach the outlet of the first column of zone III. Too short time leads to contamination of the extract stream with the intermediate component and too long time to contamination of the intermediate component's outlet stream with the most retained component. For the duration of the second step, the effects are opposite (Mata and Rodrigues, 2001).

Lee and Wankat (2010) have compared the operation of the JO process to that of the steady-state recycling process and a cascade of four-zone SMB processes (four or eight columns in one unit) in the case of linear isotherms (fractionation of nadolol isomers). The JO process was found to give higher productivity than the other two process options. In addition, desorbent to feed ratios with the JO process was approximately half of the values obtained with the recycling process (Lee and Wankat, 2010). In addition, Agrawal and Kawajiri (2012) have shown that in ternary fractionations the JO process clearly outperforms multizone SMB units (five and eight zones) and SMB cascade of two four-zone SMB units with respect to both productivity and eluent consumption.

3.4. Adsorption

Besides chromatography, also adsorption can be used to purify lignocellulose hydrolysates. The main difference to fixed-bed chromatography is that the process is operated with relatively long loading and regeneration steps, separated by a washing step if needed. Long loading steps are achieved by utilizing stationary phase materials with high capacity and strong interactions with the impurity compounds. Regeneration of such materials usually requires special regeneration agents (e.g., organic solvents, acids, or bases) and/or higher temperature to be economical. Otherwise, the operation and design of adsorption separation is similar to fixed-bed chromatography.

4. FUNDAMENTAL PHENOMENA IN CHROMATOGRAPHIC FRACTIONATION OF LIGNOCELLULOSIC HYDROLYSATES

Implementation of the fractionation of acidic lignocellulosic hydrolysates depends on the type of the hydrolysate. With concentrated acid hydrolysates, the main issue is the separation of the monosaccharides from the hydrolysis acid and recovery of the acid for reuse. In addition, the by-products formed in the hydrolysis (acetic acid, furfural, and HMF) must be separated from the monosaccharides. Chromatographic separation with strong acid cation-exchange resin can be used to fractionate concentrated acid hydrolysates in a single step; however, it has been observed that it would be more economical to do the fractionation in two steps (Heinonen and Sainio, 2010; Heinonen et al., 2012a; Sainio et al., 2011; Sun et al., 2011; Xie et al., 2005). In the first step, the sugar degradation products (and possible phenolic compounds) are removed using adsorption (Chandel et al., 2007; Nilvebrant et al., 2001;

Ranjan et al., 2009; Sainio et al., 2011; Sun et al., 2011; Villarreal et al., 2006; Weil et al., 2002). In the second step, sulfuric acid and acetic acid are separated from the monosaccharides chromatographically using strong acid cation-exchange resins as adsorbents (Farone and Cuzens, 1993, 1995; Heinonen and Sainio, 2010, 2012; Hester and Farina, 1997; Nanguneri and Hester, 1990; Neuman et al., 1987; Sprinfield and Hester, 1999, 2001; Sun et al., 2011; Wooley et al., 1998; Xie et al., 2005).

With dilute acid hydrolysates, the recovery of hydrolysis acid is not essential due to low concentration of the acid in the hydrolysates (Gurgel et al., 2012; Hamelinck et al., 2005; Karimi et al., 2006; Lee et al., 1999; Takerzadeh and Karimi, 2007; Taherzadeh and Karimi, 2008; Xiang et al., 2004). These hydrolysates, however, contain many valuable by-products formed in hydrolysis which have to be removed prior to down-stream processing (fermentation to ethanol, etc.) (Galbe and Zacchi, 2002; Lee et al., 1999; Lenihan et al., 2010; Takerzadeh and Karimi, 2007; Taherzadeh and Karimi, 2008). The fractionation of dilute acid hydrolysates can be accomplished in one step similarly as the fractionation of concentrated acid hydrolysates (adsorption and chromatography).

Lignocellulosic hydrolysates contain strong (sulfuric acid) and weak electrolytes (mainly acetic acid, but may also contain formic acid and levulinic acid), and nonelectrolytes (monosaccharides, furfural, HMF). The concentrations of these components can be very different, for example, in concentrated acid hydrolysates, the sulfuric acid concentration can be onefold higher than the concentrations of the other components (Heinonen and Sainio, 2010).

Many different separation mechanisms can be observed in the adsorption/chromatographic fractionation of lignocellulosic hydrolysates. Chromatographic fractionation of (concentrated acid) lignocellulosic hydrolysates with strong acid cation-exchange resins as adsorbents is based on electrolyte exclusion (or ion exclusion) chromatography (Glód, 1997; Heinonen and Sainio, 2010; Laatikainen et al., 2011; Nanguneri and Hester, 1990; Neuman et al., 1987; Sprinfield and Hester, 1999). In this method, first reported by Wheaton and Bauman (1953), the strong electrolytes are excluded from the resin phase completely or partially due to electrical repulsion caused by the fixed ionic groups in the resin (Dorfner, 1991; Glód, 1997; Helfferich, 1995; Nanguneri and Hester, 1990; Neuman et al., 1987; Sprinfield and Hester, 1999; Wheaton and Bauman, 1953). As a consequence, the strong electrolytes (sulfuric acid, etc.) propagate fast through the column: breakthrough volume of the strong electrolytes corresponds to the dead volume of the resin bed (interstitial volume) due to zero sorption at infinite dilution (Glód, 1997; Heinonen and Sainio, 2012; Wheaton and Bauman, 1953). Weak electrolytes (acetic acid, etc.) and nonelectrolytes (monosaccharides, furfural, HMF, etc.) are not affected by the electrolyte exclusion and propagate slower through the column due to different interactions between these components and the adsorbent, and are, therefore, separated from the strong electrolytes. As a consequence, the strong electrolytes are separated from the weak electrolytes and nonelectrolytes. No ion exchange occurs in the electrolyte exclusion chromatography and, therefore, no regeneration of the resin due to ion exchange is required (Dorfner, 1991; Helfferich, 1995; Wheaton and Bauman, 1953). In the sugar industry, electrolyte exclusion chromatography is used for the separation of sucrose and betaine from molasses using sequential SMB system (see, e.g., Heikkilä et al., 1989).

In the following chapters, the separation mechanisms observed in chromatographic separation and adsorption processes are elucidated. Afterwards, a simple model, consisting of phase equilibrium models and dynamic column model, for the chromatographic fractionation of lignocellulosic hydrolysates is presented.

4.1. Sorption mechanisms

4.1.1 Electrolyte exclusion

Chromatographic separation of strong electrolytes from weak electrolytes and nonelectrolytes is based on the electrolyte exclusion. Strong electrolytes are excluded from the resin phase (i.e., the solution phase inside the resin pores) due to the electrolyte exclusion, but the weak electrolytes and nonelectrolytes are not affected and are, therefore, adsorbed to the resin. As a consequence, the strong electrolyte propagates through the column faster than the weak electrolytes and nonelectrolytes and separation occurs. No actual ion exchange occurs in the electrolyte exclusion (Dorfner, 1991; Helfferich, 1995; Wheaton and Bauman, 1953).

The electrolyte exclusion can be explained in terms of the Donnan potential as follows. When a strong acid cation-exchange resin is put into a dilute solution of a strong electrolyte, there exists a large concentration difference between the phases. Cation (counterion, i.e., exchangeable ion) concentration is higher in the resin phase, whereas the mobile anion (coion, i.e., ion with same sign of charge as the fixed ionic group in the resin) concentration is higher in the bulk solution phase. Diffusion of cations from the resin phase to bulk solution phase and the opposite movement of anions from the bulk solution phase to the resin phase would erode these concentration differences. However, because these species are ionic, the diffusion would interfere with electroneutrality. The first few ions that try to diffuse create a potential difference between the phases known as Donnan potential which draws the cations back to the resin phase and anions back to the bulk solution phase (Dorfner, 1991; Helfferich, 1995; Wheaton and Bauman, 1953). Now that the anions of a strong electrolyte cannot diffuse into the resin phase, neither can the cations of the electrolyte because electroneutrality condition must hold in the bulk solution phase. As a consequence, the strong electrolyte is excluded from the resin phase.

The ions' tendency to level the concentration differences between the phases is counterbalanced by the effects of the electrical field and equilibrium, called Donnan equilibrium, is formed. Donnan equilibrium means that the electrical potential of each ion is equal in the bulk solution and resin phases (Dorfner, 1991; Helfferich, 1995). The consequence of this equilibrium is that the counterion (i.e., exchangeable ion in the resin) concentration in the ion exchanger remains higher and coion concentration lower in the resin phase than in the bulk solution phase. The situation is analogous with both cation and anion exchangers, but the sign of the Donnan potential is different: for cation-exchange resins, Donnan potential is negative, and for anion-exchange resins, it is positive (Dorfner, 1991).

Donnan potential, and therefore also electrolyte exclusion, depends on many factors. First of all, it decreases when the electrolyte concentration in the bulk solution increases (Dorfner, 1991; Helfferich, 1995; Wheaton and Bauman, 1953). Complete electrolyte exclusion can be achieved only at infinite dilution. As the electrolyte concentration increases, the exclusion effect diminishes and eventually vanishes completely. The sorption isotherms of strong electrolytes are unfavorable (concave upward) when electrolyte exclusion is the main sorption mechanism. In addition, the slopes of the sorption isotherms are zero at infinite dilution due to complete exclusion (Heinonen and Sainio, 2012; Heinonen et al., 2012b; Helfferich, 1995; Laatikainen et al., 2011; Wheaton and Bauman, 1953).

Although all kinds of ion exchangers (weak and strong, and anionic and cationic) can be used in the electrolyte exclusion chromatography, the separation efficiency is favored by high ionization level of the functional groups. Therefore, strong ion exchangers perform better (Helfferich, 1995; Wheaton and Bauman, 1953). High ion-exchange capacity (counterion concentration) and high degree of cross-linking (i.e., high charge density) increase the electrolyte exclusion (Dorfner, 1991; Helfferich, 1995).

The Electrolyte exclusion is higher for low valence of the counterion (exchangeable ion in the resin) due to the fact that the electrostatic force that affects the ion is proportional to the charge of the ion. However, high coion

(ion with same sign of charge as the fixed ionic groups in the resin) valence leads to stronger exclusion of electrolyte (Helfferich, 1995).

Size of the ions also affects the electrolyte exclusion: large ions are excluded from highly cross-linked resin due to the steric effects (Helfferich, 1995). Ion association or formation of ionic pairs between the counterions and the ionic groups in the resin localizes the counterions and decreases the effectiveness of the electrolyte exclusion. For example, when weak cation-exchange resins are only partially ionized (depends on solution pH), they do not exclude acids: acids are adsorbed to the ion-exchange resin like nonelectrolytes.

Complex formation between the ions in bulk solution and counterions in the resin can also weaken the effect of the electrolyte exclusion. This mechanism is known as site sharing (Helfferich, 1995). For example, sulfuric acid is sorbed to strong base anion-exchange resin in sulfate form because the sulfate counterions can form HSO_4^- complexes with the protons of sulfuric acid which is in the solution. This means that one sulfate counterion (valence of -2) is replaced by two HSO_4^- counterions (valence of -1) resulting in sorption of sulfuric acid. Sorption of strong electrolytes is also enhanced by increasing temperature and by various other interactions between electrolyte and ion-exchange resin (Dorfner, 1991; Helfferich, 1995).

The effect of electrolyte exclusion on the sorption of weak electrolytes (i.e., acetic acid in lignocellulosic hydrolysates) is very small. This is because of the low or nonexistent dissociation of weak electrolytes in the presence of the counterions in the resin (Dorfner, 1991; Helfferich, 1995; Wheaton and Bauman, 1953). In addition, for example, with acidic lignocellulosic hydrolysates (especially, concentrated acid hydrolysates), the high sulfuric acid concentration in the solution prevents dissociation of acetic acid (Heinonen et al., 2012b; Laatikainen et al., 2011). Weak electrolytes are adsorbed to the resin like nonelectrolytes (Helfferich, 1995).

Weak electrolytes and nonelectrolytes do not have a direct effect on the electrolyte exclusion (Helfferich, 1995; Wheaton and Bauman, 1953). However, these components can affect the electrolyte exclusion indirectly if they affect resin swelling/shrinking through activity of water. Resin swelling/shrinking affects electrolyte exclusion because it changes the charge density in the resin phase.

4.1.2 van der Waals interactions

Adsorption of nonelectrolytes on an ion-exchange resin with a hydrocarbon matrix is affected by van der Waals forces (London interactions and dipole–dipole interactions) (Dorfner, 1991; Helfferich, 1995; Rieman, 1961). These interactions favor local adsorption of hydrocarbon groups of solutes to the resin matrix and enhance the sorption of nonelectrolytes (and associated electrolytes). London interactions between the solute molecules and resin matrix are weak interactions. Dipole–dipole interactions between polar solvent molecules and between solvent and solute molecules are strong interactions (Dorfner, 1991; Helfferich, 1995). As a consequence of dipole–dipole interactions, organic solutes tend to coagulate or to be squeezed out of the polar solution onto the resin–solution phase boundary (Helfferich, 1995).

London interactions depend on the molecular structure of the hydrocarbon chains of the solutes and the matrix: if these are of similar type, stronger sorption occurs (Helfferich, 1995). For example, aromatic hydrocarbons or other hydrocarbons containing ring structures are more strongly adsorbed on a PS–DVB resin than aliphatic hydrocarbons.

With ion-exchange resins, interactions between polar solute groups and fixed ionic groups and counterions may be the strongest factor enhancing or reducing adsorption. Due to the ionic groups, ion-exchange resins are strongly polar adsorbents and tend to adsorb polar substances more strongly than nonpolar (Dorfner, 1991; Rieman, 1961). Repulsive forces may also exist between the polar fixed ionic groups and nonelectrolytes (Rieman, 1961). In addition, solvation of fixed ionic groups with the solute molecules can be one adsorption mechanism with ion-exchange resins (Section 4.1.5) (Dorfner, 1991).

Naturally, the concentration of adsorbing component in the bulk solution phase affects sorption: the higher the concentration, the higher the adsorbed amount. However, with components having favorable (concave downward shaped) isotherms, the distribution coefficients become smaller when their concentration increases due to the fact that a limited amount of substance can be adsorbed to a certain volume/mass of adsorbent. With components having unfavorable isotherms, the distribution coefficients increase with increasing concentration at least in the concentration range that is normally used. Usually, the adsorption isotherms of nonelectrolytes are favorable at least to some extent (Dorfner, 1991; Helfferich, 1995); however, exceptions can be found. For example, sorption isotherms of monosaccharides on a strong acid cation-exchange resin are slightly unfavorable. This is due to nonidealities in the liquid phase (Ching et al., 1987; Nowak et al., 2007, 2010; Vente et al., 2005), which in turn result from the large size of the solvated monosaccharide molecules (size exclusion, see below) (LÄHDE).

4.1.3 Size exclusion

Molecular size of the adsorbing components also affects sorption. Large molecules might not be adsorbed because their size prevents their diffusion into the pores of the adsorbent (Helfferich, 1995; Wankat, 1990). Resin pore size distribution is not uniform, and therefore, there are no clear limits for how large molecules can fit into the pores of an adsorbent (Helfferich, 1995). With resins that swell and shrink considerably, due to changes in the solution-phase composition, the pore size, and therefore also the effect of size exclusion to sorption, depends on resin swelling. Those components that cause strong swelling (increase in pore size) are more easily adsorbed (Dorfner, 1991).

In addition to the size exclusion effect, degree of resin cross-linking affects the sorption of solute molecules of different size also through swelling pressure (see Section 4.2). Inside a swollen resin is a rather large swelling pressure that tends to squeeze solvent and solute molecules out of the resin phase. Large molecules are more strongly affected than small ones. The importance of the swelling pressure effect increases as the degree of resin cross-linking and solute molecule size increases. Therefore, with resins having a moderate degree of cross-linking, the effect of swelling pressure on the sorption of small molecules is rather small (Helfferich, 1995).

4.1.4 Partition

Partition means that the chromatographic separation is based mainly on the differences between the solubilities of the solutes in the mobile and stationary phases (IUPAC, 2012). The latter means here the solution inside the pores of a porous adsorbent. In partition chromatography, ion-exchange resins are used as adsorbents and a salt (NaCl, etc.) or an organic solvent (ethanol, etc.) is added to the aqueous bulk solution phase. The role of the additive is to either increase or decrease the distribution coefficient of the adsorbing components between the bulk solution phase and the solution phase inside the resin pores (Hunter and Carta, 2001; Mirica et al., 2012; Nfor et al., 2011; Ng and Snyder, 2012; Rieman, 1961; Sargent and Rieman, 1956; Tiihonen et al., 2002b).

Salting out is a special case of partition chromatography: addition of salt to a bulk solution decreases the solubility of organic nonelectrolytes (and weak electrolytes) to the bulk solution. These components are salted out from the bulk solution phase to the solution phase inside the resin pores. In other words, the salt has a cooperative effect on the sorption of these organic nonelectrolytes. Salting out chromatography has been used, for example, in the fractionation of alcohols, aldehydes and ketones, ethers, and amines (Rieman, 1961; Sargent and Rieman, 1956).

The salting out mechanism can be explained as follows. Only "free water," that is, the water (or other solvent) that is not bound to ionic solvation shells, is available for dissolving nonelectrolytes and associated weak electrolytes (if these solutes do not displace solvation water, see below).

Inside the pores of an ion-exchange resin, the "free water" content is low due to solvation of the fixed ionic groups and the counterions. Due to this, the nonelectrolytes are salted out from the resin phase to the bulk solution phase where the "free water" content available for dissolution of the non-electrolytes is higher (Helfferich, 1995).

However, salting out works in two ways. If a strong electrolyte is added to the bulk solution, the "free water" content in the bulk solution decreases because water molecules form solvation shells with the ions of the strong electrolyte. This leads to salting out of the nonelectrolytes and associated weak electrolytes from the bulk solution phase to the resin phase (solution inside the pores of the resin) and the sorption of these species increases. This is the case, for example, when sulfuric acid is added to a solution of monosaccharides and acetic acid in the presence of a strong acid cation-exchange resin (Heinonen and Sainio, 2010, 2012).

The salting out phenomenon has been observed in the chromatographic fractionation of lignocellulosic hydrolysates with strong acid cation-exchange resins in acid form (Heinonen and Sainio, 2012; Laatikainen et al., 2011; Neuman et al., 1987; Sprinfield and Hester, 2001). In this case, the sorption of the nonelectrolytes and weak electrolytes in the hydrolysates (monosaccharides, acetic acid, furfural, and HMF) is enhanced by sulfuric acid: the higher the sulfuric acid concentration, the stronger the sorption of the organic components (Heinonen and Sainio, 2010, 2012; Laatikainen et al., 2011; Neuman et al., 1987; Sprinfield and Hester, 2001). Due to this, very peculiar elution profiles of the organic solutes have been reported in the chromatographic fractionation of concentrated acid lignocellulosic hydrolysates (see Section 5.2).

4.1.5 Complex formation

Sugars can be separated from each other using cation-exchange resins with metal ions as counterions because sugar molecules can form donor–acceptor complexes with metal counterions (Angyal, 1973; Helfferich, 1995; Howard et al., 1988; Reeves, 1949a, 1949b; Saladini et al., 2001; Stefansson and Westerlund, 1996; Tiihonen et al., 2002a, 2002b). This kind of complex formation is called ligand exchange: no ion-exchange occurs, but the resin functions solely as a solid carrier for the metal ions (Helfferich, 1995; Howard et al., 1988). Strong acid cation-exchange resins with PS–DVB matrix are usually used as stationary phase in ligand-exhange chromatography (Stefansson and Westerlund, 1996), and water as eluent (Helfferich, 1995; Howard et al., 1988; Saladini et al., 2001; Stefansson and Westerlund, 1996; Tiihonen et al., 2002a, 2002b).

Cations capable of forming complexes with ligands include Ca^{2+}, Cu^{2+}, Ni^{2+}, Pb^{2+}, Sr^{2+}, Ba^{2+}, Y^{3+}, La^{3+}, Pr^{3+}, and Ag^{2+} (Helfferich, 1995; Howard et al., 1988; Stefansson and Westerlund, 1996). Higher valence of the metal ion leads to more stable complexes (Angyal, 1973). Complexes formed by univalent cations (Na^{+}, etc.) are usually very weak, and therefore, these ions are considered noncomplexing (Tiihonen et al., 2002a). Ca^{2+} is the most suitable for sugar separation purposes. With Pb^{2+}, for example, the sugar–metal complexes are too strong, which leads to increased retention times in chromatographic separations (Stefansson and Westerlund, 1996).

Ligand exchange is a highly selective chromatographic separation method, and there is a strong driving force for the sorption of ligands because the ligands are efficiently sequestered by the metal ions. The selectivity depends usually on the solution concentration. When the solution concentration is low, ligands with higher coordinative valence are favored. At high solution concentrations, however, ligands with lower coordinative valences are favored. In addition, the metal ions prefer ligands that have the strongest tendency to form complexes, enabling the separation of different ligands. Complexes are formed already when the bulk solution-phase concentration of the organic component is very low; this property can be efficiently used in the separation of ligands and nonligands (Helfferich, 1995).

In ligand exchange, water (solvent) molecules in the hydration (solvation) spheres of metal counterions are replaced by hydroxyl groups (or other highly polar groups like carboxylic groups) of the organic molecules resulting in complex formation (Angyal, 1973; Helfferich, 1995; Howard et al., 1988; Saladini et al., 2001). The number of water molecules in the hydration sphere of the metal ion and, therefore, also the number of ligands that can complex with the metal ion depend on the coordination number of the ion and the type of the ligand (mono-, bi-, or tridentate ligand) (Helfferich, 1995; Stefansson and Westerlund, 1996). The ligands can also be replaced by other ligands. Organic compounds that can act as ligands include polyols (sugars and sugar alcohols), anions of carboxylic and amino acids, ammonia, and aliphatic amines (Helfferich, 1995).

Formation of metal–ligand complexes requires that the polar functional groups of the organic ligands are attached to adjacent carbon atoms and properly oriented (and spaced): conformation of the organic molecules (functional groups) determines their relative affinity for the metal ions, their distribution coefficients, and the stability of the complexes (Angyal, 1973; Howard et al., 1988; Reeves, 1949a, 1949b; Stefansson and Westerlund, 1996).

The conformation of the ligands determines the distance (sterically favorable) between adjacent hydroxyl groups. If the distance is close to an optimal distance for complex formation, or the energy required to move the hydroxyl groups to optimal distance from each other is small, a complex is formed. The optimal distance between the adjacent hydroxyl groups is determined by the ionic radius of the metal ion. With the axial–equatorial sequence of hydroxyl groups, the optimal ionic radius is approximately 1 Å (Ca^{2+} 0.99 Å, Na^{+} 0.97 Å, La^{3+} 1.02 Å) (Angyal, 1973; Stefansson and Westerlund, 1996; Tiihonen et al., 2002a, 2002b).

Complex formation requires at least two functional groups in adjacent positions and in right conformation: monohydric alcohols (e.g., ethanol) do not form complexes with metal ions (Angyal, 1973; Stefansson and Westerlund, 1996; Tiihonen et al., 2002a, 2002b). Complex formation between metal ion and a monosaccharide with six-carbon atom rings occurs if the hydroxyl groups are in axial–equatorial sequence (Angyal, 1973; Howard et al., 1988; Stefansson and Westerlund, 1996). For example, α- and β-glucoses can be separated from each other with Ca^{2+} form strong acid PS–DVB resin because the former has hydroxyl groups in axial–equatorial sequence but the latter does not (Howard et al., 1988). Complex formation with ligands containing five-carbon atom rings occurs if the hydroxyl groups are in *cis–cis* sequence (Angyal, 1973; Stefansson and Westerlund, 1996).

Generally, carbohydrate complexes in neutral and acidic aqueous solutions have a low stability due to the poor donor ability of the hydroxylic oxygen atoms (Saladini et al., 2001). In alkaline conditions, the carbohydrate complexes are relatively strong, because of the dissociation of carbohydrates which are weak acids (pK_a in order of 12–14) (Angyal, 1973; Stefansson and Westerlund, 1996). In addition, the selectivity of the ligand exchange for carbohydrates increases in alkaline conditions because common ligands such as amino acids, hydroxy acids, and amines stay unbound. However, increase of pH beyond the limit value of full ionization of carbohydrates leads to decrease in the complex stability due to competition for complex formation with the increasing amount of hydroxyl ions (Stefansson and Westerlund, 1996).

4.2. Resin swelling and shrinking

Ion-exchange resins sorb solvents in which they are placed and swell due to this. The swelling results from the osmotic pressure difference between the external solution and the resin phase (solution inside resin pores): the ion concentration is considerably higher in the resin than in the external solution. Solvent is taken up in order to decrease the osmotic pressure difference between the two phases, that is, to dilute the ion concentration in the resin phase (Dorfner, 1991; Helfferich, 1995). When ion-exchange resins swell, the solvent molecules form solvation shells with the ionic groups and counterions of the resin. If the resins would not be cross-linked, the solvation of the ions and ionic groups would lead to dissolution of the polymers (Helfferich, 1995). However, due to the cross-linked structure of the resins, swelling continues only until an equilibrium is attained. In this equilibrium, the solvation of the ions which causes stretching of the resin matrix (i.e., the dissolution tendency) is balanced by the elastic forces of the resin matrix (i.e., resistance for the stretching of the matrix) (Dorfner, 1991; Helfferich, 1995).

The solution inside the pores of a resin is under higher pressure than the external solution due to the contradictive forces of the resin matrix. This pressure difference between the external solution and the solution in the resin pores is known as swelling pressure. It affects the solvent uptake and therefore also the solvent activity in the resin. The swelling pressure increases when the resin swells due to an increase in the elastic forces opposing the swelling. The swelling pressure also increases when the degree of resin cross-linking increases due to stronger forces opposing the swelling. On the other hand, swelling tendency of a resin also decreases with increasing degree of cross-linking due to more rigid polymer network. Swelling pressure affects also the sorption of solutes, especially larger solutes (see Section 4.1.3) (Helfferich, 1995).

Many forces contribute to resin swelling. First of all, the formation of solvation shells for the counterions and ionic groups, and the decrease in the osmotic pressure difference. Decrease in the osmotic pressure difference occurs when ion concentration in the bulk solution increases. The electrostatic repulsion between the ionic groups in the resin contributes to resin swelling as well as the interactions between the solvent and the resin matrix especially if the solvent is organic (Dorfner, 1991; Helfferich, 1995).

Polar solvents are better swelling agents than nonpolar solvents due to stronger interactions between the solvent and the ions and ionic and polar groups of the resin. Complete ionization of ionic groups also increases swelling because of greater solvation. Ion pair formation and association of the counterions and ionic groups decrease the solvation tendency and, therefore, also the swelling. Affinity of the fixed ionic groups toward the solvent also affects the swelling: the greater the affinity, the larger is the swelling. In addition, nature of the ionic groups affects the swelling: resins containing sulfonic acid groups swell more than resins that contain acetate groups. Ion-exchange resins with high ion-exchange capacity swell more than resin with small capacity due to the higher ion concentration in the resin which results in higher osmotic pressure difference (Dorfner, 1991; Helfferich, 1995).

Increasing solute concentration decreases resin swelling (Dorfner, 1991; Helfferich, 1995). This is particularly clear with strong electrolytes which cause considerable decrease in swelling at high concentrations. This shrinking is due to decreasing osmotic pressure (activity) difference. In other words, water is drawn from the resin to solvate the electrolyte ions in the bulk solution.

The effect of the nature of the counterions to resin swelling is more complex than the effects listed above. With moderately and highly cross-linked resins, the swelling is affected by the solvation tendency and size of the counterion: ions with large hydrated ionic volume result in greater swelling. However, in highly cross-linked resins, the solvation of the ions may be incomplete (steric obstructions), and in this case, resin swelling depends on the size of the actual ion (Helfferich, 1995).

With weakly cross-linked resins (gel-type resins), the valence of the counterion is important for resin swelling. These resins contain large amounts of free water, that is, solvent not bound to the solvation shells of the ionic species in the resin. The amount of free water depends on the number of the counterions which, in turn, depends on the valence of the ions. When univalent counterions are replaced with bivalent ions, the number of counterions is cut to half which, in turn, results in lower swelling of the gel-type resins (Dorfner, 1991; Helfferich, 1995). For example, swelling of gel-type sulfonated PS–DVB resins decreases according to the sequence $Na^+ > Ca^{2+} > Al^{3+}$ (Dorfner, 1991). In addition, if the counterions are exchanged for ions with larger hydrated radius, free water is squeezed out of the resin due to increased swelling pressure (Helfferich, 1995).

The effect of temperature on swelling of common ion-exchange resins used in separation applications is not very significant. Usually, resin swelling increases with increasing temperature (Dorfner, 1991).

Resin volume changes affect diffusion coefficients and dispersion coefficients in chromatographic separation. When resin swells, the pore size of the resin increases and therefore also the diffusion rates are increased. The opposite is true when resin shrinks. Resin shrinking increases also dispersion and favors channeling especially at the column walls (Helfferich, 1995). When resin swelling or shrinking is considerable, it must be taken into account in the modeling of the chromatographic separation or ion exchange. Usually, this complicates the modeling considerably.

4.3. Phase equilibrium models

Accurate modeling of phase equilibria is essential for rigorous modeling of chromatographic separation processes. The phase equilibrium models include the sorption isotherm models and the possible resin swelling/shrinking model.

In the earliest studies regarding the chromatographic fractionation of lignocellulosic hydrolysates, the phase equilibrium was modeled simply with constant distribution coefficients (Nanguneri and Hester, 1990; Neuman et al., 1987). With these simple phase equilibrium models, the correlation between the experimental and calculated results was inadequate. Wooley et al. (1998) and Xie et al. (2005) modeled the phase equilibrium in the case of dilute acid lignocellulosic hydrolysates with linear and Langmuir isotherms depending on the adsorbent used. In the relatively narrow concentration ranges used in these studies, satisfactory correlation between the experimental and calculated results was obtained.

The modeling of the phase equilibria in the case of chromatographic fractionation of dilute acid lignocellulosic hydrolysates with gel-type ion-exchange resins is relatively easy due to low sulfuric acid concentration in these hydrolysates (Wooley et al., 1998; Xie et al., 2005). The situation is somewhat more complicated when fractionation of a concentrated acid lignocellulosic hydrolysate using a gel-type strong acid cation-exchange resin in acid (H^+) form is studied (Heinonen and Sainio, 2010, 2012; Heinonen et al., 2012a, 2012b; Laatikainen et al., 2011). In these systems, the concentration of sulfuric acid is considerably high, and it therefore affects the sorption of the other solutes present in the system (salting out, see Section 4.1.4) as well as the resin volume (resin swelling, see Section 4.2). In addition, due to high concentration and electrolyte exclusion, the sorption of sulfuric acid cannot be modeled with linear isotherm model as was done by Wooley et al. (1998) and Xie et al. (2005) in the fractionation of dilute acid hydrolysates. Heinonen and Sainio (2012) successfully used simple empirical nonlinear correlations to describe the phase equilibria in a system of concentrated acid lignocellulosic hydrolysate and gel-type strong acid cation-exchange resin in H^+ form. Simple isotherm models were derived for the main components of concentrated acid lignocellulosic hydrolysates: sulfuric acid, monosaccharides (glucose and xylose), and acetic acid (Heinonen and Sainio, 2012). This modeling approach is reviewed here in detail.

In chromatographic fractionation of lignocellulosic hydrolysates with strong acid cation-exchange resin in acid form, sulfuric acid sorption is affected by electrolyte exclusion (Dorfner, 1991; Laatikainen et al., 2011):

at infinite dilution, the sorption of sulfuric acid is zero because of complete electrolyte exclusion. As the sulfuric acid concentration increases, the acid sorption also increases as the electrolyte exclusion effect diminishes. For these reasons, the slope of the sulfuric acid isotherm model has to be zero at infinite dilution and increases with concentration. The simplest expression which can be used to describe accurately sulfuric acid sorption on a strong acid cation-exchange resin in acid form is (Heinonen and Sainio, 2012)

$$q_{H_2SO_4} = \alpha_{H_2SO_4}(C_{H_2SO_4})^{\beta_{H_2SO_4}}, \qquad [5.1]$$

where q and C are solid phase and liquid-phase concentrations, respectively, and α and β are positive constants. Physical constraints (finite volume of eluent during column regeneration) require that β is larger than unity.

Monosaccharide sorption isotherms on a strong acid cation-exchange resin in acid form are slightly unfavorably shaped (concave upward) (Heinonen and Sainio, 2012; Laatikainen et al., 2011). Similar isotherms for monosaccharides have also been obtained also earlier with strong acid cation-exchange resin in calcium form (Ching et al., 1987; Nowak et al., 2007, 2010; Vente et al., 2005) and acid form (Nanguneri and Hester, 1990). The unfavorable shapes of the monosaccharide isotherms originate from nonidealities in the liquid phase (Ching et al., 1987; Nowak et al., 2007, 2010; Vente et al., 2005).

Neuman et al. (1987) first noted that glucose sorption is enhanced by the presence of sulfuric acid. This was later verified for glucose, xylose, acetic acid, furfural, and HMF by Heinonen and Sainio (2012), Laatikainen et al. (2011), and Sainio et al. (2011). This cooperative effect is due to salting out phenomenon (see Section 4.1.4). The glucose and xylose sorption on a strong acid cation-exchange resin in H^+ form can be described with a simple anti-Langmuir type isotherm model in which the Henry constant depends linearly on liquid-phase concentration of sulfuric acid (Heinonen and Sainio, 2012)

$$q_k = \frac{(\alpha_k + \beta_k C_{H_2SO_4})C_k}{1 - \sum_{j=1}^{n} \gamma_j C_j}, \qquad [5.2]$$

where subscript k stands for glucose and xylose, γ is a positive constant, and the sum term in the nominator stands for the total monosaccharide concentration.

Acetic acid is a weak electrolyte ($K_a \approx 1.6 \times 10^{-5}$ mol/L at 50 °C (Harned and Ehlers, 1933)). In acidic conditions, for example, in concentrated acid hydrolysates and in the presence of strong acid cation-exchange resin in acid form, acetic acid is associated and can be treated as nonelectrolyte. Due to the association of acetic acid, electrolyte exclusion does not affect its sorption. Cooperative effect of sulfuric acid also on acetic acid sorption has been observed (Heinonen and Sainio, 2012; Laatikainen et al., 2011; Sainio et al., 2011). In the concentration range in which acetic acid is found on lignocellulosic hydrolysates, the sorption of acetic acid on a strong acid cation-exchange resin in acid form can be described with a linear isotherm model with the Henry constant depending on liquid-phase sulfuric acid concentration (Heinonen and Sainio, 2012)

$$q_{AcOH} = (\alpha_{AcOH} + \beta_{AcOH} C_{H_2SO_4}) C_{AcOH}. \quad [5.3]$$

The modeling of chromatographic fractionation of lignocellulosic hydrolysates with a gel-type strong acid cation-exchange resin in H^+ form is complicated due to changes in resin volume. These changes occur when the solution-phase composition changes during the separation. Accurate modeling of these volume changes is crucial in order to precisely model the aforementioned separation process.

Many approaches to model resin volume changes in adsorption processes have been applied (Durão et al., 1992; Marra and Cooney, 1973, 1978; Mazzotti et al., 1997; Sainio and Paatero, 2007; Tiihonen et al., 2002a, 2002b). In aqueous solutions, the activity of water is the main solution property of interest (Helfferich 1952). In studies concerning the modeling of chromatographic fractionation of concentrated acid lignocellulosic hydrolysates, Heinonen and Sainio (2012) demonstrated that the volume changes of the strong acid cation-exchange resin particles can be directly related to the sorbed amount of sulfuric acid if such experimental data are available. The effect of the other adsorbates present in these systems to resin volume changes is insignificant. A simple empirical correlation for the extent of resin shrinking was applied (Heinonen and Sainio, 2012)

$$\theta = \frac{V_{bed,acid}}{V_{bed,water}} = 1 - \frac{\delta_1 q_{H_2SO_4}}{\delta_2 + q_{H_2SO_4}}, \quad [5.4]$$

where θ is extent of resin shrinking, that is, the ratio of resin bed volume (V_{bed}) in acid to resin bed volume in water, $q_{H_2SO_4}$ is the adsorbed amount of sulfuric acid, and δ_1 and δ_2 are positive constants. Eq. (5.4) is scaled so that θ is unity in pure water, and decreases with increasing $q_{H_2SO_4}$.

In this chapter, a review was given of simple empirical correlations used to describe the phase equilibrium in a system of lignocellulosic hydrolysate and a gel-type strong acid cation-exchange resin. If the reader is interested in other modeling approaches which have been used for the aforementioned system, these can be found in studies by Heinonen et al. (2012b) and Laatikainen et al. (2011). These models can also be applied for other chromatographic separations of electrolytes and nonelectrolytes.

4.4. Dynamic column model for the fractionation of lignocellulosic hydrolysates

Several dynamic column models have been used in the modeling of chromatographic fractionation of lignocellulosic hydrolysates (Heinonen and Sainio, 2012; Laatikainen et al., 2011; Nanguneri and Hester, 1990; Neuman et al., 1987; Wooley et al., 1998; Xie et al., 2005). The first models used for this purpose were an equilibrium-dispersive model used by Neuman et al. (1987) and a plate model used by Nanguneri and Hester (1990). A more accurate column model, the VERSE (VErsatile Reaction and SEparation) model which includes column mass balances for both bulk solution and resin pore solution, and the possibility for chemical reaction, was used by Wooley et al. (1998) and Xie et al. (2005). The aforementioned column models are applicable for the modeling of chromatographic fractionation of dilute hydrolysates, but they are not adequate for fractionation of concentrated acid hydrolysate with gel-type strong acid cation-exchange resins. This is because the models do not take into account volume changes of the gel-type resin caused by sulfuric acid in the concentrated acid hydrolysates. Heinonen and Sainio (2012) presented a column model that takes into account the resin volume changes as changing local resin bed porosity. This model is well applicable for the modeling of chromatographic fractionation of concentrated acid hydrolysates. The column model by Heinonen and Sainio (2012) is reviewed below.

In the model by Heinonen and Sainio (2012) the resin particles are assumed to be attached to a fixed two-dimensional grid (Mazzotti et al., 1997). Change in the sulfuric acid concentration causes a change to the resin particle diameter (resin volume) and therefore also to the void volume (bed porosity).

The mass balance in the chromatographic column for component i over a small volume element of length Δz was given by Heinonen and Sainio (2012) as

$$\frac{\partial C_i}{\partial t}=-\frac{\dot{V}}{\varepsilon A_{col}}\frac{\partial C_i}{\partial z}+\frac{q_i-C_i}{\varepsilon}\frac{\partial \varepsilon}{\partial t}-\frac{1-\varepsilon}{\varepsilon}\frac{\partial q_i}{\partial t}+\frac{D_{ax,i}}{\varepsilon}\frac{\partial}{\partial z}\left(\varepsilon\frac{\partial C_i}{\partial z}\right). \quad [5.5]$$

where ε is local bed porosity, $\dot{V}$ is volumetric flow rate, A_{col} is column cross-sectional area, D_{ax} is axial dispersion coefficient, t and z are temporal and spatial coordinates, respectively. In Eq. (5.5), C_i, q_i, and ε are functions of both t and z. The second term on the right-hand side of Eq. (5.5) takes into account the influence of changing phase ratio due to resin volume changes on the liquid-phase concentration.

Danckwert's boundary conditions were used at the column inlet and outlet:

$$\frac{\partial C_i}{\partial z}=\frac{\dot{V}}{A_{col}\varepsilon}\frac{\left(C_i^{in}-C_i\right)}{D_{ax,i}},\quad z=0, \tag{5.6}$$

$$\frac{\partial C_i}{\partial z}=0,\quad z=h_{col}, \tag{5.7}$$

where C_i^{in} is the concentration at the column inlet and h_{col} is the column height. The initial condition was

$$C_i(z,t=0)=0,\quad 0\leq z\leq L. \tag{5.8}$$

Resin bed porosity ε was calculated from

$$\varepsilon=1-\theta(1-\varepsilon_{ref}), \tag{5.9}$$

where ε_{ref} is bed porosity in reference state, which is a column equilibrated with pure water in the current case (Heinonen and Sainio, 2012). The rate of bed porosity change is obtained through differentiation of Eq. (5.9) with respect to time:

$$\frac{\partial \varepsilon}{\partial t}=-\frac{\partial \theta}{\partial t}(1-\varepsilon_{ref}), \tag{5.10}$$

from which, by using chain rule and Eq. (5.4), an equation for the rate of bed porosity change depending on the rate of sulfuric acid sorption is obtained:

$$\frac{\partial \varepsilon}{\partial t}=\frac{\alpha_\theta\beta_\theta}{\left(\beta_\theta+q_{H_2SO_4}\right)^2}\frac{\partial q_{H_2SO_4}}{\partial t}(1-\varepsilon_{ref}). \tag{5.11}$$

Heinonen and Sainio (2012) took into account the effect of adsorption kinetics on the resin shrinking by relating the extent of resin shrinking to the adsorbed amount of sulfuric acid.

Solid film linear driving force model (Guiochon et al., 2006) was used to describe mass transfer between the phases:

$$\frac{\partial \bar{q}_i}{\partial t}=k_{m,i}\left(q_i^*-\bar{q}_i\right)+\frac{\bar{q}_i}{1-\varepsilon}\frac{\partial \varepsilon}{\partial t}, \tag{5.12}$$

where $k_{m,i}$ is intraparticle mass transfer coefficient, $q_i{}^*$ is solid phase concentration of i at equilibrium with liquid-phase concentration C_i, and $\bar{q}_i$ is average solid-phase concentration of i. The second term on the right-hand side of Eq. (5.12) originates from a change in the phase ratio due to resin volume change.

The intraparticle mass transfer coefficient can be calculated from (Heinonen and Sainio, 2012)

$$k_{\mathrm{m,i}} = \frac{60 D_{\mathrm{i}}^{P}}{d_{\mathrm{p}}^{2}}, \quad [5.13]$$

where $D_{\mathrm{i}}^{\mathrm{P}}$ is diffusion coefficient which depends on resin shrinking. Diffusion coefficient can be calculated from the correlation of Mackie and Meares (Laatikainen et al., 2011; Mackie and Meares, 1955):

$$D_{\mathrm{i}}^{\mathrm{P}} = D_{\mathrm{i,ref}}^{\mathrm{P}} \left[\frac{(1-\Psi_{\mathrm{p}})(1+\Psi_{\mathrm{p,ref}})}{(1+\Psi_{\mathrm{p}})(1-\Psi_{\mathrm{p,ref}})} \right]^{2}, \quad [5.14]$$

where $D_{\mathrm{i,ref}}^{\mathrm{P}}$ is diffusion coefficient in the reference state, Ψ_{p} volume fraction of the polymer, and $\Psi_{\mathrm{p,ref}}$ volume fraction of the polymer in reference state. Volume fraction of the polymer can be calculated from

$$\Psi_{\mathrm{p}} = \frac{\Psi_{\mathrm{p,ref}}}{\theta}. \quad [5.15]$$

The volume fraction of the polymer in reference state for water swollen gel-type strong acid cation-exchange resin (sulfonated PS–DVB resin, 8 wt% DVB content) in acid form (e.g., CS16GC, Finex Oy, Finland) is 0.391 (Laatikainen et al., 2011). Particle diameter d_{p} can be calculated, when taken into account that the total volume of a volume element remains constant, from

$$d_{\mathrm{p}} = d_{\mathrm{p,ref}} \left(\frac{1-\varepsilon}{1-\varepsilon_{\mathrm{ref}}} \right)^{1/3}, \quad [5.16]$$

where $d_{\mathrm{p,ref}}$ is particle diameter in the reference state (here: water swollen resin).

The partial differential equations in Eq. (5.5) in the column model presented in this chapter can be easily solved numerically using the Method of Lines (Schiesser, 1991). In the study by Heinonen and Sainio (2012), the diffusion coefficients in reference state in Eq. (5.14) and the axial dispersion coefficients (same value for all components) in Eq. (5.5) were estimated by fitting the simulated elution profiles to the experimentally obtained profiles.

5. CHROMATOGRAPHIC FRACTIONATION OF LIGNOCELLULOSIC HYDROLYSATES

Chromatographic fractionation of lignocellulosic hydrolysates has been studied with three types of adsorbents: strong acid cation-exchange resins in acid (H^+) form (Farone and Cuzens, 1993, 1995; Heinonen and Sainio, 2010, 2012; Heinonen et al., 2012a, 2012b; Nanguneri and Hester, 1990; Neuman et al., 1987; Sprinfield and Hester, 1999, 2001; Wooley et al., 1998; Xie et al., 2005), weakly basic poly-4-vinyl-pyridine resin (PVP, Reillex HP, Reilly Industries Inc.) (Xie et al., 2005), and basic anion-exchange resin (DIAION MA03SS, Mitsubishi Chemical Co.) in sulfate form (Sun et al., 2011). In all cases water was used as eluent. The strong acid cation-exchange resin has received most attention because no regeneration of the resin is needed (Farone and Cuzens, 1993, 1995; Heinonen et al., 2012a, 2012b; Heinonen and Sainio, 2010, 2012; Nanguneri and Hester, 1990; Neuman et al., 1987; Sprinfield and Hester, 1999, 2001; Wooley et al., 1998; Xie et al., 2005): the separation of the components in the hydrolysates is based on electrolyte exclusion, sorption, partition, and size exclusion.

The need for adsorbent regeneration is avoided also if anion-exchange resin in sulfate form is used as an adsorbent for the fractionation of lignocellulosic hydrolysates (Sun et al., 2011). The separation is also based on same principles as with strong acid cation-exchange resin in acid form. However, the exclusion of sulfuric acid from the resin phase is not as efficient as with strong acid cation-exchange resin in acid form: larger counterion (sulfate ion) leads to less efficient exclusion and, in addition, sulfuric acid is sorbed to the resin by site sharing mechanism (see Section 4.1.1). Use of basic anion-exchange resins for the fractionation of lignocellulosic hydrolysates (Xie et al., 2005) is discouraged if ion exchange is involved due to the need for regeneration of the adsorbents.

5.1. Phase equilibrium

The phase equilibrium in a system of concentrated acid lignocellulosic hydrolysate and a gel-type strong acid cation-exchange resin in acid (H^+) form were determined by Heinonen and Sainio (2012) and Laatikainen et al. (2011). Sorption isotherms of the main components of concentrated acid hydrolysates (sulfuric acid, glucose, xylose, acetic acid) and resin volume changes due to the changes in the liquid-phase composition are shown here.

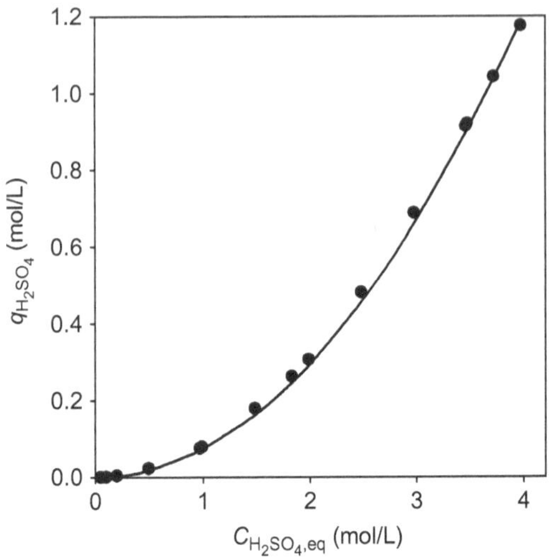

Figure 5.5 Sorption of sulfuric acid from water on a gel-type strong acid cation-exchange resin (sulfonated PS–DVB resin, 8 wt% DVB content, d_p 246 μm, H^+ capacity 1.93 meq/mL (Heinonen and Sainio, 2010)) in acid (H^+) form. Experimental details: measurement of the isotherm with pulse on a plateau method (Guiochon et al., 2006); flow rate = 0.2 cm/min, d_{col} = 2.5 cm, h_{col} = 20 cm, top–down flow, water as eluent, and T = 50 °C. Line is the fit of the empirical isotherm model ((Eq. 5.1), see Section 4.3). Model parameters are given in Table 5.2. *Reprinted with permission from Heinonen and Sainio (2012). Copyright: 2012. John Wiley and Sons.*

The sorption of sulfuric acid on a gel-type strong acid cation-exchange resin (sulfonated PS–DVB resin, 8 wt% DVB content, CS16GC, Finex Oy) in acid form is shown in Fig. 5.5. Sulfuric acid sorption is affected by electrolyte exclusion. The shape of the sulfuric acid isotherm (Fig. 5.5) is clearly unfavorable (concave upward). This is typical for strong electrolytes when electrolyte exclusion (see Section 4.1.1) affects their sorption at low concentrations. The Henry constant of the isotherm at infinite dilution is zero because complete electrolyte exclusion prevents all HSO_4^- and SO_4^{2-} ions from entering the pores of the resin. As a consequence, the whole electrolyte is excluded due to electroneutrality condition.

As the sulfuric acid concentration increases, the effect of electrolyte exclusion on sulfuric acid sorption diminishes and eventually vanishes completely. The weakening of the exclusion is seen from the increasing slope of the isotherm (Fig. 5.5). The isotherm shape becomes linear when the effect of the electrolyte exclusion vanishes: in the case presented here, this occurs when the sulfuric acid concentration in solution is higher than 3 mol/L. The correlation between the simple empirical sorption isotherm model for sulfuric acid ((Eq. 5.1), see Section 4.3; model parameters are given in Table 5.2) and the experimental results is good (Fig. 5.5).

Lignocellulosic hydrolysates contain many monosaccharides originating from cellulose and hemicelluloses. However, it has been observed that with a gel-type strong acid cation exchange resins in acid form they behave very similarly, and general sorption behavior can be characterized by studing only a few monosaccharides (Heinonen and Sainio, 2010; Xie et al., 2005). Only the sorption of glucose and xylose on a gel-type strong acid cation-exchange resin (sulfonated PS–DVB resin, 8wt% DVB content, CS16GC, Finex Oy) in acid form is presented in here.

Glucose and xylose sorption isotherms (Fig. 5.6) on strong acid PS–DVB resin are slightly unfavorable (concave upward) due to the nonidealities in the solution phase (Ching et al., 1987; Nowak et al., 2007, 2010; Vente et al., 2005). Sulfuric acid has a considerable cooperative effect on the sorption of the monosaccharides. For glucose, this kind of behavior was first observed by Neuman et al. (1987). The effect results from the salting out effect (see Section 4.1.4 for details). The monosaccharide sorption on a strong acid cation-exchange resin is almost doubled when the sorption occurs from

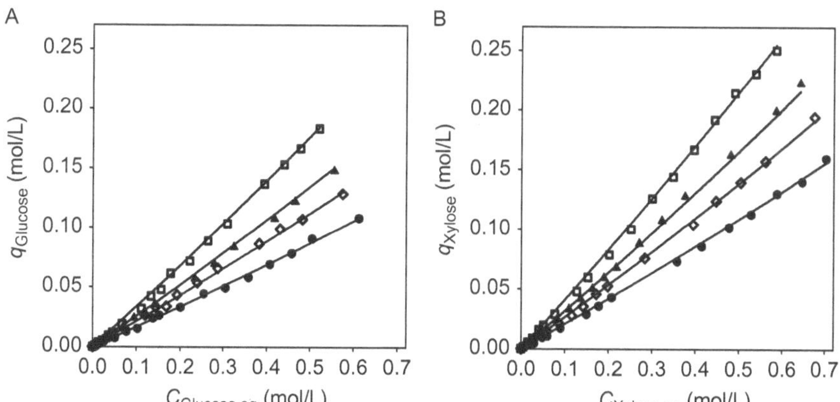

Figure 5.6 Sorption of glucose (A) and xylose (B) from water and various sulfuric acid solutions on a gel-type strong acid cation-exchange resin (sulfonated PS–DVB resin, 8 wt% DVB content, $d_p = 246\ \mu m$, H^+ capacity = 1.93 meq/mL (Heinonen and Sainio, 2010)) in acid (H^+) form. Experimental details: measurement of the isotherm with batch method (Guiochon et al., 2006), $m_{resin} = 5$ g, $V_{solution} = 7$ mL, $T = 50$ °C. Symbols: (●) water, (◇) 0.97 mol/L H_2SO_4, (▲) 1.84 mol/L H_2SO_4, and (□) 3.46 mol/L H_2SO_4. Lines are the fit of the empirical isotherm model ((Eq. 5.2), see Section 4.3). Model parameters are given in Table 5.2. *Reprinted with permission from Heinonen and Sainio (2012). Copyright: 2012. John Wiley and Sons.*

3.5 mol/L of H_2SO_4 instead of pure water. The curvatures of the monosaccharide isotherms do not change in the presence of sulfuric acid, although the Henry constant increases with increasing acid concentration.

Heinonen and Sainio (2012) described the monosaccharide sorption with a simple empirical isotherm model ((Eq. 5.2), see Section 4.3; model parameters are given in Table 5.2) which gives a good correlation between the experimental and calculated results. Linear dependency of the Henry constant on the sulfuric acid concentration is adequate to describe the co-operative sorption effect of sulfuric acid on the monosaccharides.

Sorption of acetic acid on a gel-type strong acid cation-exchange resin in acid form is also enhanced by sulfuric acid due to salting out (Fig. 5.7). Acetic acid sorption is not significantly affected by electrolyte exclusion due to the high proton concentration in the resin and in the solution which keeps acetic acid in associated form: acetic acid behaves like a nonelectrolyte. The sorption isotherm on strong acid PS–DVB resin is linear in the concentration range studied by Heinonen and Sainio (2012) and Laatikainen et al. (2011) (Fig. 5.7). The linear isotherm model presented by Heinonen and Sainio (2012)

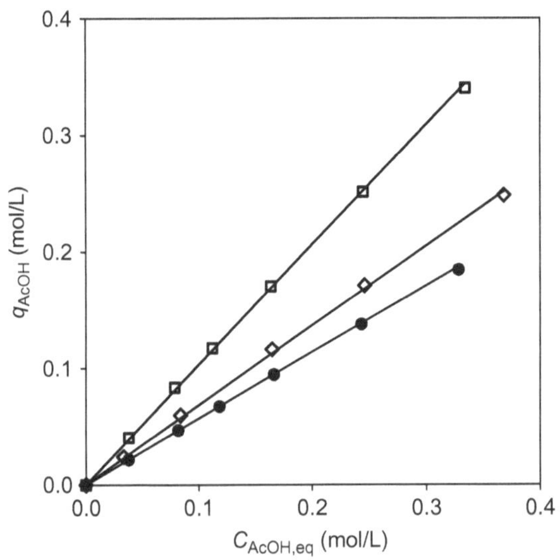

Figure 5.7 Sorption of acetic acid from water and various sulfuric acid solutions on a gel-type strong acid cation-exchange resin (sulfonated PS–DVB resin, 8 wt % DVB content, $d_p = 246$ μm, H^+ capacity = 1.93 meq/mL (Heinonen and Sainio, 2010)) in acid (H^+) form. Experimental details: measurement of the isotherm with frontal analysis method (Guiochon et al., 2006), see caption of Fig. 5.5 for other details. Symbols: (●) water, (◇) 0.96 mol/L H_2SO_4, and (□) 3.92 mol/L H_2SO_4. Lines are the fit of the empirical isotherm model ((Eq. 5.3), see Section 4.3). Model parameters are given in Table 5.2. *Reprinted with permission from Heinonen and Sainio (2012). Copyright: 2012. John Wiley and Sons.*

((Eq. 5.3), see Section 4.3; model parameters are given in Table 5.2) describes adequately the acetic acid sorption on strong acid PS–DVB resin.

Changes in the liquid-phase composition may lead to changes in the volume of an elastic adsorbent (see Section 4.2). This is especially true with gel-type ion-exchange resins, and the system presented in this chapter is no exception from this. Basically, all the components in the lignocellulosic hydrolysates can affect the volume of the gel-type strong acid cation-exchange resin used as an adsorbent. However, it has been observed by Heinonen and Sainio (2012) that resin volume changes are due to sulfuric acid in the case of concentrated acid lignocellulosic hydrolysates.

The effect of sulfuric acid on the shrinking of the gel-type resin studied by Heinonen and Sainio (2012) (sulfonated PS–DVB resin, 8 wt% DVB content, in acid form) is significant (Fig. 5.8). Obviously, such large volume change must be taken into account in the modeling of the chromatographic fractionation of concentrated acid lignocellulosic hydrolysates. As observed in Fig. 5.8, the shrinking of the resin due to sulfuric acid is well described by the empirical model ((Eq. 5.4), see Section 4.3; model parameters are given in Table 5.2).

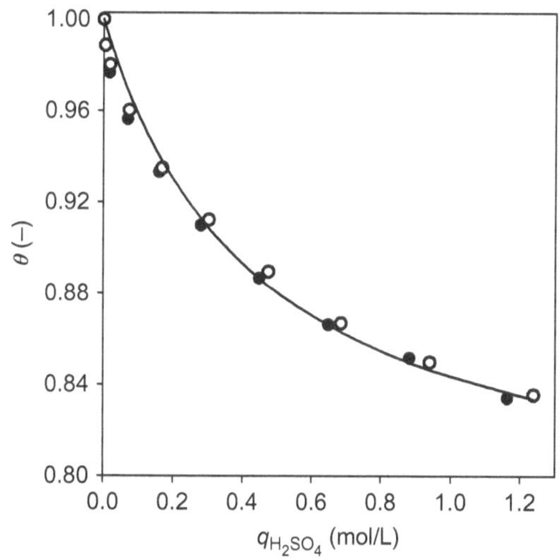

Figure 5.8 Effect of sulfuric acid loading on the shrinking of a gel-type strong acid cation-exchange resin (sulfonated PS–DVB resin, 8 wt % DVB content, $d_p = 246$ μm, H^+ capacity = 1.93 meq/mL (Heinonen and Sainio, 2010)) in acid (H^+) form. Results from two independent experiments are presented (filled and open symbols). Experimental details: equilibrium measurements in a column, $T = 50$ °C, h_{col} in pure water = 20 cm, $d_{col} = 2.5$ cm. Line is the fit of the empirical resin shrinking model ((Eq. 5.4), see Section 4.3). Model parameters are given in Table 5.2. *Reprinted with permission from Heinonen and Sainio (2012). Copyright: 2012. John Wiley and Sons.*

5.2. Fractionation using batch chromatography

Neuman et al. (1987) studied experimentally the effects of column loading and temperature on the separation of glucose (1.0 wt%) and sulfuric acid (7.7 wt%) with a sulfonated gel-type strong acid PS–DVB cation-exchange resin in acid form (Amberlite IR-118, 4.5 wt% DVB, Rohm and Haas Co.) in a batch column. Water was used as an eluent. The separation efficiency was found to increase with decreasing column loading, and the highest efficiency was obtained at a temperature of 55 or 68 °C. Neuman et al. (1987) noted that the sulfuric acid profile had a diffuse front and a shock layer at the back of the profile. Neuman et al. (1987) observed also fronting of the glucose profile: a phenomenon resulting from the cooperative effect of sulfuric acid on the monosaccharide sorption.

Nanguneri and Hester (1990) studied experimentally and numerically the chromatographic fractionation of sulfuric acid (8.0 wt%) and glucose (17.35 wt%) with a sulfonated gel-type strong acid PS–DVB cation-exchange resin in acid form (Amberlite IR-122, 10 wt% DVB content, Rohm and Haas Co.). The unsymmetrical shape of the sulfuric acid profile was clearly seen also in these experiments. Fronting of the glucose profile due to sulfuric acid was also observed (Nanguneri and Hester, 1990).

Xie et al. (2005) compared two adsorbents for the chromatographic fractionation of dilute acid (1 wt% sulfuric acid) corn stover hydrolysate at a temperature of 60 °C. The adsorbents used in this study were Dowex 99 (a sulfonated gel-type strong acid PS–DVB resin in acid form, Dow Chemicals Co.) and a weakly basic PVP (Reillex HP, poly-4-vinyl pyridine resin, Reilly Industries Inc.). Both adsorbents were able to fractionate the hydrolysate into sulfuric acid, monosaccharide, and various by-product (acetic acid, HMF, furfural) fractions. With Dowex 99, no additional regeneration was needed and the components of the hydrolysate eluted in the following order: sulfuric acid, monosaccharides, acetic acid, HMF, and furfural. PVP resin exchanged the sulfate ions of sulfuric acid to hydroxide ions, and therefore, an additional regeneration phase with a base was required to elute sulfate from the resin (Xie et al., 2005).

The aforementioned studies do not discuss chromatographic fractionation of concentrated acid lignocellulosic hydrolysates. Some patents (e.g., Farone and Cuzens, 1993, 1995; Hester and Farina, 1997) describe this particular application. Farone and Cuzens (1993, 1995) investigated the recovery of monosaccharides (23.0 wt%) from concentrated acid hydrolysate (33.6 wt% sulfuric acid) using a gel-type strong acid cation-exchange resin in acid form (CS16GC, Finex Oy). 90% pure sulfuric acid and 94% pure monosaccharide streams were obtained (Farone and Cuzens, 1993, 1995). Heinonen et al. (2012a), Heinonen and Sainio (2010, 2012), and Laatikainen et al. (2011) have investigated the chromatographic fractionation of concentrated acid lignocellulosic hydrolysates with gel-type strong acid PS–DVB cation-exchange resins in acid form (CS09GC, CS12GC, CS16GC, and Finex Oy).

Heinonen and Sainio (2012) used a synthetic solution representing concentrated acid lignocellulosic hydrolysate and a sulfonated gel-type PS–DVB resin (8 wt% DVB content, CS16GC, Finex Oy). The solution used contained sulfuric acid (20 wt%), glucose, xylose, and acetic acid (Heinonen and Sainio, 2012). HMF and furfural were not included because it was deemed uneconomical to separate these components from the monosaccharides simultaneously with sulfuric acid and acetic acid due to very strong sorption on the resin studied (Heinonen and Sainio, 2010; Sainio et al., 2011; Xie et al., 2005). Instead, a separate adsorption unit can be used for the recovery of HMF and furfural prior to the monosaccharide–acid separation (Chandel et al., 2007; Nilvebrant et al., 2001; Ranjan et al., 2009; Sainio et al., 2011; Villarreal et al., 2006; Weil et al., 2002).

Pulse elution profiles of synthetic solution containing sulfuric acid (20 wt%), glucose, xylose, and acetic acid in a column filled with a gel-type strong acid cation-exchange resin in acid form (CS16GC, Finex Oy) are shown in Fig. 5.9. In addition to the experimental profiles obtained by Heinonen and Sainio (2012), the profiles calculated with the model presented in Section 4.4 are also included.

Heinonen and Sainio (2012) obtained the values for the diffusion coefficients in reference state and dispersion coefficients by fitting the calculated column profiles to the experimental ones. The averaged value of axial dispersion coefficient was 6.0×10^{-8} m^2/s for all components. Intraparticle diffusion coefficients in reference state were 3.0×10^{-10} m^2/s, 1.5×10^{-10} m^2/s, 1.5×10^{-10}m^2/s, and 4.0×10^{-10} m^2/s for sulfuric acid, glucose, xylose, and acetic acid, respectively.

Sulfuric acid sorption on a strong acid cation-exchange resin is affected by the electrolyte exclusion: at infinite dilution, the exclusion is complete and sulfuric acid is not adsorbed to the resin. Due to this, the breakthrough of sulfuric acid is at void volume of the resin bed (Fig. 5.9). As the sulfuric acid concentration increases, the strength of the electrolyte exclusion decreases and sulfuric acid sorption increases (see sulfuric acid isotherm in Section 5.1). Each concentration is associated with a propagation velocity (Guiochon et al., 2006)

$$u_{c,i}=\frac{u}{1+F\frac{dq_i}{dC_i}}, \qquad [5.17]$$

where u is the interstitial velocity ($\dot{V}/\varepsilon A_{col}$), $u_{c,i}$ is the velocity associated to a particular concentration C_i, F is the phase ratio ($(1-\varepsilon)/\varepsilon$), and dq_i/dC_i is the slope of the isotherm at C_i. With unfavorable isotherms, the slope of the isotherm increases as the liquid-phase concentration increases. As a result, the propagation velocity of concentration C_i of component i ($u_{c,i}$) decreases. Sulfuric acid isotherm on a strong acid cation-exchange resin in acid form is unfavorable due to electrolyte exclusion. Therefore, the propagation of sulfuric acid slows down as the liquid-phase concentration increases and the diffuse front of the sulfuric acid profile is formed (Fig. 5.9). In the rear of the sulfuric acid profile, a shock layer exists: the low sulfuric acid concentrations propagate faster than the high ones, but due to physical reasons, the low concentrations cannot pass the high ones, and a shock layer is formed (Fig. 5.9).

The model derived by Heinonen and Sainio (2012) (see Section 4) can describe the elution of sulfuric acid with reasonable accuracy (Fig. 5.9). Sulfuric acid sorption appears to be not affected by the other components present in the system. This is evident from the good correlation between the experimental and calculated results: the sulfuric acid sorption isotherm model does not include interactions with other components.

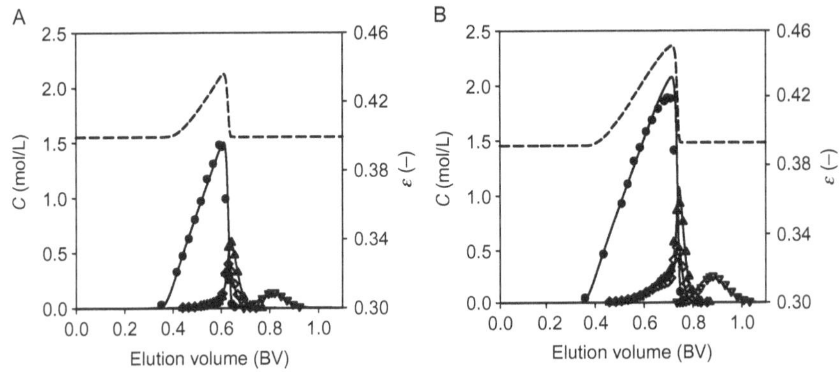

Figure 5.9 Elution profiles of 0.101 BV (BV = resin bed volume) (A) and 0.204 BV (B) pulses of synthetic hydrolysate solution with a gel-type strong acid cation-exchange resin (sulfonated PS–DVB resin, 8 wt % DVB content, $d_p = 246$ μm, H^+ capacity = 1.93 meq/mL (Heinonen and Sainio, 2010), CS16GC, Finex Oy) in acid form. Feed composition: 2.22 mol/L (20 wt%) sulfuric acid, 0.24 mol/L glucose, 0.38 mol/L xylose, and 0.16 mol/L acetic acid. Experimental details: flow rate = 2.454 mL/min, $\varepsilon = 0.395$, $d_{col} = 2.5$ cm, $h_{col} = 20$ cm, top–down flow, water as eluent, and $T = 50$ °C. Symbols: (●) sulfuric acid, (◇) glucose, (▲) xylose, and (▽) acetic acid. Lines are the results calculated with the model presented in Section 4; dotted line is the calculated ε. Isotherm model parameters are given in Table 5.2 and other parameters in the text (Section 5.2). *Reprinted with permission from Heinonen and Sainio (2012). Copyright: 2012. John Wiley and Sons.*

Heinonen and Sainio (2012) described the resin shrinking as changing bed porosity. This property is also shown in Fig. 5.9. The profile of resin bed porosity near the column outlet is similar to the profile of sulfuric acid because it is linked to the sorbed amount of the acid.

Monosaccharides (glucose and xylose) elute after sulfuric acid on a strong acid PS–DVB cation-exchange resin in acid (H^+) form (Fig. 5.9). The cooperative effect of sulfuric acid on the sorption of the monosaccharides results in interesting phenomena seen in the elution. The outlet profiles of these components have elongated fronts under the sulfuric acid profile: sulfuric acid slows down the propagation of the monosaccharides. The effect becomes stronger as the sulfuric acid concentration increases (Fig. 5.9). As a result, the front parts of the monosaccharide profiles under the sulfuric acid profile spread and become more diffuse. The phenomenon is opposite to the tag-along effect (Guiochon et al., 2006) often seen in chromatographic separations.

If no sulfuric acid would be present in the hydrolysate, the monosaccharides would elute close to the breakthrough point of the profiles (approximately 0.45 BV in Fig. 5.9). Elongation of the monosaccharide profiles reduces the economy of batchwise separation because part of the monosaccharides is lost to the sulfuric acid fraction. However, as the sulfuric acid fraction is recycled back to the hydrolysis, the monosaccharides eluting under the sulfuric acid are also recycled (see Section 5.2.2) and therefore are not lost product.

Another interesting phenomenon arising from the cooperative effect of sulfuric acid on the monosaccharide sorption is the focusing of the monosaccharides at the rear of the sulfuric acid profile (Fig. 5.9). The monosaccharide concentrations increase above the feed values instantly after the elongated fronts. This kind of focusing effect can occur when some thermodynamic variable (in the present case, sulfuric acid concentration) is changed causing a large change in the sorption of the adsorbates (Wankat, 1990). After some separation has been achieved, the front parts of the monosaccharide profiles are still slowed down by sulfuric acid. The molecules eluting behind the sulfuric acid zone, however, are not affected by the cooperative sorption and therefore propagate faster. This results in the focusing of the monosaccharides at the rear of the sulfuric acid profile (Fig. 5.9). Focusing of later eluting components is rarely seen in chromatography, but it is beneficial for the separation efficiency in this case.

The elution of the monosaccharides is also correctly predicted (Fig. 5.9) by the simple model (see Section 4) of Heinonen and Sainio (2012). The effects of sulfuric acid on the monosaccharide elution are accurately described. From the good correlation between the experimental and calculated profiles, it is clear that acetic acid does not affect the sorption of the monosaccharides (Fig. 5.9): the monosaccharide isotherm models do not include interactions with acetic acid.

Acetic acid elutes after sulfuric acid and monosaccharides with a strong acid PS–DVB cation-exchange resin in acid form as adsorbent (Fig. 5.9). The cooperative effect of sulfuric acid on the acetic acid sorption is not clearly seen in the elution profiles due to largely different sorption strengths of these components (see Figs. 5.5 and 5.7). Only the retention time of the acetic acid is slightly increased by sulfuric acid. The model presented in Section 4 predicts correctly the elution of acetic acid (Fig. 5.9).

5.2.1 Effect of resin cross-linking on separation

The degree of resin cross-linking contributes significantly to resin properties. For example, in the fractionation of lignocellulosic hydrolysates, the degree of resin cross-linking affects electrolyte exclusion and swelling as well as the sorption of components (through size exclusion and swelling pressure effects).

The effect of the cross-linking degree of a gel-type sulfonated strong acid PS–DVB cation-exchange resins on the fractionation of authentic concentrated acid (30 wt% H_2SO_4) lignocellulosic hydrolysates (spruce and birch) has been investigated by Heinonen and Sainio (2010). Three resins (Table 5.3) with different degrees of cross-linking (4.5, 6, and 8 wt% DVB content) were used.

The resin cross-linking does not affect the breakthrough of sulfuric acid because of the complete electrolyte exclusion at infinite dilution (Fig. 5.10). Otherwise the sulfuric acid profiles are different. Sulfuric acid sorption is high in resins with low degree of cross-linking, and the concentration profiles at column outlet are wider and more dilute. This originates from the less effective electrolyte exclusion (see Section 4.1.1). Although the resins shrink in

Table 5.2 Parameters of the phase equilibrium models used in the modeling of chromatographic fractionation of concentrated acid lignocellulosic hydrolysates (see Section 4.3)

	H_2SO_4	Glucose	Xylose	Acetic acid	Swelling
α	0.072	0.164	0.205	0.57	0.226
β	2.03	4.854×10^{-2}	5.812×10^{-2}	0.117	0.442
γ	–	0.114	0.114	–	–

Adsorbent: a gel-type strong acid cation-exchange resin (sulfonated PS–DVB resin, 8 wt% DVB content) in acid (H^+) form.

Reprinted with permission from Heinonen and Sainio (2012). Copyright: 2012. John Wiley and Sons.

Table 5.3 Properties of the resins (Finex Oy, Finland) used in the investigation of the effect of resin cross-linking on the chromatographic fractionation of concentrated acid lignocellulosic hydrolysates

Resin	CS09GC	CS12GC	CS16GC
Matrix	PS–DVB	PS–DVB	PS–DVB
Functional group	Sulfonic acid	Sulfonic acid	Sulfonic acid
Crosslinking density (wt%)	4.5	6.0	8.0
Particle diameter (μm)	211.0	217.0	217.5
Vol. capacity (H^+) (mequiv/mL)	1.07	1.51	1.93

the presence of sulfuric acid (which leads to increased charge density and more efficient electrolyte exclusion), the charge density is lower with less cross-linked resin. Tailing of the sulfuric acid profile also increases with decreasing cross-linking due to the increased dispersion caused by the larger changes in the height of the packed bed (Fig. 5.10).

The degree of cross-linking does not affect the width of the front part of the monosaccharide profile (Fig. 5.10). However, sorption of monosaccharides is higher in resins of low degree of cross-linking because of diminished size exclusion. As a consequence, the propagation velocity of the monosaccharides behind the sulfuric acid profile is lower and the focusing effect of the monosaccharides less strong (Fig. 5.10) when the degree of cross-linking is low.

Also, acetic acid sorption increases as the degree of resin cross-linking decreases (Fig. 5.10). The phenomenon is not very strong due to smaller molecular dimensions.

HMF and furfural are completely separated from the other components regardeless of the resin cross-linking (Fig. 5.10). The sorption of these components increases when resin cross-linking is increased from 4.5 to 6.0 wt%, but additional increase does not result in increased sorption.

The effect of resin cross-linking on monosaccharide yield is shown in Fig. 5.11 and productivity of the separation process with respect to the monosaccharides in Fig. 5.12 for authentic concentrated acid spruce and birch hydrolysates (Table 5.4). The constraints used to determine the cut points of the monosaccharide fraction were the desired sulfuric acid yield in the H_2SO_4 fraction $Y_{H_2SO_4}$, and the maximum allowed inhibitor concentration in the monosaccharide fraction $C_{AcOH}^{product}$. In practice, the only inhibitor eluting near the monosaccharides is acetic acid and the inhibitors' concentration corresponds to the acetic acid concentration (see Fig. 5.10).

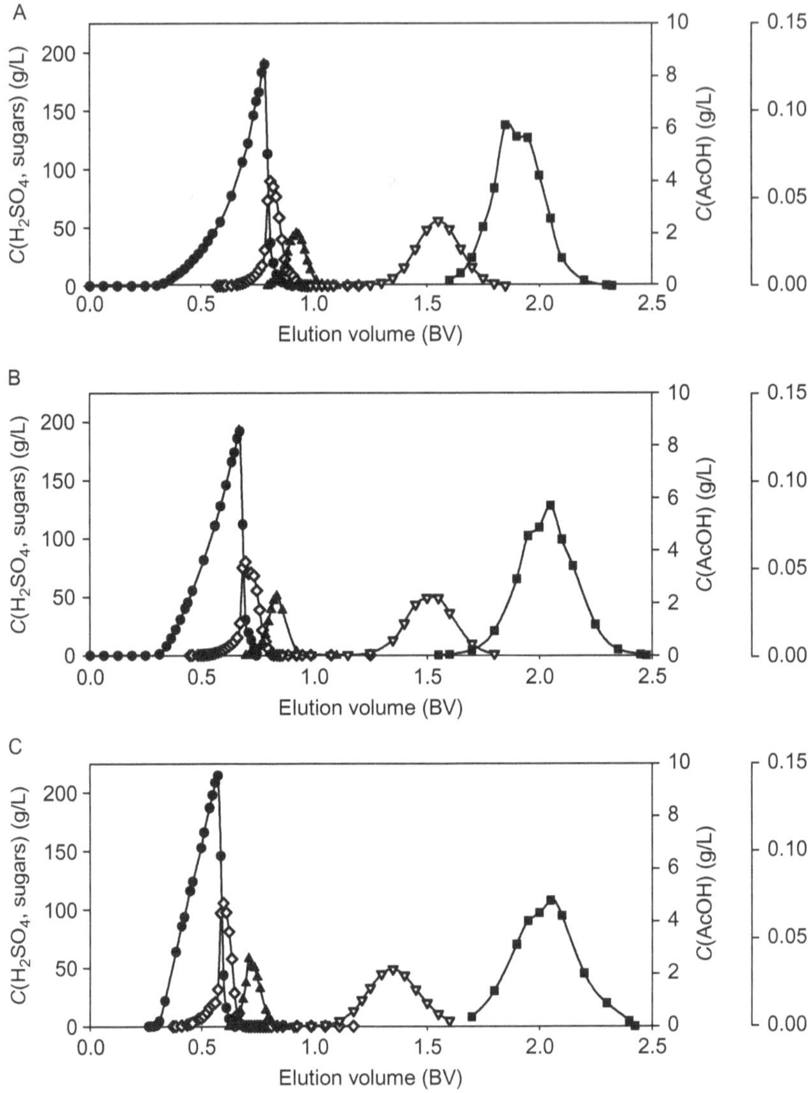

Figure 5.10 Effect of resin cross-linking on the chromatographic recovery of monosaccharides from concentrated acid spruce hydrolysate. Degree of resin cross-linking: 4.5 wt% (A), 6.0 wt% (B), and 8.0 wt% (C). All monosaccharides are shown as one pseudocomponent. Feed composition: see Table 5.4. Experimental details: flow rate = 2.655 mL/min, d_{col} = 2.6 cm, h_{col} = 20 cm, top–down flow, water as eluent, and T = 50 °C. Symbols: (●) sulfuric acid, (◇) monosaccharides, (▲) acetic acid, (▽) HMF, and (■) furfural. Lines are presented to guide the eye.

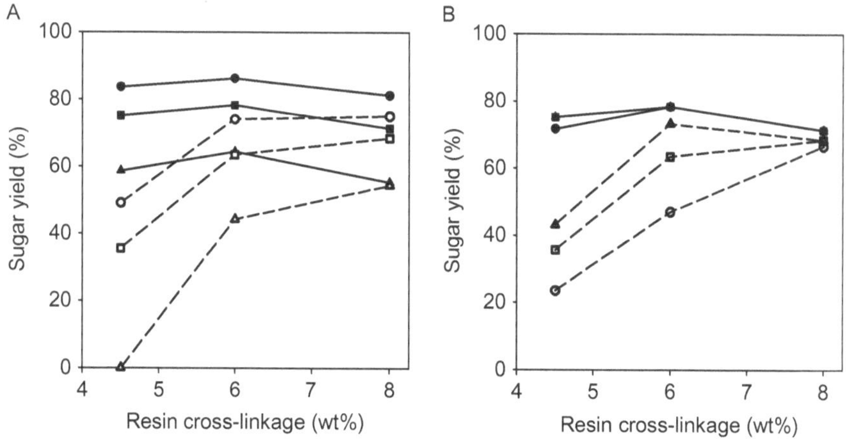

Figure 5.11 The effect of a gel-type sulfonated PS–DVB cation-exchange (in acid form) resin cross-linkage on the monosaccharide yield. (A): $Y_{H_2SO_4}$ is varied and $C_{AcOH}^{product}$ is kept constant at 1.0 g/L: (●) = 90%, (■) = 95%, and (▲) = 98% H_2SO_4 yield; (B): $C_{AcOH}^{product}$ is varied and $Y_{H_2SO_4}$ is kept constant at 95%: (●) = 0.5 g/L, (■) = 1.0 g/L, and (▲) = 1.5 g/L acetic acid. Hydrolysate compotisions: see Table 5.4. Filled symbols, spruce; open symbols, birch hydrolysate. Lines are presented to guide the eye. *Reprinted with permission from Heinonen and Sainio (2010). Copyright: 2010. American Chemical Society.*

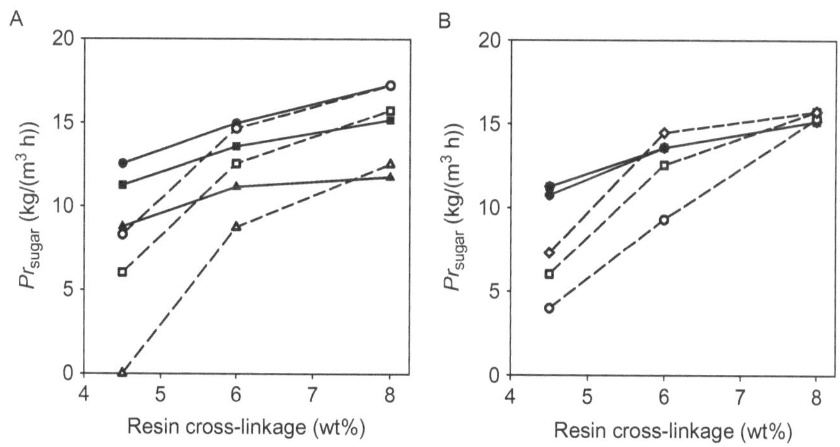

Figure 5.12 The effect of a gel-type sulfonated PS–DVB cation-exchange (in acid form) resin-cross-linkage on the productivity of the separation process with respect to the monosaccharides. (A): $Y_{H_2SO_4}$ is varied and is $C_{AcOH}^{product}$ kept constant at 1.0 g/L: (●) = 90%, (■) = 95%, and (▲) = 98% H_2SO_4 yield; (B): $C_{AcOH}^{product}$ is varied and $Y_{H_2SO_4}$ is kept constant at 95%: (●) = 0.5 g/L, (■) = 1.0 g/L, and (▲) = 1.5 g/L acetic acid. Hydrolysate compotisions: see Table 5.4. Filled symbols, spruce; open symbols, birch hydrolysate. Lines are presented to guide the eye. *Reprinted with permission from Heinonen and Sainio (2010). Copyright: 2010. American Chemical Society.*

Table 5.4 Compositions of concentrated acid spruce and birch hydrolysates used in the investigation of the effect of resin cross-linkage density on chromatographic fractionation of the hydrolysates with a gel-type sulfonated strong acid PS-DVB cation-exchange resin in acid form

	C (g/L)	
	Spruce hydrolysate	**Birch hydrolysate**
H_2SO_4	340.75	329.77
Glucose	47.16	48.73
Xylose	7.40	29.03
Galactose	4.70	1.01
Mannose	16.20	3.73
Arabinose	2.42	6.44
Total sugars	77.88	88.94
Acetic acid	2.04	9.07
HMF	0.10	0.12
Furfural	0.30	1.28

Reprinted with permission from Heinonen and Sainio (2010). Copyright: 2010. American Chemical Society.

The effect of resin cross-linking on the monosaccharide yield with spruce hydrolysate is small (Fig. 5.11), although a weak maximum is observed at 6.0 wt% cross-linking. Electrolyte exclusion becomes more effective when the cross-linking density is increased. This leads to lower sorption of sulfuric acid (see Fig. 5.10), and therefore, the separation between sulfuric acid and monosaccharides should improve. However, at the same time, the sorption of monosaccharides decreases due to stronger size exclusion. As a consequence, the monosaccharide yield obtained with the 6.0 wt% cross-linking resin is somewhat higher than with the other degrees of cross-linking. The low amount of acetic acid in the spruce hydrolysate does not significantly affect the yield (Fig. 5.11).

In the fractionation of birch hydrolysate, the monosaccharide yield is more dependent on the resin cross-linking due to higher acetic acid concentration (Fig. 5.11). Sorption of acetic acid increases with decreasing cross-linking although it is not as much affected as the sorption of the monosaccharides (see Fig. 5.10). Due to this, the monosaccharide–acetic acid separation efficiency as well as the monosaccharide yield increases when the degree of cross-linking is increased (Fig. 5.11). With both hydrolysates, the yield of monosaccharides decreases as $Y_{H_2SO_4}$ increases (at constant $C_{AcOH}^{product}$) or as $C_{AcOH}^{product}$ decreases (at constant $Y_{H_2SO_4}$) (Fig. 5.11).

The productivity of the batchwise chromatographic separation process with respect to the monosaccharides (Pr_{sugar}) can be calculated from (Heinonen and Sainio, 2012)

$$Pr_{sugar} = \frac{Y_{sugar} C_{sugar}^{feed} V^{feed}}{V_{bed} t_{cycle}}, \qquad [5.18]$$

where Y_{sugar} is monosaccharide yield, C_{sugar}^{feed} is monosaccharide concentration in the feed, V^{feed} is feed (injection) volume, V_{bed} is resin bed volume, and t_{cycle} is cycle time.

The productivity of the separation process with respect to the monosaccharides increases with increasing resin cross-linking due to decreasing cycle time (Fig. 5.12). Decrease in the cycle time results from the weaker sorption of the components. Heinonen and Sainio (2010) obtained highest productivity with 8.0 wt% cross-linked resin (Fig. 5.12). The effect of resin cross-linking on the productivity is more significant with the birch hydrolysate than with the spruce hydrolysate due to differences in the monosaccharide yields. The productivity of the batchwise chromatographic separation process decreases as $Y_{H_2SO_4}$ increases (at constant $C_{AcOH}^{product}$) or as $C_{AcOH}^{product}$ decreases (at constant $Y_{H_2SO_4}$) due to the lower monosaccharide yield (Fig. 5.12).

5.2.2 Process performance

Economical evaluations of the chromatographic fractionation of authentic concentrated acid hydrolysates have not been done in academic literature. However, Nanguneri and Hester (1990) made a cost analysis based on the binary fractionation of sulfuric acid and glucose. Recovery of sulfuric acid using chromatography was calculated to be 40% cheaper than $Ca(OH)_2$ neutralization (Nanguneri and Hester, 1990). This naturally depends on the way the separation step is designed and operated. The sensitivity of the process performance to operating parameters (column loading), the concentration of sulfuric acid in the hydrolysis reactor, as well as coupling of streams within the process (recycling of sulfuric acid) were investigated by Heinonen and Sainio (2012).

a. *Stand-alone separation process.* Heinonen and Sainio (2012) investigated the process performance of a stand-alone batchwise chromatographic fractionation of concentrated acid lignocellulosic hydrolysate using the model presented in Section 4. Stand-alone means here that recycling of fractions from the separation step to the hydrolysis reactor was not considered. HMF and furfural were omitted from the investigation: these were assumed to be removed prior to the monosaccharide–acid separation. The column loading (vol% of resin bed volume) was varied from 0.1 vol% to 20 vol%.

The cut points were defined as follows. The first cut point (i.e., the beginning of the sulfuric acid fraction) was set to the point of sulfuric acid breakthrough. Beginning of the monosaccharide fraction (end of the sulfuric acid fraction, second cut point) was defined such that the ratio of the total amount of monosaccharides to the amount of sulfuric acid in moles ($n_{sugar}/n_{H_2SO_4}$) in the monosaccharide fraction was at least 8.0 while simultaneously maximizing the productivity. Due to this, the obtained $n_{sugar}/n_{H_2SO_4}$ was always close to eight. The monosaccharide yield was not taken as a constraint for the cut points as the monosaccharides eluting under the sulfuric acid profile are collected to the sulfuric acid fraction and recycled back to the reactor. Therefore, these monosaccharides are not lost product (Heinonen and Sainio, 2012).

End point of the monosaccharide fraction (beginning of the acetic acid fraction, third cut point) was defined by Heinonen and Sainio (2012) by setting a maximum for the allowed acetic acid concentration in the monosaccharide fraction: 6.7 mmol/L (0.4 g/L). Due to this, a small fraction of the monosaccharides ends up to the acetic acid fraction and can be regarded as lost product. The last cut point (end point of the acetic acid fraction) was set to the point where acetic acid concentration decreases below 0.2 mmol/L (0.01 g/L). The cycle time of the process was defined as a difference between the first and the last cut point (Heinonen and Sainio, 2012).

The productivity of the batchwise monosaccharide recovery process with respect to the monosaccharides (Pr_{sugar}) was calculated using Eq. (5.18). Eluent consumption (EC) for the batchwise chromatographic separation process was calculated from (Heinonen and Sainio, 2012)

$$EC = \frac{t_{cycle}\dot{V} - V^{feed}}{Y_{sugar} C_{sugar}^{feed} V^{feed}}. \qquad [5.19]$$

Pr_{sugar} as a function of column loading goes through a maximum (Fig. 5.13A). Maximum Pr_{sugar} is obtained for 10, 20, and 30 wt% hydrolysis acid concentrations with loadings 15.1, 11.5, and 5.5 vol%, respectively, with the hydrolysate composition used by Heinonen and Sainio (2012). Pr_{sugar} is not significantly affected by the loading in a wide range of loadings around the loading giving the maximum value (Fig. 5.13A).

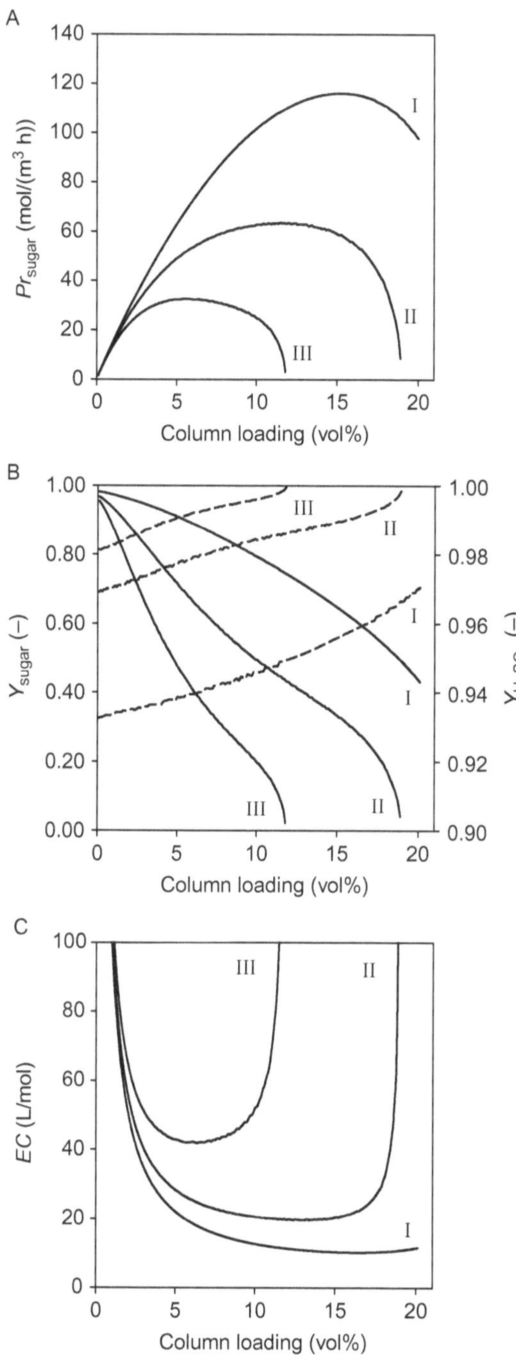
A
140
120
100
80
60
40
20
0
Pr_{sugar} (mol/(m^3 h))
0 5 10 15 20
Column loading (vol%)
I
II
III
B
1.00
0.80
0.60
0.40
0.20
0.00
Y_{sugar} (–)
1.00
0.98
0.96
0.94
0.92
0.90
$Y_{H_2SO_4}$ (–)
0 5 10 15 20
Column loading (vol%)
III
II
I
I
III
II
C
100
80
60
40
20
0
EC (L/mol)
0 5 10 15 20
Column loading (vol%)
III
II
I

Although Y_{sugar} decreases with increasing loading (Fig. 5.13B), the amount of monosaccharides treated during one cycle increases, and therefore also Pr_{sugar} increases until a certain limit. Eventually, with higher column loadings, Y_{sugar} becomes so low that Pr_{sugar} begins to decrease rapidly. With column loadings smaller than a certain limit, Pr_{sugar} is lower than the maximum value, due to small amount of monosaccharides treated per cycle, although Y_{sugar} is higher than with the loading giving the maximum productivity (Fig. 5.13B). Interestingly, the maximum productivity for the monosaccharides is achieved at relatively low yield.

Hydrolysis acid concentration has a significant effect on Pr_{sugar} (Fig. 5.13A) due to the overlapping of the monosaccharide and sulfuric acid profiles and due to the cooperative effect of sulfuric acid on the monosaccharide sorption. Higher productivity of monosaccharides can be achieved with lower feed concentration of sulfuric acid. As sulfuric acid concentration is increased, larger fraction of the monosaccharides elutes under the sulfuric acid profile and end up to the sulfuric acid fraction. This lowers Y_{sugar} (Fig. 5.13B), and, therefore, also Pr_{sugar}. Increase in cycle time with increasing column loading and feed concentration of sulfuric acid has also a negative effect on the productivity, but its influence is overshadowed by that of the column loading and feed concentration.

The dependence of $Y_{\mathrm{H_2SO_4}}$ on the column loading is opposite to that of Y_{sugar} (Fig. 5.13B). With increasing column loading, the overlapping of monosaccharide and sulfuric acid profiles increases and the second cut point has to be moved to the right so that the limiting $n_{\mathrm{sugar}}/n_{\mathrm{H_2SO_4}}$ value in monosaccharide fraction is achieved. The sulfuric acid yield increases as a larger part of the sulfuric acid shock layer is taken into the sulfuric acid fraction. High sulfuric acid yield is beneficial for the process economy because lower amount of fresh sulfuric acid is needed for the hydrolysis of lignocellulosic biomass.

Eluent consumption decreases first as the column loading increases as the amount of monosaccharides collected in the monosaccharide fraction

Figure 5.13 The effects of column loading and sulfuric acid feed concentration on Pr_{sugar} (A), Y_{sugar} (solid) and $Y_{\mathrm{H_2SO_4}}$ (dashed) in their target fractions (B), and *EC* (C). Feed composition: 10 wt% (I), 20 wt% (II), or 30 wt% (III) sulfuric acid, and 0.3 mol/L glucose, 0.3 mol/L xylose, and 0.16 mol/L acetic acid. Operational conditions: $h_{\mathrm{col}}=20$ cm, $d_{\mathrm{col}}=2.5$ cm, flow rate = 2.454 mL/min. Model parameters are given in Table 5.2 (isotherm parameters, Section 5.1) and Section 5.2 (D_p and D_{ax} values). For fractionation constaints, see the text (Section 5.2.2). *Reprinted with permission from Heinonen and Sainio (2012). Copyright: 2012. John Wiley and Sons.*

increases (Fig. 5.13C). However, the eluent consumption begins to increase rapidly as the monosaccharide yield approaches zero (Fig. 5.13C). A decrease in the feed concentration of sulfuric acid has a positive effect also on the eluent consumption, due to higher amount of monosaccharides recovered. Around the minimum value of the eluent consumption, it is only slightly affected by column loading: *EC* at the column loading giving the maximum productivity is only a little higher than the minimum value (Fig. 5.13C).

Sulfuric acid is diluted during the elution regardless of the column loading due to the thermodynamics. However, this unwanted dilution decreases with decreasing feed concentration of sulfuric acid and increasing column loading (Fig. 5.14).

At low column loadings, dilution of the monosaccharide fraction also occurs (Fig. 5.14). However, as column loading is increased, the monosaccharide concentration increases and eventually becomes higher than the feed

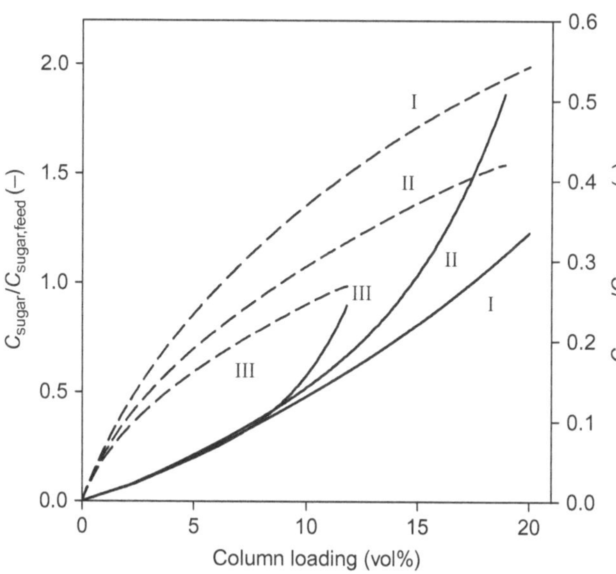

Figure 5.14 The effects of column loading and sulfuric acid feed concentration on the average concentrations of sulfuric acid (solid) and monosaccharides (dash) in their target fractions. Feed composition: 10 wt% (I), 20 wt% (II), or 30 wt% (III) sulfuric acid, and 0.3 mol/L glucose, 0.3 mol/L xylose, and 0.16 mol/L acetic acid. Operational conditions: $h_{col}=20$ cm, $d_{col}=2.5$ cm, flow rate $=2.454$ mL/min. Model parameters are given in Table 5.2 (isotherm parameters, Section 5.1) and Section 5.2 (D_p and D_{ax} values). For fractionation constaints, see the text (Section 5.2.2).

concentration. This is due to the cooperative effect of sulfuric acid on the monosaccharide sorption (focusing effect, see Section 5.2).

b. *Chromatographic separation coupled with biomass hydrolysis and recycling of sulfuric acid.* Investigation of the stand-alone separation process rarely gives an adequately realistic picture of the separation performance if recycling from the separation unit to preceding unit operations exists. These additional unit operations should be included to the process performance evaluation and the evaluation should be done at steady-state conditions. The process scheme investigated by Heinonen and Sainio (2012) is shown in Fig. 5.15.

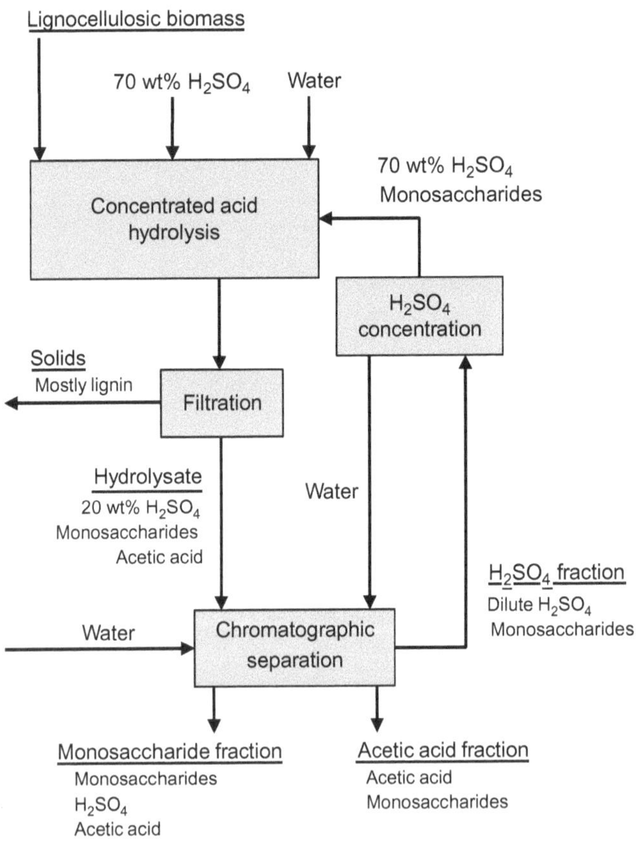

Figure 5.15 The streams and main unit operations of a process consisting of two-step concentrated acid hydrolysis, batchwise chromatographic separation, and concentration and recycling of sulfuric acid fraction. *Reprinted with permission from Heinonen and Sainio (2012). Copyright: 2012. John Wiley and Sons.*

As seen in Fig. 5.15, the concentrated acid hydrolysis is conducted in two steps (see also Farone and Cuzens, 1993, 1995; Hoshino et al., 2007). In the first step, the biomass is treated with 70 wt% sulfuric acid (decrystallization of cellulose). After the pretreatment, the hydrolysis acid is diluted to 20 wt% and the actual hydrolysis is conducted. Filtration is used to separate the concentrated acid hydrolysate from the remaining solids after the hydrolysis. The hydrolysate is led to the chromatographic separation step and fractionated into sulfuric acid, monosaccharide, and acetic acid fractions. The monosaccharide and acetic acid fractions are taken to downstream processing, but the sulfuric acid fraction (with all the other components in this fraction) is recycled back to the hydrolysis reactor through a concentration step. In the concentration step, the sulfuric acid is concentrated to 70 wt% (Farone and Cuzens, 1993, 1995), for example, by using multiple effect evaporator or vapor compression distillation (Gsell, 2004).

Heinonen and Sainio (2012) evaluated the process performance using the following parameters. In the acid hydrolysis, 1 ton of wood chips and 70 wt% of sulfuric acid were mixed in a way that after the second hydrolysis step the wood to liquid mass ratio was 1:6. In the second hydrolysis step, sulfuric acid was diluted to 20 wt% concentration with water. Wood was assumed to contain 70 wt% polysaccharides (40% cellulose and 30% hemicelluloses) (Sun et al. 2002; Taherzadeh and Karimi, 2008; Hamelinck et al., 2005). In the hydrolysis, polysaccharides were presumed to be cleaved only into glucose from cellulose and xylose from hemicelluloses with 80 wt% yield (Farone and Cuzens, 1993, 1995; Hoshino et al., 2007). This is a realistic simplification because practically all monosaccharides would behave as these two in the chromatographic fractionation. In addition to the monosaccharides, acetic acid was also formed in the hydrolysis in such amounts that the concentration of acetic acid after the hydrolysis was 0.16 mol/L (Heinonen and Sainio, 2010). The formation of other by-products (mainly furfural and HMF) was neglected in the calculations because their amounts were expected to be low (Heinonen and Sainio, 2010), and they can be effectively removed in a separate adsorption unit (Chandel et al., 2007; Nilvebrant et al., 2001; Ranjan et al., 2009; Sainio et al., 2011; Villarreal et al., 2006; Weil et al., 2002). The volume of the hydrolysate was calculated with the density of 20 wt% aqueous H_2SO_4. After the hydrolysis, the solids were separated from the hydrolysate with filtration: 6.5 vol% of the hydrolysate was lost, concentrations remained unchanged. The amount of hydrolysate after the filtration was 5000 L and remained constant throughout the process.

The batchwise chromatographic separation was simulated using 4-m high column and a flow rate of 4 BV/h. The flow rate and column were chosen only for demonstration purposes, and no optimization was conducted. Heinonen and Sainio (2012) defined the cut points for the separation process in same manner as in the investigation of the stand-alone chromatographic separation (see Section 5.2.2.a).

Six cycles were required for the process shown in Fig. 5.15 to reach steady-state regardless of the loading (Heinonen and Sainio, 2012). With the recycling of the sulfuric acid fraction, the column loading giving the maximum productivity (18.2 vol%) is considerably higher than in the single pass operation (13.0 vol%). In addition, in the case studied by Heinonen and Sainio (2012), recycling of the sulfuric acid fraction increases the productivity about 160% from the maximum obtained without the recycling (Table 5.5). The large increase in the productivity is due to the increased amount of monosaccharides in the feed to the chromatographic separation unit (Table 5.6). The additional monosaccharides are recycled with sulfuric acid from the chromatographic separation back to the hydrolysis reactor.

Without the recycling of sulfuric acid fraction, the productivity is close to the maximum value in a wide range of loadings. With the recycling, maximizing the productivity requires more careful optimization. Change in the column loading from 18.2 vol% quickly decreases the productivity considerably (Fig. 5.16).

The sulfuric acid is concentrated in the concentration step to 70 wt% (11.5 mol/L, Table 5.6) by removing water. Approximately 90% of the water in the sulfuric acid fraction has to be removed. This water can be reused as

Table 5.5 Performance of the chromatographic separation in a process shown in Fig. 5.15 without H_2SO_4 fraction recycling and with H_2SO_4 fraction recycling in steady state

	Separation with a single pass	**Separation with recycling**
Pr_{sugar} (mol/(m^3 h))	212.89	546.31
Y_{sugar} (%)	32.0	90.6
$Y_{H_2SO_4}$ (%)	99.0	97.0
EC (L/mol)	13.91	5.42

Model parameters are given in Table 5.2 (isotherm parameters, Section 5.1) and Section 5.2 (D_p and D_{ax} values). For fractionation constaints, see the text (Section 5.2.2).

Reprinted with permission from Heinonen and Sainio (2012). Copyright: 2012. John Wiley and Sons.

Table 5.6 Volumes and concentrations in a lignocellulosic biomass conversion and purification process (see Fig. 5.15) without H_2SO_4 fraction recycling and with H_2SO_4 fraction recycling in steady state

	Feed to separation column	H_2SO_4 fraction	H_2SO_4 fraction after concentration	Monosaccharide fraction	Acetic acid fraction
Separation with a single pass					
V (L)	5000	9897	1000	1259	8096
$C_{H_2SO_4}$ (mol/L)	2.379	1.162	11.494	0.099	0
$C_{Glucose}$ (mol/L)	0.332	0.121	1.2	0.338	4.2×10^{-3}
C_{Xylose} (mol/L)	0.299	0.081	0.8	0.476	11.7×10^{-3}
C_{AcOH} (mol/L)	0.16	0	0	6.6×10^{-3}	97.6×10^{-3}
Separation with H_2SO_4 fraction recycling in steady state					
V (L)	5000	9767	982	1434	8050
$C_{H_2SO_4}$ (mol/L)	2.379	1.156	11.494	0.228	0
$C_{Glucose}$ (mol/L)	0.742	0.224	2.229	0.978	13.7×10^{-3}
C_{Xylose} (mol/L)	0.474	0.096	0.953	0.854	25.9×10^{-3}
C_{AcOH} (mol/L)	0.16	0	0	6.6×10^{-3}	98.2×10^{-3}

Model parameters are given in Table 5.2 (isotherm parameters, Section 5.1) and Section 5.2 (D_p and D_{ax} values). For fractionation constaints see Section 5.2.2.

Reprinted with permission from Heinonen and (2012). Copyright: 2012. John Wiley and Sons.

eluent in the chromatographic separation: the fresh water consumption is decreased by approximately 61% in the case investigated by Heinonen and Sainio (2012). Through the acid recycling, 91.3% of the acid needed in the hydrolysis is obtained.

Sulfuric acid concentration in the feed to the separation unit remained constant with the recycling (Table 5.6) because of a constant amount of sulfuric acid used in the hydrolysis on each cycle. However, the amount of monosaccharides in the feed is increased (Table 5.6) due to the recycling of the sulfuric acid fraction which contains also monosaccharides. The increase of the monosaccharide amount in the feed affects the position of the second cut point which is defined as the ratio of monosaccharides to sulfuric acid in the middle fraction. The second point is shifted slightly toward smaller value. Due to this, the sulfuric acid yield and concentration in the

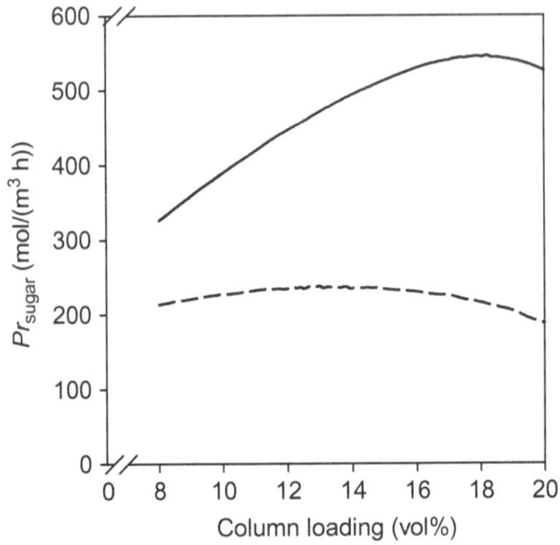

Figure 5.16 Pr_{sugar} in the chromatographic separation in a process shown in Fig. 5.15 as a function of column loading: without sulfuric acid fraction recycling (dashed) and in steady state with sulfuric acid fraction recycling (solid). Feed composition: see Table 5.6. Operational conditions: $h_{col}=20$ cm, $d_{col}=2.5$ cm, flow rate $=2.454$ mL/min. Model parameters are given in Table 5.2 (isotherm parameters, Section 5.1) and Section 5.2 (D_p and D_{ax} values). For fractionation constaints, see the text (Section 5.2.2). *Reprinted with permission from Heinonen and Sainio (2012). Copyright: 2012. John Wiley and Sons.*

acid fraction decreases slightly while the sulfuric acid concentration in the monosaccharide fraction is doubled (Table 5.6).

Recycling of the sulfuric acid fraction increases also the monosaccharide yield (Table 5.5). In the case studied by Heinonen and Sainio (2012), recycling enables 90.6% of the monosaccharides that are fed to the separation column to be taken out for the downstream processing. The rest of the monosaccharides are taken out from the process in the acetic acid fraction. The monosaccharide yield from the whole process (yield in the hydrolysis: 80%) with the recycling is 72.5%.

Recycling of the sulfuric acid fraction increases the glucose and xylose concentrations in the product fraction approximately by a factor of 2.9 and 1.8, respectively (Table 5.6). This kind of increase in the monosaccharide concentrations would be highly beneficial for the downstream processing.

Acetic acid is removed completely from the system in the monosaccharide and acetic acid fractions, and therefore, its amount in the feed to the separation column was not affected by the recycling.

5.2.3 Fermentation of chromatographically purified lignocellulosic hydrolysates

Fermentative production of ethanol is a good way to test the effectiveness of the chromatographic fractionation of concentrated acid lignocellulosic hydrolysates. Xie et al. (2005) investigated the efficiency of chromatographic purification of dilute acid corn stover hydrolysate by fermentations using *Saccharomyces cerevisiae* yeast. Hydrolysates purified with a gel-type strong acid PS–DVB cation-exchange resin (Dowex 99, Dow Chemicals Co.) in acid form and with a weak base PVP (poly-4-vinyl pyridine, Reillex HP, Reilly Industries Inc.) resin were used as substrates. The highest monosaccharide consumption rate was obtained with the PVP-purified hydrolysate: 89% of the monosaccharides were utilized in 24 h. According to Xie et al. (2005), this is close to the fermentability of a synthetic solution of pure monosaccharides: 90% in 24 h. With the Dowex-purified hydrolysate, 52% of the monosaccharides were utilized in the same time (48 h was required to utilize 85% of the monosaccharides). $Ca(OH)_2$-neutralized hydrolysate was used as reference: 48% of the monosaccharides were utilized in 24 h (64% in 48 h) (Xie et al., 2005).

Sun et al. (2011) investigated the fermentation of chromatographically purified concentrated sulfuric acid bamboo hydrolysates using *S. cerevisiae* yeast. Glucose was well converted into ethanol by *S. cerevisiae*, but the yeast was unable to consume xylose properly. From glucose, 92% ethanol yield was obtained in continuous fermentation with a tower-type reactor. The ethanol productivity was 8.2 g/L/h (Sun et al., 2011).

Also, Heinonen et al. (2012a) have studied the fermentative ethanol production from chromatographically purified concentrated acid lignocellulosic hydrolysates (Table 5.7). *S. cerevisiae* (genetically modified to utilize pentoses) and *Pichia stipitis* (naturally pentose utilizing) yeasts were used for the fermentation. Fermentation of $Ca(OH)_2$-neutralized hydrolysate was studied as a reference (Heinonen et al., 2012a).

According to Heinonen et al. (2012a), ethanol productivity from chromatographically purified and $Ca(OH)_2$-neutralized concentrated acid hydrolysates were comparable (Fig. 5.17A). The chromatographic purification removes all HMF and furfural from the hydrolysates, and only a slight amount of acetic acid remains in the purified solution (Table 5.7). This is not the case with $Ca(OH)_2$ neutralization: in particular, acetic acid is unaffected (Chandel et al., 2007; Palmqvist and Hahn-Hägerdahl, 2000a). This, however, has a very limited effect on the fermentation of the hydrolysates with *S. cerevisiae* (Fig. 5.17A). *S. cerevisiae* can tolerate inhibitory components rather well.

Table 5.7 Compositions of concentrated acid spruce hydrolysate prior and after chromatographic purification (Heinonen et al., 2012a)

	Component	C (g/L)
Unpurified hydrolysate	H_2SO_4	340.8
	Glucose	47.2
	X+G+M+A	30.7
	AcOH	2
	HMF	0.1
	Furfural	0.3
Purified hydrolysate	H_2SO_4	14.7
	Glucose	33.7
	X+G+M+A	22.3
	AcOH	0.3
	HMF	0
	Furfural	0

X+G+M+A=combined amount of xylose, galactose, mannose, and arabinose.

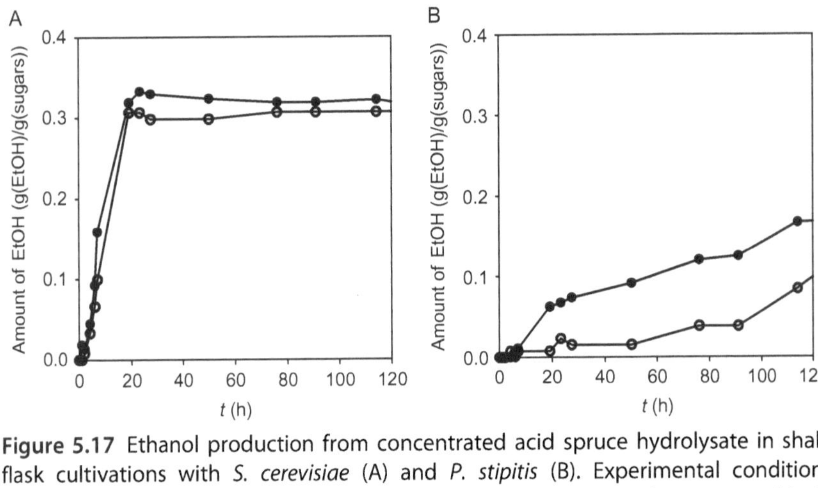

Figure 5.17 Ethanol production from concentrated acid spruce hydrolysate in shake flask cultivations with *S. cerevisiae* (A) and *P. stipitis* (B). Experimental conditions: $c_{yeast}=1.0$ g/L, $V_{solution}=50$ mL, $T=30$ °C, $pH_0=5.0$. Symbols: (●) chromatographically purified hydrolysate, final pH adjustment with $Ca(OH)_2$; (○) $Ca(OH)_2$-neutralized hydrolysate. Concentrations are average values from two parallel experiments. Lines are presented to guide the eye. *Reprinted with permission from Heinonen et al. (2012a). Copyright: 2012. John Wiley and Sons.*

P. stipitis strain gave substantially lower ethanol productivity than the *S. cerevisiae* (Fig. 5.17B). This is due to its higher sensitivity of the yeast for inhibitory compounds (Du et al., 2010; Heinonen et al., 2012a; Ladisch and Svarczkopf, 1991; Matsushika et al., 2009). This is clearly seen when ethanol productivity from the chromatographically purified and $Ca(OH)_2$-neutralized hydrolysates are compared (Fig. 5.17B): $Ca(OH)_2$ treatment is not sufficient for inhibitor removal. The chromatographically purified hydrolysate contained only a small amount of acetic acid, whereas the $Ca(OH)_2$-neutralized hydrolysate contained HMF, furfural, and high amounts of acetic acid. Similar results have been observed by Chandel et al. (2007) in fermentation experiments with *Candida shehatae*. *P. stipitis* is also known to be sensitive also to ethanol (Du et al., 2010; Ladisch and Svarczkopf, 1991; Matsushika et al., 2009) which also explains the slow ethanol productivity obtained.

Chromatographic purification of lignocellulosic hydrolysates provides a significant increase in the ethanol yield over $Ca(OH)_2$ neutralization although the ethanol concentrations achieved with both purification methods are comparable (Table 5.8).

5.3. Fractionation using steady-state recycling chromatography

As discussed above, the purity of monosaccharides in a concentrated acid hydrolysate could be increased from as low as 20% up to 80% in a single step using fixed-bed chromatography. Higher purities are certainly possible, but the yield and thus productivity would decrease significantly. The use of steady-state recycling chromatography (see Section 3.2) to increase productivity while maintaining high purity (95%) of the monosaccharide fraction has been investigated by Hellstén and Sainio (2012) using a binary model system of sulfuric acid and glucose.

Table 5.8 Ethanol yields from total sugars and ethanol concentrations from purified spruce hydrolysate after 120-h fermentation

	Chromatographically purified	$Ca(OH)_2$-neutralized
Yeast		
S. cerevisiae		
Y(EtOH) (%)	74.3	61.3
C(EtOH) (g/L)	14.9	14.8
P. stipitis		
Y(EtOH) (%)	34.1	19.4
C(EtOH) (g/L)	7.1	5.1

Yields are expressed as percentage of theoretical ethanol yield: 0.51 g(EtOH)/g(monosaccharides) (Takerzadeh and Karimi, 2007). Dilution of the samples due to the addition of the yeasts to the solution is taken into account in the calculation.

Reprinted with permission from Heinonen et al. (2012a). Copyright: 2012. John Wiley and Sons.

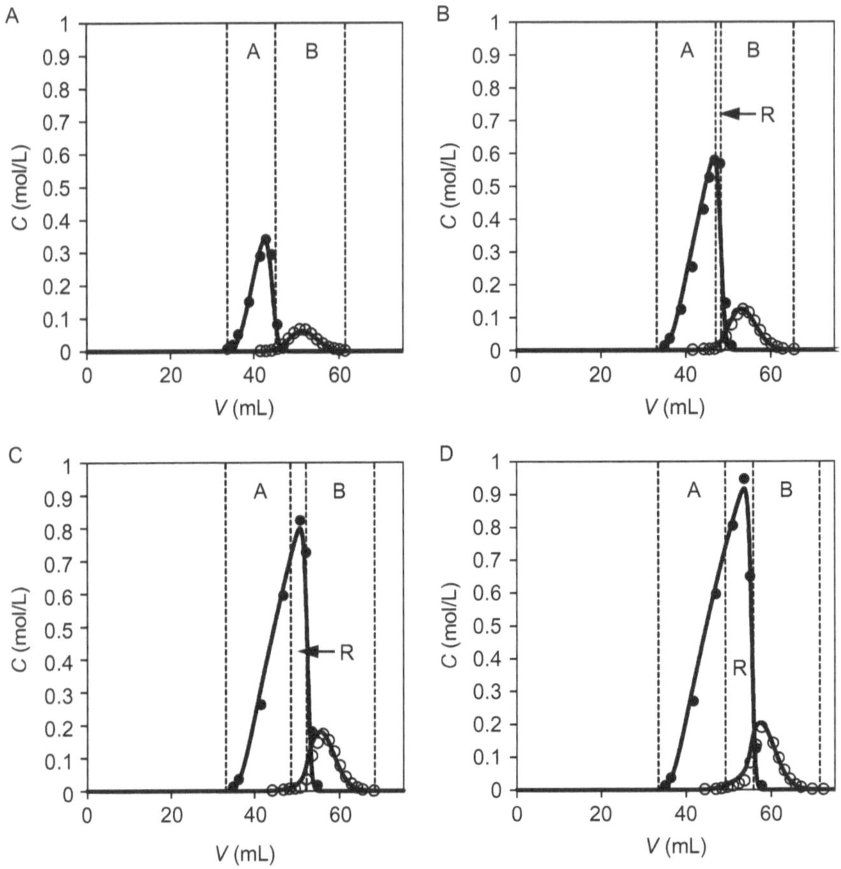

Figure 5.18 Chromatographic separation of H_2SO_4 and glucose in batch mode (A) and SSR mode (B–D) with a gel-type strong acid cation-exchange resin (sulfonated PS-DVB resin, 8 wt% DVB content, CS16GC, Finex Oy) in acid form. Injection volume: (A) 0.9 mL, (B) 3 mL, (C) 6 mL, (D) 9.0 mL. Flow rate: 2.655 mL/min. $h_{col}=20$ cm, $d_{col}=2.5$ cm. Filled circles, H_2SO_4; open circles, glucose. Solid lines represent simulation results. Dashed vertical lines show the fraction cuts (A and B are product fractions, R is the recycle fraction). *Reprinted with permission from Hellstén and Sainio (2012). Copyright: 2012. Taylor and Francis.*

Experimental and calculated chromatograms of batch separation and SSR at steady state are presented in Fig. 5.18. The figure shows also the cut times applied in the experiments. The cut times were determined beforehand using a short-cut design method (Kaspereit and Sainio, 2011). The cycle time of the batch process was approximately 28 min, and the cut time between the sulfuric acid fraction (fraction A) and the glucose fraction (fraction B) occurred at 45.2 min. In the SSR mode, the steady-state was reached after 6 cycles.

As seen in Fig. 5.18, significantly higher product concentrations can be obtained when SSR is used instead of batch chromatography. For example, the concentration of glucose in product fraction B is 3.6 times higher when SSR with the injection width of 9 mL is used instead of the batch separation with injection width of 0.9 mL. The optimal injection volume is 6 mL, yieding a productivity 0.41 mol/L_{bed}/h with the eluent consumption of 3.3 L/mol. The increase in productivity compared to the batch process was 59% (Hellstén and Sainio, 2012).

5.4. Fractionation using SMB chromatography

The potential of continuous chromatography has not been fully explored for fractionation of lignocellulose hydrolysates. Sprinfield and Hester (1999) investigated experimentally continuous separation of sulfuric acid (10 wt%) and glucose (10 wt%). A standard four-zone SMB with 18 columns was used. The fractionations were conducted using two different gel-type strong acid cation-exchange resins (sulfonated PS–DVB resin) in acid form: Dowex monosphere 99 (6 wt% DVB content, Dow Chemicals Co.) and U.S. Filter SM38 (4 wt% DVB content U.S. Filter Co.). Water was used as eluent. Dowex resin was found to be better suited for the glucose–sulfuric acid separation due to higher degree of cross-linking (affects resin volume changes, electrolyte exclusion, and size exclusion as discussed in Section 4). Using the Dowex resin and a column setup 5:4:6:3, 98% of sulfuric acid was collected to the raffinate stream and 96% of the glucose to the extract stream. Weak focusing effect was also noted for glucose in the experiments. Also, a nine-column setup (combining two columns as one) was investigated, and it was found to be sufficient for the separation (Sprinfield and Hester, 1999).

Sprinfield and Hester (2001) also studied modeling of the aforementioned continuous acid–sugar separation system using a plate model (mixing cell model). The sorption of glucose and sulfuric acid was described using distribution coefficients with values depending on the degree of resin shrinking. The value of the glucose distribution coefficient also depended on the sulfuric acid concentration, that is, the cooperative effect of sulfuric acid on the glucose sorption was included in the model. A reasonable correlation between the experimental and calculated results was obtained.

Continuous fractionation of an authentic dilute acid yellow poplar hydrolysate (sulfuric acid, glucose, xylose, and acetic acid) using a nine-zone SMB system of 20 columns has been investigated by Wooley et al. (1998). Dowex 99 (sulfonated PS–DVB resin, 6 wt% DVB content, Dow Chemicals Co.) in acid form was used as adsorbent and water as eluent.

The nine-zone SMB was constructed by coupling a four-zone SMB with a five-zone SMB. In the first unit, the hydrolysate feed was splitted into three streams. Sulfuric acid was taken out from the system through raffinate stream. Sugars and most of the acetic acid were led through the bypass stream into the second unit. Part of the acetic acid was taken out from the first SMB unit in the extract stream. In the second unit, the sugar–acetic acid mixture was separated into two fractions. The sugars were taken out from the system in raffinate stream of the second unit, and acetic acid in extract stream of the second unit. Good separation efficiency was achieved with the nine-zone SMB: recoveries of glucose and xylose in the raffinate stream of the second unit were 88% and 83%, respectively. The purity of sugar stream was 100% (Wooley et al., 1998).

Sun et al. (2011) investigated the use of Intermittent SMB (I-SMB) (see Section 3.3.2) for the fractionation of concentrated sulfuric acid (27 wt%) hydrolysates of bamboo chips. In this case, only a binary fractionation of the hydrolysate was required: the hydrolysate contained mainly sulfuric acid and monosaccharides (glucose and xylose). Anion-exchange resin (DIAION MA03SS, Mitsubishi Chemical Co.) in sulfate form was used as adsorbent and water as eluent (Sun et al., 2011). With the I-SMB process, Sun et al. (2011) obtained 90.5% pure sulfuric acid and 98.4% pure monosaccharide streams in steady-state conditions.

Japan Organo process (see Section 3.3.2) is an attractive process option for the continuous chromatographic fractionation of concentrated acid lignocellulosic hydrolysates. The JO process can be used for ternary (or more complex) fractionations and, unlike the nine-zone SMB or other multizone SMB systems, it can be run with fewer columns (Masuda et al., 1993; Mata and Rodrigues, 2001).

Previously unpublished simulation results for separation of sulfuric acid (20 wt% or 2.32 mol/L), glucose (0.35 mol/L), xylose (0.35 mol/L), and acetic acid (0.16 mol/L) with the JO process are presented below. The model used is presented in Section 4. With strong acid PS–DVB cation-exchange resin in acid form as adsorbent and water as eluent, sulfuric acid as the least adsorbed component is collected from the raffinate outlet, the monosaccharides as the intermediately adsorbed components are collected from the intermediate outlet, and acetic acid as the most adsorbed component is collected from the extract stream (see Fig. 5.4).

The same purity constraints were used in the calculations as in the process performance evaluation of the batchwise fractionation process (see Section 5.2.2): purity constraints were given only for the monosaccharide outlet stream.

Investigation of eluent consumption was excluded from the JO process performance study, and internal recycling stream was not utilized. Due to these reasons, the JO process was simulated without zone IV. Due to this, sulfuric acid (in raffinate) was collected also in the first step (compare to Fig. 5.4). Five columns were connected in configuration 1:2:2. The inlet flow rates to the system in steps one and two were defined as the maximum possible flow rates when the pressure loss from system inlet to system outlet Δp was 4 bar. The flow rates were calculated using the simple Kozeny–Carman equation (McCabe et al., 2005):

$$\frac{\Delta p}{h_{\mathrm{col}}} = \frac{150 v \mu}{\Phi^2 d_{\mathrm{p}}^2} \frac{(1-\varepsilon)^2}{\varepsilon^3}, \qquad [5.20]$$

where v is superficial flow velocity (i.e., fluid velocity through an empty column), μ is fluid viscosity, and Φ is particle sphericity (one for spherical resin beads).

In step one of the JO process, two inlet streams exist: eluent stream and feed stream. The flow rates of the outlet streams in the first step were the same as the inlet flow rates (calculated with Eq. (5.20)) because no recycle stream was utilized. In the second step, the only inlet stream is the eluent stream (see Fig. 5.4). Two outlet streams exist in step two (extract and raffinate). The effect of the flow rates of the outlet streams in the second step on the separation performance was evaluated by changing the ratio of these streams $\varphi\left(\dot{V}_{\mathrm{extract}}/\dot{V}_{\mathrm{raffinate}}\right)$. Due to the dependence of the flow rates from the maximum pressure drop, the eluent flow rate in step two changed slightly as the value of φ was changed. Also, the durations of the first step and one switch in the second step, t_1 (1.0–2.2 min) and t_2 (3.0–3.7 min), respectively, were varied.

The effects of the outlet flow rate ratio in the second step and the durations of the first and one switch in the second step on the productivity of the JO process with respect to the monosaccharides were investigated. The productivity of the process with respect to the monosaccharides was calculated from

$$Pr_{\mathrm{sugar}} = \frac{n_{\mathrm{sugar}}^{\mathrm{out,I}}}{n_{\mathrm{col}} V_{\mathrm{col}} (1-\varepsilon)(t_1 + n_{\mathrm{sub,2}} t_2)}, \qquad [5.21]$$

where $n_{\mathrm{sugar}}^{\mathrm{out,I}}$ is the amount of monosaccharides in the intermediate outlet stream (I) (see Fig. 5.4), n_{col} is the number of columns in the SMB train (five in this case), and $n_{\mathrm{sub,2}}$ is the number of SMB port switches in the second step. $n_{\mathrm{sub,2}}$ was set to five in this study, that is, in the second step, one full SMB cycle was performed, that is, the duration of step two is $n_{\mathrm{sub,2}} t_2$. Other investigated performance parameters were the purities Pu and recovery yields Re of the components in their target outlet streams. Purity of component i in outlet stream X can be calculated with

$$Pu_X = \frac{C_{i,X}}{\sum_{j=1}^{n} C_{j,X}}, \quad [5.22]$$

where the summation in the denominator stands for the total concentration in the outlet stream X. Recovery yield of acetic acid in the extract outlet stream (E) is calculated from

$$Re_{AcOH,E} = \frac{C_{AcOH,E}^{step\ 2}\dot{V}_{eluent,2}\varphi n_{sub,2}t_2}{C_{AcOH}^{feed}\dot{V}_{feed,1}t_1}100\%, \quad [5.23]$$

recovery yield of the monosaccharides in the intermediate outlet stream from

$$Re_{sugar,I} = \frac{C_{sugar,I}^{step\ 1}\dot{V}_{eluent,1}}{C_{sugar}^{feed}\dot{V}_{feed,1}}100\%, \quad [5.24]$$

and recovery yield of sulfuric acid in the raffinate outlet stream from

$$Re_{H_2SO_4,R} = \frac{C_{H_2SO_4,R}^{step\ 1}\dot{V}_{feed,1}t_1 + C_{H_2SO_4,R}^{step\ 2}(1-\varphi)\dot{V}_{eluent,2}n_{sub,2}t_2}{C_{H_2SO_4}^{feed}\dot{V}_{feed,1}t_1}100\%, \quad [5.25]$$

where $C_{i,X}^{stepk}$ is the average concentration of component i in the outlet stream X in step k, C_i^{feed} is the feed concentration of component i, $\dot{V}_{feed,1}$ is the feed flow rate in step 1, and $\dot{V}_{eluent,k}$ is the eluent flow rate in step k.

15 cycles were needed for the system to reach steady-state. The spatial profiles of the fractionation of concentrated acid lignocellulosic hydrolysate with the Japan Organo SMB in steady-state at the end of the first and second (number of switches equal to $n_{sub,2}$) steps are shown in Fig. 5.19.

The spatial sulfuric acid profile in the Japan Organo SMB system (Fig. 5.19) is similar to the outlet profile of sulfuric acid in batch chromatography system with diffuse front and rear shock layer (see Fig. 5.9). Due to the cooperative effect of sulfuric acid on their sorption, the monosaccharides have extensive fronts eluting under the sulfuric acid profile. In addition, focusing of the monosaccharides can be noted at the rear of the sulfuric acid profiles (Fig. 5.19): the maximum concentrations are approximately three times larger than those in the feed solution. Acetic acid is well separated from sulfuric acid and the monosaccharides also in the Japan Organo SMB system (Fig. 5.19).

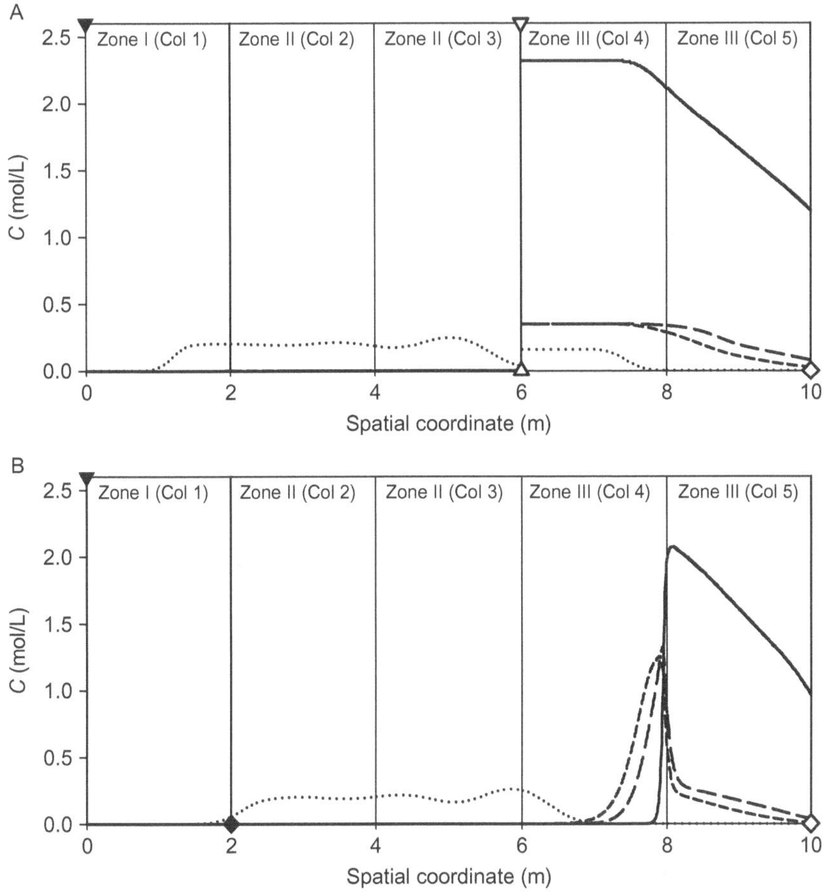

Figure 5.19 Calculated spatial steady-state profiles of sulfuric acid, glucose, xylose, and acetic acid inside a five-column (1:2:2 setup) Japan Organo SMB process without recycle stream at the end of the first step (A) and at the end of the second step (B). Feed composition: 2.32 mol/L sulfuric acid, 0.35 mol/L glucose, 0.35 mol/L xylose, and 0.16 mol/L acetic acid. Simulation parameters: $h_{col}=2$ m, $d_{col}=2$ m, $t_1=2.0$ min, $t_2=3.59$ min, feed flow rate in step 1 = 123.1 m^3/h, eluent flow rate in step 1 = 82.1 m^3/h, eluent flow rate in step 2 = 73.5 m^3/h, and $\varphi=0.216$. Symbols: ▼ and ▽ indicate the inlet positions of eluent and feed streams, respectively, and ◆, ◇, and △ indicate the outlet positions of extract, raffinate, and intermediate streams, respectively. The model used is presented in Section 4. Isotherm parameters are given in Table 5.2, and other parameters in Section 5.2.

The productivity of the JO process with respect to the monosaccharides was found to increase as the duration of the first step increased until t_1 was so large that the front of the acetic acid profile reached the second column of zone III. Increasing the value of t_1 beyond this threshold quickly decreased Pr_{sugar} to zero with the given purity constraints. This is because the amount of acetic acid in the monosaccharide outlet stream becomes too large when the acetic acid front reaches the second column of zone III.

Maximum Pr_{sugar} (296 mol/m^3 h) was found with the duration of the first step as 2.0 min. By varying t_2 and the ratio of the outlet flow rates in the second step, φ, it was found that Pr_{sugar} stays on a relatively constant level as φ is increased (i.e., the flow rate through zones II and III decreases) and t_2 is increased. The maximum Pr_{sugar} was obtained with t_2 value of 3.37 min and φ value of 0.01 (i.e., 1% of the eluent stream is taken out in the extract stream). With batchwise fractionation using a column with similar dimensions as in the SMB calculations, maximum Pr_{sugar} obtained was 50% smaller than the one obtained with the JO process. This clearly indicates that the JO process is more effective for the fractionation of lignocellulosic hydrolysates than batch process.

With the t_1, t_2, and φ values giving the maximum Pr_{sugar}, the recovery yields of sulfuric acid, monosaccharide, and acetic acid outlet streams were 97.4%, 60.9%, and 4.6%, respectively. The recovery yield of acetic acid was very low, and therefore, the t_2 and φ giving the maximum productivity is not optimal for the recovery yield of acetic acid: most of acetic acid circulates to the sulfuric acid outlet stream because of the low φ value. The purities of the sulfuric acid, monosaccharide, and acetic acid outlet streams were 84.0%, 89.5%, and 100%, respectively. The relatively low purity of sulfuric acid stream is due to the coelution of monosaccharides and acetic acid in this stream.

Recovery yield of acetic acid can be increased considerably without significant impact on Pr_{sugar} by increasing the values of t_2 and φ. For example, with t_2 and φ values of 3.59 min and 0.216 (Fig. 5.19), respectively, acetic acid recovery yield is 93.5% while Pr_{sugar} (283 mol/m^3 h) remains close to the maximum value (296 mol/m^3 h). At this point, the recovery yields of sulfuric acid and monosaccharides were 97.3% and 61.7%, respectively. Most of these components ended up to their target outlet streams. Those monosaccharides that elute under the sulfuric acid profile (38.3% with t_1 2.0 min, t_2 3.59 min, and φ 0.216) are taken out from the system in the sulfuric acid outlet stream. Similarly, a small portion of sulfuric acid (2.7% with t_1 2.0 min, t_2 3.59 min, and φ 0.216) elutes in the monosaccharide outlet stream.

Purities of outlet streams with t_1 2.0 min, t_2 3.59 min, and φ 0.216 were 89.0%, 88.7%, and 100% for sulfuric acid, monosaccharides, and acetic acid, respectively. In this case, the relatively low purity of the sulfuric acid stream is only due to the monosaccharides eluting under the sulfuric acid profile. On the other hand, sulfuric acid eluting in the monosaccharide outlet stream decreases the purity of the monosaccharide stream. Acetic acid is easily separated from the other components and elutes pure from the JO process.

Table 5.9 Average concentrations in steady state in the outlet streams of the five-column Japan Organo SMB process

	Average concentrations in steady state (mol/L)		
	Extract	**Intermediate**	**Raffinate**
H_2SO_4	0	0.08	0.44
Sugars	0	0.65	0.05
Acetic acid	0.13	3.36×10^{-3}	1.6×10^{-3}

For simulation parameters see caption of Fig. 5.19.

Average concentrations in each outlet streams in steady state are given in Table 5.9. All components are diluted in their target outlet streams: the most diluted component is the sulfuric acid for which the outlet concentration is approximately 20% of the feed concentration (Table 5.9). For the monosaccharides and acetic acid, the dilution is much smaller: the average total monosaccharide outlet concentration is only approximately 7% smaller than the feed concentration. Monosaccharide dilution is decreased by the focusing effect caused by sulfuric acid (Fig. 5.19).

6. ADSORPTIVE REMOVAL OF INHIBITORY COMPOUNDS

While concentrated acid hydrolysis produces harmful by-products in smaller quantities than dilute acid hydrolysis (Sun and Cheng, 2002; Takerzadeh and Karimi, 2007), harmful compounds should still be removed before further processing such as fermentation. As demonstrated in Section 5.2 (see Fig. 5.10), the sugar decomposition products that act as fermentation inhibitors can be separated from the monosaccharides using batch chromatography. Although such a process would consist of only a single step, it may not be economical because of long cycle times caused by strong affinity of furans to polystyrene-based ion-exchange resins.

Several techniques have been proposed for the hydrolyzate detoxification, including overliming (Larsson et al., 1999), adsorption (Nilvebrant 2001; Villarreal et al., 2006; Weil et al., 2002; Xie et al., 2005), evaporation (Larsson 1999), and enzyme or microorganism treatment (Larsson 1999). Effectiveness of the detoxification methods depends on the type of hydrolyzate and the concentration of inhibitors. A comprehensive review of detoxification methods was recently given by Wang and Feng (2010).

The literature on detoxification of biomass hydrolyzates for bioethanol production mostly considers solutions produced with the dilute acid process. Sainio et al. (2011) compared three adsorbents for removal of furfural and HMF from a concentrated acid hydrolysate in a fixed-bed adsorption column. The adsorbents were a granulated activated carbon (GAC), a neutral styrenic polymer (XAD–16), and a strong acid cation-exchange resin (Finex CS16GC) in acid (H^+) form.

According to experimental adsorption equilibrium isotherm data measured in the presence of 20 wt% H_2SO_4, monosaccharides are not significantly adsorbed on any of these adsorbents (Sainio et al., 2011). Furfural and HMF showed very high affinities toward GAC and XAD–16 (distribution coefficient at low concentration in the order of several hundreds or thousands) and moderate affinity toward the ion-exchange resin (distribution coefficient in the order of 3). It was therefore expected that GAC and XAD–16 would allow processing a larger amount of feed before regeneration is required. This was indeed observed in the adsorption column experiments that are summarized in Table 5.10.

Unfortunately, long breakthrough or saturation times are associated with extensive regeneration times, and regeneration with water is not an economically feasible option. Considerably reduced regeneration times were obtained by using an aqueous ethanol solution instead (Sainio et al., 2011).

Table 5.10 Characteristic properties of column outlet profiles during loading, regeneration with water, and regeneration with 50 wt% EtOH(aq)

Adsorbent	Compound	*t*(Breakthrough)	*t*(Saturation)	*t*(reg.) water	*t*(reg.) EtOH	max(C/C^{feed}) EtOH
CS16GC	Glucose	1.7	6.5	1.9	3.7	1.05
	Acetic acid	2.2	9	2.2	4.1	1
	HMF	4.1	8.6	4.4	4.7	2.7
	Furfural	6.0	10.3	7.1	4.7	5.5
XAD–16	Glucose	0.7	1.9	3	2.5	1
	Acetic acid	1.2	12.5	3.2	2.5	1
	HMF	3.8	13.5	15	4	3.9
	Furfural	8.5	14	53	6.5	6.8
GAC	Glucose	1.5	4	8	5	1
	Acetic acid	3.1	5.5	10	5	1
	HMF	39	75	210	35	8.4
	Furfural	42	75	>300	40	8.7

t(Breakthrough) evaluated at $C/C^{feed} = 0.1$; *t*(reg.) evaluated at $C/C^{feed} = 0.03$. Times are dimensionless.
Reprinted with permission from Sainio et al. (2011). Copyright: 2011. Elsevier.

As observed in Table 5.10, an almost 10-fold decrease in cycle time is obtained with 50 wt% EtOH(aq) in the case of activated carbon adsorbent. Decreaseding adsorption in the presence of ethanol was reported also by Weil et al. (2002). They studied the short-chain aliphatic alcohols (methanol to *n*-butanol) for desorbing furfural from XAD-4, but no quantitative data were given. They recommended ethanol for desorption because of the possibility of significant adsorption of *n*-propanol and *n*-butanol on the hydrophobic adsorbent.

Sainio et al. (2011) calculated the productivity of the adsorption process based on experimental data by considering the total length of a cycle consisting of loading, regeneration, and washing steps. The influence of the required extent of inhibitor removal on productivity is shown in Fig. 5.20. Productivity is here scaled in a way that the value corresponds to the duration of the loading step relative to the complete loading–regeneration–washing cycle.

When no organic desorbent is used, an ion-exchange resin yields the highest productivity (Fig. 5.20A). This is because it is faster to regenerate with water than the more hydrophobic adsorbents. The process is not very cost–effective, however, because the adsorption column spends more time in regeneration than in operation if more than 60% of furfural and HMF are to be removed. Using an organic desorbent shortens the cycle time in all

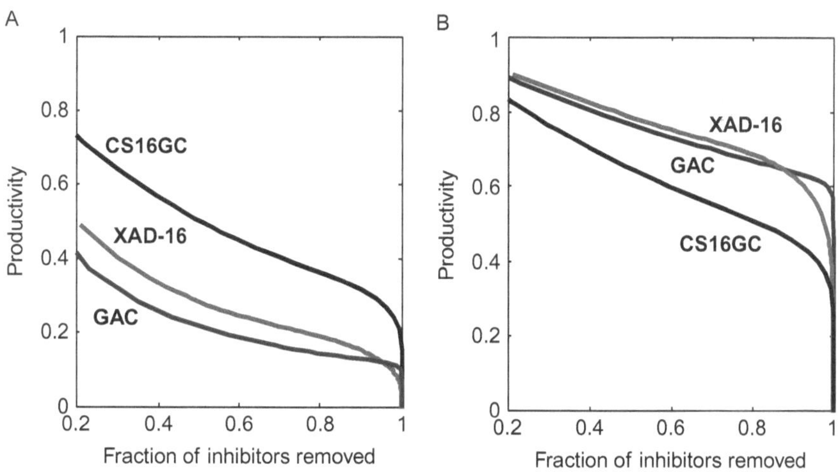

Figure 5.20 Influence of required product purity and type of adsorbent on the productivity of fixed-bed adsorptive detoxification of a (synthetic) concentrated acid of lignocellulose. (A) Desorption with water; (B) desorption with 50 wt% EtOH(aq). *Reprinted with permission from Sainio et al. (2011). Copyright: 2011. Elsevier.*

cases (Fig. 5.20B), but the effect is the strongest for GAC and XAD–16. The hydrophobic adsorbents yield nearly identical productivity, unless very high extent of detoxification is required. The activated carbon was found to be the best option for almost complete removal of furfural and HMF.

7. SELECTIVE SEPARATION OF MONOSACCHARIDES

Separation of individual monosaccharides from their mixture using chromatography is an established technology. Sulfonated gel-type strong acid cation-exchange PS–DVB resins in metal ion form are used as adsorbents (Bart et al., 1996; Farone and Cuzens, 1993, 1995; Howard et al., 1988; Lee, 2003; Saari and Hurme, 2011; Schoenrock, 1991). The fractionation of monosaccharides is based on the complex formation between the monosaccharides and the metal counterions of the resin. In other words, the mechanism is predominantly ligand exchange (see Section 4.1.5). Larger sugars (e.g., sucrose, oligosaccharides) are separated from each other mainly due to size exclusion: larger molecules are excluded totally or partially from the resin and interact only very weakly with the metal counterions of the resin (Howard et al., 1988; Stefansson and Westerlund, 1996).

Metals from groups I and II can be used as complexing counterions in chromatographic fractionation of sugars. Most commonly, Ca^{2+} ions are used as counterions in sugar separations, although at least Na^{+}, K^{+}, Pb^{2+} ions can also be used (Farone and Cuzens, 1993, 1995; Howard et al., 1988). The monosaccharides are separated from each other because the complexes formed with the metal counterions vary in strength due to different molecular structures of the monosaccharides. In purification of lignocellulosic hydrolysates (hydrolysis acid removal), complex formation between the monosaccharides and protons in the resin does not occur, because protons do not form complexes (Angyal, 1973; Helfferich, 1995; Howard et al., 1988; Reeves, 1949a, 1949b; Saladini et al., 2001; Stefansson and Westerlund, 1996; Tiihonen et al., 2002a, 2002b).

In the sweetener (sugar) industry, chromatography is widely used for the separation of individual sugars from sugar syrups derived from sugar cane or sugar beet (Howard et al., 1988). The most common chromatographic separation process in the sugar industry is the enrichment of fructose in fructose–glucose syrups using SMB chromatography. This syrup is a product from hydrolysis of sucrose to fructose and glucose. Sulfonated gel-type strong acid PS–DVB cation-exchange resin in calcium form is used as an adsorbent in the separation (Bart et al., 1996; Howard et al., 1988; Lee, 2003; Saari and Hurme, 2011; Schoenrock, 1991).

Other chromatographic sugar separations in the sweetener industry include, for example, the separation of galactose from lactose hydrolysates; the separation of lactulose from lactose and galactose (Dendene et al., 1995; Saari et al., 2010; Saari and Hurme, 2011); and the separation of fructose, sorbitol, and mannitol (Bart et al., 1996). The first mentioned separation can be accomplished with a sulfonated PS–DVB resin in calcium form (Saari et al., 2010; Saari and Hurme, 2011). In the second separation mentioned above, two steps are needed. First, galactose is separated form lactulose–lactose mixture with a resin in potassium or sodium form at high temperature. In the second step, lactulose and lactose are separated with a resin in calcium form at low temperature (Dendene et al., 1995; Saari and Hurme, 2011). When fructose is hydrogenated in the presence of a heterogeneous catalyst, sorbitol and mannitol are produced in equimolar amounts. Chromatography with sulfonated PS–DVB resin in calcium form as adsorbent can be used to fractionate fructose, sorbitol, and mannitol (Bart et al., 1996).

Farone and Fatigati (2003) have patented a process in which glucose and xylose from purified lignocellulosic hydrolysates are separated using chromatography. This separation is also done using sulfonated gel-type strong acid PS–DVB cation-exchange resin in Ca^{2+} form as an adsorbent. Most preferably, the separation of glucose and xylose is done using SMB chromatography: 94% pure xylose stream (86.4% recovery yield) and 76% pure glucose stream (88.3% recovery yield) can be produced with a feed containing 19% glucose and 39% xylose (Farone and Fatigati, 2003).

The recovery of xylose from lignocellulosic hydrolysates can also be accomplished using strong basic PS–DVB anion-exchange resin in SO_4^{2-} form (Saari et al., 2010; Saari and Hurme, 2011).

8. CONCLUSIONS

Acid-catalyzed hydrolysis of lignocellulosic biomasses is an efficient way to obtain monosaccharides for use as precursors for biofuel or chemicals production. Recovery of monosaccharides can be efficiently accomplished using chromatographic separation and adsorption.

Using chromatography, lignocellulosic hydrolysates can be fractionated selectively to obtain sulfuric acid, monosaccharides, acetic acid, and other by-product fractions. The sulfuric acid can then be recycled to be reused in the hydrolysis process, and the other fractions can be led to downstream processing.

Gel-type strong acid PS–DVB resin in acid (H^+) form is preferred as an adsorbent because no regeneration of the resin is required. Sulfuric acid has a cooperative effect on the sorption of the other components in the lignocellulosic hydrolysates which brings about some interesting phenomena: the fronts of the monosaccharide profiles are elongated, and at the rear of the sulfuric acid profile so-called focusing of the monosaccharides occurs. These phenomena have both negative (elongated fronts) and positive effects (focusing) on the separation performance.

The process performance of a stand-alone chromatographic fractionation unit for concentrated acid lignocellulosic hydrolysates seems not very promising because of low monosaccharide yield. However, since the sulfuric acid fraction (and hence the monosaccharides in this fraction) can be recycled back to the hydrolysis reactor, the entire process should be considered when carrying out a performance evaluation. When recycling is properly included in the analysis, it is concluded that chromatographic separation becomes an efficient method for the fractionation of lignocellulosic hydrolysates.

There are several process options for chromatographic fractionation of lignocellulosic hydrolysates: batch chromatography, steady-state recycling chromatography, or continuous SMB chromatography. A basic four-zone SMB is inadequate for this fractionation due to the fact that more than two fractions have to be obtained, but there are more advanced SMB variants (JO, I-SMB, etc.) that can be used.

Also, removal of by-products formed during acid hydrolysis can be accomplished using chromatographic separation. However, due to strong sorption of these components on common PS–DVB resins, it would be more economical to separate them in a separate adsorption unit prior to the chromatographic acid–monosaccharide separation. Adsorption using activated carbon, zeolites, polymeric adsorbents, or ion-exchange resins as adsorbents can be used for the recovery of the valuable by-products.

REFERENCES

Agrawal G, Kawajiri Y: Comparison of various ternary simulated moving bed separation schemes by multi-objective optimization, *J Chromatogr A* 1238:105–113, 2012.

Airaksinen J, Heikkilä H, Lewandowski J, Laiho K: Separation process, U.S. Patent 2010/0212662 A1, August 26, 2010.

Alén R, Sjöström E, Vaskikari P: Carbon dioxide precipitation of lignin from alkaline pulping liquors, *Cell Chem Technol* 19:537–541, 1985.

Ando M, Tanimura M, Tamura M: Method for chromatographic separation, U.S. Patent 4970002, November 13, 1990.

Angyal SJ: Complex formation between sugars and metal ions, *Pure Appl Chem* 35:132–146, 1973.

Arslan Y, Takac S, Eken-Saracoglu N: Kinetic study of hemicellulosic sugar production from hazelnut shells, *Chem Eng J* 185–186:23–28, 2012.

Baniel A, Eyal A: Process for the production of HCl gas from chloride salts and for the production of carbohydrates, U.S. Patent No. 2011/0178290 A1, July 21, 2011.

Barreto AG Jr., da Silva IJ Jr., dos Santos MAG, de Veredas V, Santana CC: Modeling, simulation, and experimental evaluation of continuous chromatographic separation of ketamine enantiomers on MCTA, *J Liquid Chromatogr Relat Technol* 31:3057–3076, 2008.

Bart HJ, Messenböck RC, Byers CH, Prior A, Wolfgang J: Continuous chromatographic separation of fructose, mannitol and sorbitol, *Chem Eng Process* 35:459–471, 1996.

Binder JB, Blank JJ, Cefali AV, Raines RT: Synthesis of furfural from xylose and xylan, *ChemSusChem* 3:1268–1272, 2010.

Blommel PG, Cortright RD: Production of conventional liquid fuels from sugars, 2008, www.biofuelstp.eu/downloads/Virent_Technology_Whitepaper.pdf, 21.8.2012.

Broughton D, Gerhold CG: Continuous sorption process employing fixed bed of sorbent and moving inlets and outlets, U.S. Patent 2985589, May 23, 1961.

Bubnik Z, Pour V, Guberova A, Starhova H, Hinkova A, Kadlec P: Application of continuous chromatographic separation in sugar processing, *J Food Eng* 61:509–513, 2004.

Burk MJ: Sustainable production of industrial chemicals from sugars, *Int Sugar J* 112:30–35, 2010.

Camacho F, González-Tello P, Jurado E, Robles A: Microcrystalline-cellulose hydrolysis with concentrated sulfuric acid, *J Chem Tech Biotechnol* 67:350–356, 1996.

Chandel AK, Rajeev KK, Singh A, Kuhad RC: Detoxification of sugarcane bagasse hydrolysate improves ethanol production by *Candida Shehatae* NCIM 3501, *Bioresour Technol* 98:1947–1950, 2007.

Ching CB, Ho C, Hidajat K, Ruthven DM: Experimental study of a simulated countercurrent adsorption system-V. Comparison of resin and zeolite absorbents for fructose–glucose separation at high concentration, *Chem Eng Sci* 42:2547–2555, 1987.

Clark TA, Mackie KL: Fermentation inhibitors in wood hydrolysates derived from the softwood *Pinus radiata*, *J Chem Technol Biotechnol* 34B:101–110, 1984.

Clausen EC, Caddy JL: Concentrated sulfuric acid process for converting lignocellulosic materials to sugars, U.S. Patent 5118673, February 23, 1993.

Colmenares JC, Magdziarz A, Bielejewska A: High-value added chemicals obtained from selective photo-oxidation of glucose in the presence of nanostructured titanium photocatalysts, *Bioresour Technol* 102:11254–11257, 2011.

Delgenes JP, Moletta R, Navarro JM: Effect of lignocellulose degradation products on ethanol fermentations of glucose and xylose by *Saccharomyces cerevisiae*, *Zymomonas mobilis*, *Pichia stipitis*, and *Candida shehatae*, *Enzyme Microb Technol* 19:220–225, 1996.

Dendene K, Guihard L, Balannec B, Bariou B: Study of the separation of lactose, lactulose and galactose by liquid chromatography using cationic ion-exchange resin columns, *Chromatographia* 41:561–567, 1995.

Dorfner K: Introduction to ion exchange and ion exchangers. In Dorfner K, editor: *Ion Exhangers*, Berlin, Germany, 1991, Walter de Gruyter, pp 7–189.

Du L, Li S, Zhao H: Discovery and characterization of novel D-xylose-specific transporters from *Neurospora crassa* and *Pichia stipitis*, *Mol Biosyst* 6:1–7, 2010.

Duff SJB, Murray WD: Bioconversion of forest products industry waste cellulosics to fuel ethanol: a review, *Bioresour Technol* 55:1–33, 1996.

Durão MI, Costa CAV, Rodrigues AE: Saturation and regeneration of ion exchangers with volume changes, *Ind Eng Chem Res* 31:2564–2572, 1992.

Eken-Saracoglu N, Arslan Y: Comparison of different pretreatments in ethanol fermentation using corn cob hemicellulosic hydrolysate with *Pichia Stipitis* and *Candida Shehatae*, *Biotechnol Lett* 22:855–858, 2000.

Ezeji TC, Blaschek HP: Butanol production from lignocellulosic biomass. In Blaschek HP, Ezeji TC, Scheffran J, editors: *Biofuels from agricultural wastes and by-products*, Ames, Iowa, United States, 2010, Blackwell Publishing, pp 19–37.

Farone WA, Cuzens JE: Method of producing sugars using strong acid hydrolysis of cellulosic and hemicellulosic materials, U.S. Patent No. 5562777, March 26, 1993.

Farone WA, Cuzens JE: Strong acid hydrolysis of cellulosic and hemicellulosic materials, U.S. Patent No. 5597714, June 7, 1995.

Farone W, Fatigati MA: Separation of xylose and glucose, International Patent WO 03/010339 A1, February 6, 2003.

Fuhrmann A, Kleen M, Koljonen K, Österberg M, Stenius P: Precipitation of lignin and extractives from bleaching filtrates – effects on pulp properties, TAPPI international bleaching conference, 2002.

Galbe M, Zacchi G: A review of the production of ethanol from softwood, *Appl Microbiol Biotechnol* 59:618–628, 2002.

Giacobello S, Storti G, Tola G: Design of a simulated moving bed unit for sucrose–betaine separations, *J Chromatogr A* 872:23–35, 2000.

Glód BK: Ion exclusion chromatography: parameters influencing retention, *Neurochem Res* 22:1237–1248, 1997.

Goncalves CV, Carpes MJS, Correia CRD, Santana CC: Purification of *n*-boc-rolipram racemate on chiral stationary phase using simulated moving bed chromatography under linear conditions, *Biochem Eng J* 40:526–536, 2008.

Gonzáles G, López-Santín J, Caminal G, Solá C: Dilute acid hydrolysis of wheat straw hemicelluloses at moderate temperature: a simplified kinetic model, *Biotechnol Bioeng* 28:288–293, 1986.

Gsell G: Water systems utilizing multiple effect and vapor compression technologies compared, *Pharm Eng* 24:1–7, 2004.

Guiochon G, Shirazi SG, Katti AM: *Fundamentals of preparative and non-linear chromatography*, San Diego, United States, 2006, Elsevier Academic Press.

Gurgel LVA, Marabezi K, Zambon MD, Curvelo AAS: Dilute acid hydrolysis of sugarcane bagasse at high temperatures – a kinetic study of cellulose saccharification and glucose decomposition. Part II: sulfuric acid as catalyst, *Ind Eng Chem Res* 51:1173–1185, 2012.

Hamelinck CN, von Hooijdonk G, Faaij APC: Ethanol from lignocellulosic biomass: techno-economic performance in short-, middle- and long-term, *Biomass Bioenergy* 28:384–410, 2005.

Harned HS, Ehlers RW: The dissociation constant of acetic acid from 0 to 60 °C centigrade, *J Am Chem Soc* 55:652–656, 1933.

Heikkilä H, Lewandowski J, Kuisma J: Chromatographic separation method, U.S. Patent, 7229558 B2, June 12, 2007.

Heikkilä H, Hyöky G, Kuisma J: Method for the recovery of betaine from molasses, European patent 0345511 A2, December 13, 1989.

Heikkilä H, Hyöky G, Kuisma J: A method for the fractionation of molasses, International patent WO 94/17213, August 4, 1994.

Heikkilä H, Hyöky G, Kuisma J: Method for fractionation of a solution by a chromatographic simulated moving bed process, International patent WO 97/45185, December 4, 1997.

Heikkilä H, Hyöky G, Kuisma J: Method for the fractionation of molasses, U.S. Patent 6093326, July 25, 2000.

Heinonen J, Sainio T: Chromatographic recovery of monosaccharides for the production of bioethanol from wood, *Ind Eng Chem Res* 49:3713–3729, 2010.

Heinonen J, Sainio T: Modelling and performance evaluation of chromatographic monosaccharide recovery from concentrated acid lignocellulosic hydrolysates, *J Chem Technol Biotechnol* 87:1676–1686, 2012.

Heinonen J, Tamminen A, Uusitalo J, Sainio T: Ethanol production from wood via concentrated acid hydrolysis, chromatographic separation, and fermentation, *J Chem Technol Biotechnol* 87:689–696, 2012a.

Heinonen J, Rubiera Landa HO, Sainio T, Seidel-Morgenstern A: Use of adsorbed solution theory to model competitive and co-operative sorption on elastic ion exchange resins, *Sep Purif Technol* 95:235–247, 2012b.

Helfferich F: *Ion Exchange,* Mineola, New York, United States, 1995, Dower Publications Inc.

Hellstén S, Sainio T: Steady state recycling chromatography in acid–sugar separation on an ion-exchange resin, *Sep Sci Technol* 47:2358–2365, 2012.

Hester RD, Farina GE: Exclusion chromatographic separation of ionic from nonionic solutes, U.S. Patent 5667693, September 16, 1997.

Hoshino C, Yamada T, Taneda D, Nagata Y, Fujii T, Mase T, Ueno Y: Method for producing monosaccharides from biomass and monosaccharide producing device, U.S. Patent 0148750, June 28, 2007.

Howard AJ, Carta G, Byers CH: Separation of sugars by continuous annular chromatography, *Ind Eng Chem Res* 27:1873–1882, 1988.

Hunter AK, Carta G: Protein adsorption on novel acryloamido-based polymeric ion exchangers III. Salt concentration effects and elution behavior, *J Chromatogr A* 930:79–93, 2001.

Iranmahboob J, Nadim F, Monemi S: Optimizing acid-hydrolysis: a critical step for production of ethanol from mixed wood chips, *Biomass Bioenergy* 22:401–404, 2002.

IUPAC Goldbook, goldbook.iupac.org/P04436.html, 21.8.2012.

Janga KK, Hägg M-B, Moe ST: Influence of acid concentration, temperature, and time on decrystallization in two-stage concentrated sulfuric acid hydrolysis of pinewood and aspenwood: a statistical approach, *Bioresources* 7:391–411, 2012.

Jermann S, Katsuo S, Mazzotti M: Intermittent simulated moving bed processes for chromatographic three-fraction separation, *Org Process Res Dev* 16:311–322, 2012.

Juza M, Mazzotti M, Morbidelli M: Simulated moving-bed chromatography and its application to chirotechnology, *Trends Biotechnol* 18:108–118, 2000.

Karimi K, Kheradmandinia S, Takerzadeh MJ: Conversion of rice straw to sugars by dilute-acid hydrolysis, *Biomass Bioenergy* 30:247–253, 2006.

Kaspereit M: Advanced operating concepts for simulated moving bed processes. In Grushka E, Grinberg N, editors: *Advances in Chromatography*, vol. 47, Boca Raton, United States, 2009, CRC Press, pp 165–192.

Kaspereit M, Sainio T: Simplified design of steady-state recycling chromatography under ideal and nonideal conditions, *Chem Eng Sci* 66:5428–5438, 2011.

Katsuo S, Mazzotti M: Intermittent simulated moving bed chromatography: 1. Design criteria and cyclic steady-state, *J Chromatogr A* 1217:1354–1361, 2010.

Kessler LC, Seidel-Morgenstern A: Theoretical study of continuous countercurrent chromatography based on connected 4-zone units, *J Chromatogr A* 1126:323–337, 2006.

Koljonen K, Österberg M, Kleen M, Fuhrmann A, Stenius P: Precipitation of lignin and extractives on kraft pulp: effect on surface chemistry, surface morphology and paper strength, *Cellulose* 11:209–224, 2004.

Kristensen JB, Felby C, Jørgensen H: Yield-determining factors in high-solids enzymatic hydrolysis of lignocellulose, *Biotechnol Biofuels* 2, 2009.

Kunkes EL, Simonetti DA, West RM, Serrano-Ruiz JC, Gärtner CA, Dumesic JA: Catalytic conversion of biomass to monofunctional hydrocarbons and targeted liquid-fuel classes, *Science* 333:417–421, 2008.

Kurup AS, Hidajat K, Ray AK: Optimal operation of a Pseudo-SMB process for ternary separation under non-ideal conditions, *Sep Purif Technol* 51:387–403, 2006.

Laatikainen M, Heinonen J, Sainio T: Modeling of chromatographic separation of concentrated-acid hydrolysates, *Sep Purif Technol* 80:610–619, 2011.

Ladisch M, Svarczkopf J: Ethanol production and the cost of fermentable sugars from biomass, *Bioresour Technol* 36:83–95, 1991.

Larsson S, Reimann A, Nilvebrant N-O, Jönsson LJ: Comparison of different methods for the detoxification of lignocellulose hydrolyzates of spruce, *Appl Biochem Biotechnol* 77–79:91–103, 1999.

Lee KW: Continuous separation of glucose and fructose at high concentration using two-section simulated moving bed process, *Korean J Chem Eng* 20:532–537, 2003.

Lee JW, Wankat PC: Design of pseudo-simulated moving bed process with multi-objective optimization for the separation of a ternary mixture: linear isotherms, *J Chromatogr A* 1217:3418–3426, 2010.

Lee YY, Iyer P, Torget RW: Dilute-acid hydrolysis of lignocellulosic biomass, *Adv Biochem Eng Biotechnol* 65:93–115, 1999.

Lee E, Kim JM, Kim WS, Kim IH: Simulation and operation performance of simulated moving bed for enantioseparation of mandelic acid, *Biotechnol Bioprocess Eng* 15:103–109, 2010.

Lehoucq S, Verhève D, Wouwer AV, Cavoy E: SMB enantioseparation: process development, modeling, and operating conditions, *AIChE J* 46:247–256, 2000.

Lenihan P, Orozco A, O'Neill E, Ahmad MNM, Rooney DW, Walker GM: Dilute acid hydrolysis of lignocellulosic biomass, *Chem Eng J* 156:395–403, 2010.

Lenihan P, Orozco A, O'Neill E, et al: Kinetic modelling of dilute acid hydrolysis of lignocellulosic biomass. In dos Santos Bernardes MA, editor: *Biofuels production – recent developments and prospects*, Rijeka, Croatia, 2012, InTech, pp 293–308.

Lewkowski J: Synthesis, chemistry and applications of 5-hydroxymethylfurfural and its derivatives, *Arkivoc* i:17–54, 2001.

Lima S, Antunes MM, Fernandes A, Pillinger M, Ribeiro MF, Valente AA: Acid-catalysed conversion of saccharides into furanic aldehydes in the presence of three-dimensional mesoporous Al-TUD-1, *Molecules* 15:3863–3877, 2010.

Mackie JS, Meares P: The diffusion of electrolytes in a cation-exchange resin membrane. I. Theoretical, *Proc R Soc A* 232:498–509, 1955.

Marra RA, Cooney DO: Equilibrium theory for sorption accompanied by sorbent bed shrinking or swelling, *AIChE J* 19:181–183, 1973.

Marra RA, Cooney DO: Multicomponent sorption operations: bed shrinking and swelling in an ion exclusion case, *Chem Eng Sci* 33:1597–1601, 1978.

Masuda T, Sonobe T, Matsuda F, Horie M, Process for fractional separation of multi-component fluid mixture, U.S. Patent 5198120, March 30, 1993.

Mata VG, Rodrigues AE: Separation of ternary mixtures by pseudo-simulated moving bed chromatography, *J Chromatogr A* 939:23–40, 2001.

Matsushika A, Inoue H, Murakami K, Takimuru O, Sawayama S: Bioethanol production performance of five recombinant strains of laboratory and industrial xylose-fermenting Saccharomyces cerevisiae, *Bioresour Technol* 100:2392–2398, 2009.

Mazzotti M, Neri B, Gelosa D, Morbidelli M: Dynamics of a chromatographic reactor: esterification catalyzed by acidic resins, *Ind Eng Chem Res* 36:3163–3172, 1997.

McCabe WL, Smith JC, Harriot P: *Unit operations of chemical engineering*, New York, United States, 2005, McGraw-Hill, pp 63–165.

McMurry J: *Organic chemistry*, London, United Kingdom, 2008, Thomson Learning, pp 973–1015.

Mirica KA, Lockett MR, Snyder PW, Shapiro ND, Mack ET, Nam S, Whitesides GM: Selective precipitation and purification of monovalent proteins using oligovalent ligands and ammonium sulfate, *Bioconjug Chem* 23:293–299, 2012.

Mun S: Effect of a partial-feeding application on product purities and throughput of a five-zone simulated moving bed process for the separation of a ternary nucleoside mixture, *Process Biochem* 46:977–986, 2011.

Mussatto SI, Roberto IC: Alternatives for detoxification of diluted-acid lignocellulosic hydrolyzates for use in fermentative processes: a review, *Bioresour Technol* 93:1–10, 2004.

Nanguneri SR, Hester RD: Acid/sugar separation using ion exclusion resins: a process analysis and design, *Sep Sci Technol* 25:1829–1842, 1990.

Neuman RP, Rudge SR, Ladisch MR: Sulfuric acid–sugar separation by ion exclusion, *React Polym* 5:55–61, 1987.

Nfor BK, Hylkema NN, Wiedhaup KR, Verhaert PDEM, van der Wielen LAM, Ottens M: High-throughput protein precipitation and hydrophobic interaction chromatography: salt effect and thermodynamic interrelation, *J Chromatogr A* 1218:8958–8973, 2011.

Ng PK, Snyder MA: pH-based cation exchange chromatography in the capture and elution of monoclonal antibodies, *J Sep Sci* 35:29–35, 2012.

Nghiem NP, Hicks KB, Johnston DB: Integration of succinic acid and ethanol production with potential application in a corn barley biorefinery, *Appl Biochem Biotechnol* 162:1915–1928, 2010.

Nilvebrant N-O, Reimann A, Larsson S, Jönsson LJ: Detoxification of lignocellulose hydrolysates with ion-exchange resins, *Appl Biochem Biotechnol* 91–93:35–49, 2001.

Nowak J, Gedicke K, Antos D, Piatkowski W, Seidel-Morgenstern A: Synergistic effects in competitive adsorption of carbohydrates on an ion-exchange resin, *J Chromatogr A* 1164:224–234, 2007.

Nowak J, Poplewska I, Antos D, Seidel-Morgenstern A: Adsorption behaviour of sugars versus their activity in single and multicomponent liquid solutions, *J Chromatogr A* 1216:8697–8704, 2010.

Pais LS, Loureiro JM, Rodrigues AE: Modeling strategies for enantiomers separation by SMB chromatography, *AIChE J* 44:561–569, 1998.

Palmqvist E, Hahn-Hägerdahl B: Fermentation of lignocellulosic hydrolysates. I: inhibition and detoxification, *Bioresour Technol* 74:17–24, 2000a.

Palmqvist E, Hahn-Hägerdahl B: Fermentation of lignocellulosic hydrolysates. II: inhibitors and mechanisms of inhibition, *Bioresour Technol* 74:25–33, 2000b.

Paredes G, Abel S, Mazzotti M, Morbidelli M, Stadler J: Analysis of a simulated moving bed operation for three-fraction separations (3F-SMB), *Ind Eng Chem Res* 43:6157–6167, 2004.

Parikka M: Global biomass fuel resources, *Biomass Bioenergy* 27:613–620, 2004.

Peng P, Peng F, Bian J, Xu F, Sun R: Studies on the starch and hemicelluloses fractionated by graded ethanol precipitation from Bamboo *Phyllostachys bambusoides* f. shouzhu Yi, *J Agric Food Chem* 59:2680–2688, 2011.

Pedersen S: Chemicals from sugars, 2008, www.topsoe.com/sitecore/shell/Applications/~/media/PDF%20files/Topsoe_Catalysis_Forum/2008/Pedersen.ashx, 21.8.2012.

Rafiqul ISM, Sakinah AMM: Kinetic studies on acid hydrolysis of *Meranti* wood sawdust for xylose production, *Chem Eng Sci* 71:431–437, 2012.

Ranjan R, Thust S, Gounaris CE, et al: Adsorption of fermentation inhibitors from lignocellulosic biomass hydrolyzates for improved ethanol yield and value-added product recovery, *Micropor Mesopor Mater* 122:143–148, 2009.

Reeves RE: Cuprammonium–glycoside complexes. II. The angle between hydroxyl groups on adjacent carbon atoms, *J Am Chem Soc* 71:212–214, 1949a.

Reeves RE: Cuprammonium–glycoside complexes. III. The conformation of the D-glucopyranoside ring in solution, *J Am Chem Soc* 71:215–217, 1949b.

Ribeiro AE, Gomes PS, Pais LS, Rodrigues AE: Chiral separation of Ketoprofen enantiomers by preparative and simulated moving bed chromatography, *Sep Sci Technol* 46:1726–1739, 2011.

Rieman W III: Salting-out chromatography: a review, *J Chem Educ* 38:338–343, 1961.
Rinaldi R, Schüth F: Acid hydrolysis of cellulose as the entry point into biorefinery schemes, *ChemSusChem* 2:1096–1107, 2009.
Saari P, Hurme M: Process synthesis principles in the chromatographic separation of sugars from biomass hydrolysates, *Chem Eng Technol* 34:282–288, 2011.
Saari P, Häkkä K, Jumppanen J, Heikkilä H, Hurme M: Study on industrial scale chromatographic separation methods of galactose from biomass hydrolysates, *Chem Eng Technol* 33:137–144, 2010.
Saddler JN, Yu EKC, Mes-Hartree M, Levitin N, Brownell HH: Utilization of enzymatically hydrolyzed wood hemicelluloses by microorganisms for production of liquid fuels, *Appl Environ Mirobiol* 45:153–160, 1983.
Sainio T, Kaspereit M: Analysis of steady state recycling chromatography using equilibrium theory, *Sep Purif Technol* 66:9–18, 2009.
Sainio T, Paatero E: Mass coordinates for dynamic simulation of column operations involving dimensional changes of packing material, *Comput Chem Eng* 31:374–383, 2007.
Sainio T, Turku I, Heinonen J: Adsorptive removal of fermentation inhibitors from concentrated acid hydrolysates of lignocellulosic biomass, *Bioresour Technol* 102:6048–6057, 2011.
Saladini M, Menabue L, Ferrari E: Sugar complexes with metal^{2+} ions: thermodynamic parameters of associations of Ca^{2+}, Mg^{2+} and Zn^{2+} with galactaric acid, *Carbohydr Res* 336:55–61, 2001.
Sargent R, Rieman W III: Salting-out chromatography, *J Phys Chem* 61:354–358, 1956.
Schiesser WE: *The numerical method of lines,* San Diego, United States, 1991, Academic Press.
Schoenrock KWR: Ion exchangers in the sweetener industry. In Dorfner K, editor: *Ion exchangers*, Berlin, 1991, Walter de Gruyter, pp 949–979.
Schulte M, Strube J: Preparative enantioseparation by simulated moving bed chromatography, *J Chromatogr A* 906:399–416, 2001.
Shapovalov OI, Ashkinazi LA: Biobutanol: biofuel of second generation, *Russ J Appl Chem* 81:2232–2236, 2008.
Sprinfield RM, Hester RD: Continuous ion-exclusion chromatography system for acid/sugar separation, *Sep Sci Technol* 34:1217–1241, 1999.
Sprinfield RM, Hester RD: Development and modelling of a continuous simulated moving bed ion exclusion process for the separation of acid and sugar, *Sep Sci Technol* 36:911–930, 2001.
Stefansson M, Westerlund D: Ligand-exchange chromatography of carbohydrates and glycoconjugates, *J Chromatogr A* 720:127–136, 1996.
Sun Y, Cheng J: Hydrolysis of lignocellulosic materials for ethanol production: a review, *Bioresour Technol* 83:1–11, 2002.
Sun Z-Y, Tang Y-Q, Iwanaga T, Sho T, Kida K: Production of fuel ethanol from bamboo by concentrated sulfuric acid hydrolysis followed by continuous ethanol fermentation, *Bioresour Technol* 102:10929–10935, 2011.
Taherzadeh MJ, Karimi K: Bioethanol: market and production processes. In Nag A, editor: *Biofuels refining and performance*, Fairfield, United States, 2008, McGraw-Hill, pp 69–106.
Takerzadeh MJ, Karimi K: Acid-based hydrolysis processes for ethanol from lignocellulosic materials: a review, *Bioresources* 2:472–499, 2007.
Tiihonen J, Markkanen I, Paatero E: Complex stability of sugars and sugar alcohols with Na^{+}, Ca^{2+}, and La^{3+} in chromatographic separations using poly(styrene-*co*-divinylbenzene) resins and aqueous organic eluents, *Chem Eng Commun* 189:995–1008, 2002a.
Tiihonen J, Sainio T, Kärki A, Paatero E: Co-eluent effect in partition chromatograhy. Rhamnose–xylose separation with strong and weak cation-exchangers in aqueous ethanol, *J Chromatogr A* 982:69–84, 2002b.
Torget RW, Padukone N, Hatzis C, Wyman CE: Hydrolysis and fractionation of lignocellulosic biomass, U.S. Patent 6022419, February 8, 2000.

Uçar G, Balaban M: Hydrolysis of polysaccharides with 77% sulfuric acid for quantitative saccharification, *Turk J Agric For* 27:361–365, 2003.

van Leeuwen BNM, van der Wulp AM, Duijnstee I, van Maris AJA, Straathof AJ: Fermentative production of isobutene, *Appl Microbiol Biotechnol* 93:1377–1387, 2012.

Vente JA, Bosch H, de Haan AB, Bussmann PJT: Evaluation of sugar sorption isotherm by frontal analysis under industrial processing conditions, *J Chromatogr A* 1066:71–79, 2005.

Villarreal MLM, Prata AMR, Felipe MGA, Almeida E, Silva JB: Detoxification procedures of eucalyptus hemicellulose hydrolysate for xylitol production by *Candida Quilliermondii*, *Enzyme Microb Technol* 40:17–24, 2006.

Wang B, Feng H: Detoxification of lignocellulosic hydrolysates. In Blaschek HP, Ezeji TC, Scheffran J, editors: *Biofuels from agricultural wastes and by-products*, Ames, Iowa, United States, 2010, Blackwell Publishing, pp 233–250.

Wankat PC: *Rate-controlled separations,* Essex, Great Britain, 1990, Elsevier Science Publishers Ltd.

Weber C, Farwick A, Benisch F, et al: Trends and challenges in the microbial production of lignocellulosic bioalcohol fuels, *Appl Microbiol Biotechnol* 87:1303–1315, 2010.

Weil JR, Dien B, Bothast R, Hendrickson R, Mosie NS, Ladisch MR: Removal of fermentation inhibitors formed during pretreatment of biomass by polymeric adsorbents, *Ind Eng Chem Res* 41:6132–6138, 2002.

Wheaton RM, Bauman WC: Ion exclusion – a unit operation utilizing ion exchange materials, *Ind Eng Chem* 45:228–233, 1953.

Wooley R, Ma Z, Wang N-HL: A nine-zone simulating moving bed for the recovery of glucose and xylose from biomass hydrolyzate, *Ind Eng Chem Res* 37:3699–3709, 1998.

Xiang Q, Kim JS, Lee YY: A comprehensive kinetic model for dilute-acid hydrolysis of cellulose, *Appl Biochem Biotechnol* 105–108:337–352, 2003a.

Xiang Q, Lee YY, Pettersson PO, Torget RW: Heterogeneous aspects of acid hydrolysis of α-cellulose, *Appl Biochem Biotechnol* 105–108:505–514, 2003b.

Xiang Q, Lee YY, Torget RW: Kinetics of decomposition during dilute-acid hydrolysis of lignocellulosic biomass, *Appl Biochem Biotechnol* 113–116:1127–1138, 2004.

Xie Y, Phelps D, Lee C-H, Sedlak M, Ho N, Wang N-HL: Comparison of two adsorbents for sugar recovery from biomass hydrolysate, *Ind Eng Chem Res* 44:6816–6823, 2005.

SUBJECT INDEX

Note: Page numbers followed by "*f*" indicate figures, and "*t*" indicate tables.

A

B

C

N

O

P

R

V

W

X

CONTENTS OF VOLUMES IN THIS SERIAL

Volume 7 (1968)

Robert S. Brown, Ralph Anderson, and Larry J. Shannon, *Ignition and Combustion of Solid Rocket Propellants*

Knud Østergaard, *Gas–Liquid–Particle Operations in Chemical Reaction Engineering*

J. M. Prausnilz, *Thermodynamics of Fluid–Phase Equilibria at High Pressures*

Robert V. Macbeth, *The Burn-Out Phenomenon in Forced-Convection Boiling*

William Resnick and Benjamin Gal-Or, *Gas–Liquid Dispersions*

Volume 8 (1970)

C. E. Lapple, *Electrostatic Phenomena with Particulates*

J. R. Kittrell, *Mathematical Modeling of Chemical Reactions*

W. P. Ledet and D. M. Himmelblau, *Decomposition Procedures foe the Solving of Large Scale Systems*

R. Kumar and N. R. Kuloor, *The Formation of Bubbles and Drops*

Volume 9 (1974)

Renato G. Bautista, *Hydrometallurgy*

Kishan B. Mathur and Norman Epstein, *Dynamics of Spouted Beds*

W. C. Reynolds, *Recent Advances in the Computation of Turbulent Flows*

R. E. Peck and D. T. Wasan, *Drying of Solid Particles and Sheets*

Volume 10 (1978)

G. E. O'Connor and T. W. F. Russell, *Heat Transfer in Tubular Fluid–Fluid Systems*

P. C. Kapur, *Balling and Granulation*

Richard S. H. Mah and Mordechai Shacham, *Pipeline Network Design and Synthesis*

J. Robert Selman and Charles W. Tobias, *Mass-Transfer Measurements by the Limiting-Current Technique*

Volume 11 (1981)

Jean-Claude Charpentier, *Mass-Transfer Rates in Gas–Liquid Absorbers and Reactors*

Dee H. Barker and C. R. Mitra, *The Indian Chemical Industry—Its Development and Needs*

Lawrence L. Tavlarides and Michael Stamatoudis, *The Analysis of Interphase Reactions and Mass Transfer in Liquid–Liquid Dispersions*

Terukatsu Miyauchi, Shintaro Furusaki, Shigeharu Morooka, and Yoneichi Ikeda, *Transport Phenomena and Reaction in Fluidized Catalyst Beds*

Volume 12 (1983)

C. D. Prater, J, Wei, V. W. Weekman, Jr., and B. Gross, *A Reaction Engineering Case History: Coke Burning in Thermofor Catalytic Cracking Regenerators*

Costel D. Denson, *Stripping Operations in Polymer Processing*

Robert C. Reid, *Rapid Phase Transitions from Liquid to Vapor*

John H. Seinfeld, *Atmospheric Diffusion Theory*

Volume 13 (1987)

Edward G. Jefferson, *Future Opportunities in Chemical Engineering*

Eli Ruckenstein, *Analysis of Transport Phenomena Using Scaling and Physical Models*

Rohit Khanna and John H. Seinfeld, *Mathematical Modeling of Packed Bed Reactors: Numerical Solutions and Control Model Development*

Michael P. Ramage, Kenneth R. Graziano, Paul H. Schipper, Frederick J. Krambeck, and Byung C. Choi, *KINPTR (Mobil's Kinetic Reforming Model): A Review of Mobil's Industrial Process Modeling Philosophy*

Venkat Ganesan and Glenn H. Fredrickson, *Fluctuation Effects in Microemulsion Reaction Media*
David B. Graves and Cameron F. Abrams, *Molecular Dynamics Simulations of Ion–Surface Interactions with Applications to Plasma Processing*
Christian M. Lastoskie and Keith E, Gubbins, *Characterization of Porous Materials Using Molecular Theory and Simulation*
Dimitrios Maroudas, *Modeling of Radical-Surface Interactions in the Plasma-Enhanced Chemical Vapor Deposition of Silicon Thin Films*
Sanat Kumar, M. Antonio Floriano, and Athanassiors Z. Panagiotopoulos, *Nanostructured Formation and Phase Separation in Surfactant Solutions*
Stanley I. Sandler, Amadeu K. Sum, and Shiang-Tai Lin, *Some Chemical Engineering Applications of Quantum Chemical Calculations*
Bernhardt L. Trout, *Car-Parrinello Methods in Chemical Engineering: Their Scope and potential*
R. A. van Santen and X. Rozanska, *Theory of Zeolite Catalysis*
Zhen-Gang Wang, *Morphology, Fluctuation, Metastability and Kinetics in Ordered Block*
Copolymers

Volume 29 (2004)
Michael V. Sefton, *The New Biomaterials*
Kristi S. Anseth and Kristyn S. Masters, *Cell–Material Interactions*
Surya K. Mallapragada and Jennifer B. Recknor, *Polymeric Biomaterias for Nerve Regeneration*
Anthony M. Lowman, Thomas D. Dziubla, Petr Bures, and Nicholas A. Peppas, *Structural and Dynamic Response of Neutral and Intelligent Networks in Biomedical Environments*
F. Kurtis Kasper and Antonios G. Mikos, *Biomaterials and Gene Therapy*
Balaji Narasimhan and Matt J. Kipper, *Surface-Erodible Biomaterials for Drug Delivery*

Volume 30 (2005)
Dionisio Vlachos, *A Review of Multiscale Analysis: Examples from System Biology, Materials Engineering, and Other Fluids-Surface Interacting Systems*
Lynn F. Gladden, M.D. Mantle and A.J. Sederman, *Quantifying Physics and Chemistry at Multiple Length-Scales using Magnetic Resonance Techniques*
Juraj Kosek, Frantisek Steěpánek, and Miloš Marek, Modelling of Transport and Transformation
Processes in Porous and Multiphase Bodies
Vemuri Balakotaiah and Saikat Chakraborty, *Spatially Averaged Multiscale Models for Chemical Reactors*

Volume 31 (2006)
Yang Ge and Liang-Shih Fan, *3-D Direct Numerical Simulation of Gas–Liquid and Gas–Liquid–Solid Flow Systems Using the Level-Set and Immersed-Boundary Methods*
M.A. van der Hoef, M. Ye, M. van Sint Annaland, A.T. Andrews IV, S. Sundaresan, and J.A.M. Kuipers, *Multiscale Modeling of Gas-Fluidized Beds*
Harry E.A. Van den Akker, *The Details of Turbulent Mixing Process and their Simulation*
Rodney O. Fox, *CFD Models for Analysis and Design of Chemical Reactors*
Anthony G. Dixon, Michiel Nijemeisland, and E. Hugh Stitt, *Packed Tubular Reactor Modeling and Catalyst Design Using Computational Fluid Dynamics*

Volume 32 (2007)
William H. Green, Jr., *Predictive Kinetics: A New Approach for the 21st Century*
Mario Dente, Giulia Bozzano, Tiziano Faravelli, Alessandro Marongiu, Sauro Pierucci and Eliseo Ranzi, *Kinetic Modelling of Pyrolysis Processes in Gas and Condensed Phase*
Mikhail Sinev, Vladimir Arutyunov and Andrey Romanets, *Kinetic Models of C1–C4 Alkane Oxidation as Applied to Processing of Hydrocarbon Gases: Principles, Approaches and Developments*
Pierre Galtier, *Kinetic Methods in Petroleum Process Engineering*

Volume 33 (2007)

Shinichi Matsumoto and Hirofumi Shinjoh, *Dynamic Behavior and Characterization of Automobile Catalysts*

Mehrdad Ahmadinejad, Maya R. Desai, Timothy C. Watling and Andrew P.E. York, *Simulation of Automotive Emission Control Systems*

Anke Güthenke, Daniel Chatterjee, Michel Weibel, Bernd Krutzsch, Petr Kočí, Miloš Marek, Isabella Nova and Enrico Tronconi, *Current Status of Modeling Lean Exhaust Gas Aftertreatment Catalysts*

Athanasios G. Konstandopoulos, Margaritis Kostoglou, Nickolas Vlachos and Evdoxia Kladopoulou, *Advances in the Science and Technology of Diesel Particulate Filter Simulation*

Volume 34 (2008)

C.J. van Duijn, Andro Mikelić, I.S. Pop, and Carole Rosier, *Effective Dispersion Equations for Reactive Flows with Dominant Peclet and Damkohler Numbers*

Mark Z. Lazman and Gregory S. Yablonsky, *Overall Reaction Rate Equation of Single-Route Complex Catalytic Reaction in Terms of Hypergeometric Series*

A.N. Gorban and O. Radulescu, *Dynamic and Static Limitation in Multiscale Reaction Networks, Revisited*

Liqiu Wang, Mingtian Xu, and Xiaohao Wei, *Multiscale Theorems*

Volume 35 (2009)

Rudy J. Koopmans and Anton P.J. Middelberg, *Engineering Materials from the Bottom Up – Overview*

Robert P.W. Davies, Amalia Aggeli, Neville Boden, Tom C.B. McLeish, Irena A. Nyrkova, and Alexander N. Semenov, *Mechanisms and Principles of 1 D Self-Assembly of Peptides into β*-Sheet Tapes

Paul van der Schoot, *Nucleation and Co-Operativity in Supramolecular Polymers*

Michael J. McPherson, Kier James, Stuart Kyle, Stephen Parsons, and Jessica Riley, *Recombinant Production of Self-Assembling Peptides*

Boxun Leng, Lei Huang, and Zhengzhong Shao, *Inspiration from Natural Silks and Their Proteins*

Sally L. Gras, *Surface- and Solution-Based Assembly of Amyloid Fibrils for Biomedical and Nanotechnology Applications*

Conan J. Fee, *Hybrid Systems Engineering: Polymer-Peptide Conjugates*

Volume 36 (2009)

Vincenzo Augugliaro, Sedat Yurdakal, Vittorio Loddo, Giovanni Palmisano, and Leonardo Palmisano, *Determination of Photoadsorption Capacity of Polychrystalline TiO2 Catalyst in Irradiated Slurry*

Marta I. Litter, *Treatment of Chromium, Mercury, Lead, Uranium, and Arsenic in Water by Heterogeneous Photocatalysis*

Aaron Ortiz-Gomez, Benito Serrano-Rosales, Jesus Moreira-del-Rio, and Hugo de-Lasa, Mineralization of Phenol in an Improved Photocatalytic Process Assisted with Ferric Ions: *Reaction Network and Kinetic Modeling*

R.M. Navarro, F. del Valle, J.A. Villoria de la Mano, M.C. Alvarez-Galván, and J.L.G. Fierro, *Photocatalytic Water Splitting Under Visible Light: Concept and Catalysts Development*

Ajay K. Ray, *Photocatalytic Reactor Configurations for Water Purification: Experimentation and Modeling*

Camilo A. Arancibia-Bulnes, Antonio E. Jiménez, and Claudio A. Estrada, *Development and Modeling of Solar Photocatalytic Reactors*

Orlando M. Alfano and Alberto E. Cassano, *Scaling-Up of Photoreactors: Applications to Advanced Oxidation Processes*

Yaron Paz, *Photocatalytic Treatment of Air: From Basic Aspects to Reactors*

Volume 37 (2009)

S. Roberto Gonzalez A., Yuichi Murai, and Yasushi Takeda, *Ultrasound-Based Gas–Liquid Interface Detection in Gas–Liquid Two-Phase Flows*

Z. Zhang, J. D. Stenson, and C. R. Thomas, *Micromanipulation in Mechanical Characterisation of Single Particles*

Feng-Chen Li and Koichi Hishida, *Particle Image Velocimetry Techniques and Its Applications in Multiphase Systems*
J. P. K. Seville, A. Ingram, X. Fan, and D. J. Parker, *Positron Emission Imaging in Chemical Engineering*
Fei Wang, Qussai Marashdeh, Liang-Shih Fan, and Richard A. Williams, *Electrical Capacitance, Electrical Resistance, and Positron Emission Tomography Techniques and Their Applications in Multi-Phase Flow Systems*
Alfred Leipertz and Roland Sommer, *Time-Resolved Laser-Induced Incandescence*

Volume 38 (2009)
Arata Aota and Takehiko Kitamori, *Microunit Operations and Continuous Flow Chemical Processing*
Anıl Ağıral and Han J.G.E. Gardeniers, *Microreactors with Electrical Fields*
Charlotte Wiles and Paul Watts, *High-Throughput Organic Synthesis in Microreactors*
S. Krishnadasan, A. Yashina, A.J. deMello and J.C. deMello, *Microfluidic Reactors for Nanomaterial Synthesis*

Volume 39 (2010)
B.M. Kaganovich, A.V. Keiko and V.A. Shamansky, *Equilibrium Thermodynamic Modeling of Dissipative Macroscopic Systems*
Miroslav Grmela, *Multiscale Equilibrium and Nonequilibrium Thermodynamics in Chemical Engineering*
Prasanna K. Jog, Valeriy V. Ginzburg, Rakesh Srivastava, Jeffrey D. Weinhold, Shekhar Jain, and Walter G. Chapman, *Application of Mesoscale Field-Based Models to Predict Stability of Particle Dispersions in Polymer Melts*
Semion Kuchanov, *Principles of Statistical Chemistry as Applied to Kinetic Modeling of Polymer-Obtaining Processes*

Volume 40 (2011)
Wei Wang, Wei Ge, Ning Yang and Jinghai Li, *Meso-Scale Modeling—The Key to Multi-Scale CFD Simulation*
Pil Seung Chung, Myung S. Jhon and Lorenz T. Biegler, *The Holistic Strategy in Multi-Scale Modeling*
Milo D. Meixell Jr., Boyd Gochenour and Chau-Chyun Chen, *Industrial Applications of Plant-Wide Equation-Oriented Process Modeling—2010*
Honglai Liu, Ying Hu, Xueqian Chen, Xingqing Xiao and Yongmin Huang, *Molecular Thermodynamic Models for Fluids of Chain-Like Molecules, Applications in Phase Equilibria and Micro-Phase Separation in Bulk and at Interface*

Volume 41 (2012)
Torsten Kaltschmitt and Olaf Deutschmann, *Fuel Processing for Fuel Cells*
Adam Z.Weber, Sivagaminathan Balasubramanian, and Prodip K. Das, *Proton Exchange Membrane Fuel Cells*
Keith Scott and Lei Xing, *Direct Methanol Fuel Cells*
Su Zhou and Fengxiang Chen, *PEMFC System Modeling and Control*
François Lapicque, Caroline Bonnet, Bo Tao Huang, and Yohann Chatillon, *Analysis and Evaluation of Aging Phenomena in PEMFCs*
Robert J. Kee, Huayang Zhu, Robert J. Braun, and Tyrone L. Vincent, *Modeling the Steady-State and Dynamic Characteristics of Solid-Oxide Fuel Cells*
Robert J. Braun, Tyrone L. Vincent, Huayang Zhu, and Robert J. Kee, *Analysis, Optimization, and Control of Solid-Oxide Fuel Cell Systems*

Volume 42 (2013)

T. Riitonen, V. Eta, S. Hyvärinen, L.J. Jönsson, and J.P. Mikkola, *Engineering Aspects of Bioethanol Synthesis*
R.W. Nachenius, F. Ronsse, R.H. Venderbosch, and W. Prins, *Biomass Pyrolysis*
David Kubička and Vratislav Tukač, *Hydrotreating of Triglyceride-Based Feedstocks in Refineries*
Tapio Salmi, *Chemical Reaction Engineering of Biomass Conversion*
Jari Heinonen and Tuomo Sainio, *Chromatographic Fractionation of Lignocellulosic Hydrolysates*

Kolibri 360 GmbH
Ferdinand-Jühlke-Straße 7
99095 Erfurt, Deutschland
produktsicherheit@kolibri360.de